Ground freezing in practice

J. S. Harris

🕂 Thomas Telford

Published by Thomas Telford Services Ltd, Thomas Telford House, 1 Heron Quay, London E14 4JD

First published 1995

Distributors for Thomas Telford books are
USA: American Society of Civil Engineers, Publications Sales Department, 345 East 47th Street, New York, NY 10017-2398
Japan: Maruzen Co Ltd, Book Department, 3–10 Nihonbashi 2-chome, Chuo-ku, Tokyo 103
Australia: DA Books and Journals, 648 Whitehorse Road, Mitcham 3132, Victoria

A catalogue record for this book is available from the British Library

Classification
Availability: Unrestricted
Content: Review of best current practice and sourcebook
Status: Refereed
User: Ground engineers

ISBN: 0 7277 1995 5

© J. S. Harris, 1995, except Chapter 3 © R. H. Jones, 1995, and Chapters 4 and 6 © F. A. Auld and J. S. Harris, 1995, Chapter 8 © J. S. Harris and A. Clough, 1995, and Chapter 9 © J. S. Harris and D. Jackson, 1995.

All rights, including translation reserved. Except for fair copying, no part of this publication may be reproduced, stored in a retrieval system or transmitted in any form or by any means, electronic, mechanical, photocopying or otherwise, without the prior written permission of the Publisher: Books, Publications Division, Thomas Telford Services Ltd, Thomas Telford House, 1 Heron Quay, London E14 4JD.

The book is published on the understanding that the authors are solely responsible for the statements made and opinions expressed in it and that its publication does not necessarily imply that such statements and or opinions are or reflect the views or opinions of the publishers.

Typeset in Great Britain by MHL Typesetting Ltd, Coventry
Printed and bound in Great Britain by Redwood Books, Trowbridge, Wiltshire

To my son, David, with my sincere
thanks for all his support

Contents

Acknowledgements · ix

Preface · xi

Notation · xiii

Abbreviations · xv

Legend to figures · xvii

Introduction · xix

1. **Geotechnical processes** · 1
   Range of options, 1
      Physical support, 1
      Groundwater lowering, 2
      Electro-osmosis, 3
      Injection methods, 3
      Compressed air, 4
      Artificial ground freezing (AGF), 4
   References, 15

2. **Description, sampling and testing frozen soils** · 16
   Description of frozen soils, 16
   Sampling, 17
      Field frozen samples, 17
      Laboratory frozen samples, 18
      Preparation of undisturbed specimens, 18
      Preparation of remoulded specimens, 18
   Testing, 19
      Laboratory testing, 19
      In situ testing, 23
   References, 24

3. **Properties of freezing, frozen and thawed soils** · 26
   by R. H. Jones
   Thermal considerations, 26
      Thermal conductivity, 26
      Heat capacity, 28
      Thermal diffusivity, 29
   Frost heave, 30
      Capillary model, 31

The frozen fringe and Miller's model, 32
Mathematical modelling of heave, 33
Adsorbed film models, 36
Segregation potential, 36
Discrete ice lens theory, 37
Postscript on hydrodynamic models, 38
Thermo-mechanical theory, 38
Frost susceptibility of soils, 38
Effects of salinity on freezing, 39
The effects of thawing, 41
Thaw settlement, 41
Theory of thaw consolidation, 42
Residual stress, 44
Thaw weakening, 46
Implications for artificial ground freezing, 46
Changes in properties of soils after freezing and thawing, 47
Structural changes in clay soils due to freeze−thaw, 47
Permeability and consolidation characteristics, 48
Strength, 50
Frost heave, 51
Liquid limits and suction characteristics, 51
Mechanical properties of frozen soils, 52
Stress−strain relations, 53
Creep, 54
Models of creep behaviour, 56
Time dependent creep strength, 60
Temperature effect on strain rate and strength, 62
Postscript on models of creep behaviour for AGF, 62
Effects of salinity on mechanical properties, 64
References, 64

4. **Engineering design of frozen ground works**　　70
by F. A. Auld & J. S. Harris
Ground information, 70
Structural design, 71
Shafts and pits, 72
Tunnels and drifts, 80
Underpinning, 82
Thermal design, 82
Other design considerations, 83
Water movement, 83
Low moisture content, 85
High voids ratio, 85
References, 85

5. **Refrigeration systems**   87
   Refrigeration methods, 87
   On-site mechanical plant, 87
   Off-site produced liquefied gas, 91
   Distribution systems, 92
   Brine systems, 92
   Cryogenic systems, 94
   References, 95

6. **Construction using AGF**   96
   Freeze-tube placement, 96
   Vertical and angled freeze-tubes, 96
   Horizontal and sub-horizontal freeze-tubes, 99
   Accuracy of freeze-tube placement, 100
   Freeze-tube survey, 100
   Monitoring, 102
   Refrigeration system, 102
   Ice-wall growth and integrity, 103
   Side-effects, 105
   Groundwater, 105
   Thermal factors, 108
   The permanent works, 113
   Excavation and lining, 113
   Concreting against frozen ground, 114
   Back-wall grouting, 117
   Freeze-tube recovery or abandonment, 117
   References, 118

7. **Case examples**   121
   Shafts, pits and open excavations, 121
   As the chosen design method of temporary works, 121
   As an aid to recovery, 140
   Tunnels, 144
   As the chosen design method of temporary works, 144
   As an aid to recovery, 155
   Miscellaneous, 164
   References, 170

8. **Standards and safety**   173
   Health and Safety at Work etc. Act 1974, 173
   Control of Substances Hazardous to Health (COSHH) 1988, 174
   Regulations, Standards and Codes of Practice, 175
   Selected refrigerants, 176

Ammonia, 176
Liquid nitrogen, 177
References, 178

9. **Contractual, cost and risk evaluation considerations** — 179
   Contractual arrangements, 179
   Cost, 182
   Evaluation and selection of technique, 184
   Objective, 184
   Definitions, 184
   Assessment, 185
   Reaction, 189
   References, 190

Appendices — 192
   A Symposia/Society proceedings, 192
   International Symposia on Ground Freezing (ISGF), 192
   National Symposia on Ground Freezing (NSGF), 193
   International Conferences on Permafrost (ICP), 193
   B Conversion factors, 195
   C Summary of ground freezing projects, 196
   D Evaluation models, 204

Bibliography — 211

Index — 261

Acknowledgements

Dr F. A. Auld, Director, Ian Farmer Associates; previously Chief Engineer, Cementation Mining Ltd

C. B. Chapman, Professor of Management Science, University of Southampton

A. Clough, Safety Consultant, formerly Safety Officer with British Drilling & Freezing Co Ltd

A. R. Dawson, Lecturer, Department of Civil Engineering, University of Nottingham

R. G. Futcher (retired), formerly with the Loughborough University of Technology

S. J. Harvey, Director, British Drilling & Freezing Co Ltd

D. I. Harris, Associate Director, Geotechnical Consulting Group

Dr J. T. Holden, Senior Lecturer, Department of Theoretical Mechanics, University of Nottingham

D. Jackson, Director, Jackson-Reed Associates; previously Chief QS, British Drilling & Freezing Co Ltd

Dr. R. H. Jones, Senior Lecturer, Department of Civil Engineering, University of Nottingham

I. M. Macfarlane, Senior Consultant, Trafalgar House Technology

D. Maishman, Vice President, FreezeWALL Inc (USA); previously Director, Foraky Ltd

Dr M. J. Mawdesley, Senior Lecturer, Department of Civil Engineering, University of Nottingham

Emeritis Professor P. J. Pell, formerly Head of Department of Civil Engineering, University of Nottingham

Dr K. H. Whittles, Publisher, previously Technical Editor, Blackie & Son

W. M. Wild (deceased), formerly Chairman & Managing Director, Foraky Ltd

Dr A. J. Wills, Senior Consultant, Trafalgar House Technology

Preface

Prior to my early retirement from full-time employment I was being 'bullied' to fill a perceived gap in the geotechnical literature by writing a book on the ground freezing technique. I could then plead lack of time.

The change of circumstances provided that time, whilst the approval of my former employer of 23 years, together with their prompting to continue to lecture on the subject, and much encouragement from many academic and professional colleagues, gave me the incentive.

Without

- the teachings of Mr W. M. (Bill) Wild
- the encouragement of Dr Keith H. Whittles and acceptance by the publishers
- the technical reviews or comments by Professor Chris B. Chapman, Andrew R. Dawson, Dr John T. Holden, Ian M. Macfarlane, Stephen J. Harvey, Derek Maishman, Dr Mick J. Mawdesley, Professor Peter J. Pell and Dr Alan J. Wills
- the contributions of Dr F. Alan Auld, Alan Clough, David Jackson and Dr Ron H. Jones as sub-authors
- the provision of photographs by many, and
- the patience and forbearance of my wife, Pamela, as proofreader and Ron Futcher as technical proofreader

it would not have been possible. I gratefully extend my sincere thanks to each of them.

A common complaint during the 50s and 60s was a general shortage of informative literature on the ground freezing method. An 'awakening' in the 70s led to an international meeting in Germany, hosted by the University of Bochum, the Proceedings of which were published. This was followed by the formation of an *ad hoc* International Committee which has since been active in promoting further symposia at regular intervals (listed in the appendices), whose proceedings already constitute a significant source of reference and further reading. In the UK this attitude to advancement of knowledge led to the formation of the British Ground Freezing Society, a body which hosted the 5th ISGF in Nottingham and has since enabled the preparation and publication of a series of Technical Memoranda devoted to ground freezing matters.

I set out to produce a text which is both informative and easily read, covering the many aspects of the art and practice of ground freezing likely to be of interest to users and practitioners alike. It is based on my many years experience with the then Foraky Limited (renamed British Drilling & Freezing Co Ltd in 1985), known world-wide as a specialist drilling and ground freezing contractor, but influenced also by prior geotechnical service with the railways and a major civil engineering consultant, and a period of personal research followed by joint industrial/academic supervision of a frozen ground research programme at the University of Nottingham jointly funded by the SERC (now EPSRC) and my company. I trust that readers find the result both interesting and useful.

John S. Harris

Notation*

(Other parameters or other uses of symbols, are defined in the text)

		Units
C	compressive force	N
C	volumetric heat capacity	MJ/m^{3K}
E	Young's modulus unfrozen	
H (or z)	depth below ground level	m
I_L	liquidity index	W/mK
I_p	plasticity index	W/mK
q (or K)	unconfined compressive strength	kPa or MPa
L	latent heat of fusion	kJ/kg
M	bending moment	Nm
N	normal force	N
Re	Reynold's number	
S_r	degree of saturation	%
SP	segregation potential	m^2/hK or mm^2/hK
T	temperature	K or °C
T	temperature relative to actual freezing point	°C or K
V	velocity	m/s or mm/s
K	thermal conductivity	W/mK
c	cohesion	kPa or MPa
c	specific heat capacity	J/kgK
c_v	coefficient of consolidation	m^2/year
d	diameter	m
e	engineering strain	ratio or %
h	depth below groundwater level	m
h	suction	
k	coefficient of permeability/ hydraulic conductivity	m/s or m/d
n (or θ)	porosity	dimensionless ratio
p	pressure	kPa or MPa
p_i	internal pressure on ice-wall	kPa or MPa
p_o	overburden pressure	kPa or MPa
p_v	vertical pressure	kPa or MPa
p_w	external water pressure	kPa or MPa

* Where a letter or symbol can have one of two meanings, the text defines which in each case.

NOTATION

q	uniaxial compressive strength	kPa or MPa
q_x	quantity of heat flow/unit area	J/m^2
r	radius	m
r_e	radius of excavation ($=r_i$)	m
r_{ft} (or r_s)	radius of freeze-tube circle	m
r_i (or r_a)	internal radius of ice-wall	m
r_o (or r_b)	external radius of ice-wall	m
t	time	h, days, weeks or years
t	thickness	m
u	pore pressure	kPa or MPa
w	water content	%
w_L	liquid limit	
α	thermal diffusivity	m^2/s
δ	deformation	mm
γ	unit weight	kg/m^3 or Mg/m^3
γ'	unit weight submerged	kg/m^3 or Mg/m^3
ϵ	true strain	% or ratio
λ	coefficient of soil pressure	
λ_γ	$(1 - \sin \phi)^2$	
ρ_d	dry density	kg/m^3 or Mg/m^3
ρ_w	density of water	kg/m^3 or Mg/m^3
σ	stress	
τ	tensile force	N
ν	Poisson's ratio	
ϕ	angle of internal friction	°

Subscripts

L	
N	Normalized, e.g. with respect to failure
a	air
e	excavation
f	failure (stress)
f	frozen
ff	frozen fringe
i	inner/internal
i	ice
m	failure (creep)
o	outer/external
s	soil
t	time
u	unfrozen
v	vertical
w	water (liquid)

MPa = N/mm^2

Abbreviations

ACFEL	(Predecessor of CRREL)
AGF	Artificial ground freezing
API	American Petroleum Institute
ASCE	American Society of Civil Engineers
ASME	American Society of Mining Engineers
ASTM	American Society for Testing and Materials
BCC	British Cryogenics Council
BCGA	British Compressed Gases Association
BDA	British Drilling Association
BGFS	British Ground Freezing Society
BRS	Building Research Station (UK)
BS	British Standard
CACA	Cement and Concrete Association (UK)
CBR	California Bearing Ratio
Cndn	Canadian
CP	Code of Practice
CRREL	Cold Regions Research and Engineering Laboratory (USA)
DBR	Division of Building Research (Canada)
DTp	Department of Transport (UK)
EC	European Conference
Eur	Europe(an)
FE(M)	Finite element (Method)
HSE	Health and Safety Executive
IAEG	International Association of Engineering Geologists
ICE	Institution of Civil Engineers
ICEC	International Cryogenic Engineering Conference
IME	Institution of Mining Engineers
IMM	Institution of Mining and Metallurgy
Inst	Institution/Institute
IC/IS	International Conference/Congress/Symposium
IoRef	Institute of Refrigeration
ISGF	International Symposium on Ground Freezing
J.	Journal
LN	Liquid nitrogen
LNG/LPG	Liquefied Natural/Petroleum Gas
Mem. Vol.	Memorial Volume
NATM	National Association of Testing Methods (USA)
NCE	*New Civil Engineer* (magazine)
NRCC	National Research Council of Canada
NSGF	National Symposium on Ground Freezing (recorded meetings of the BGFS)

ABBREVIATIONS

NTIS	National Technical Information Service (USA Dept of Commerce)
POAC	Port and Ocean engineering under Arctic Conditions
Proc	Proceedings/transactions
QJEG	Quarterly Journal of Engineering Geology
RETC	Rapid Excavation and Tunneling Congress (USA)
Rpt	Report
Soc.	Society
SM & FE	Soil Mechanics and Foundation Engineering
Tech	Technical/Technology
Temp	Temperature
TM	Technical memorandum
Transl.	Translation
Trnsptn	Transportation
ULS	Uniaxial long-term strength
Univ.	University
(U)	Unpublished
Wksp	Workshop

Legend to figures

☐	Unfrozen permeable soil, weak and/or saturated, e.g. sand
☰	Unfrozen soil, strong and impermeable, e.g. clay
▨	Frozen soil/ice-wall
🧱	Ancillary structures/masonry
⧄	Lining construction facilitated by freezing — in situ concrete, concrete or cast iron segments, etc.
〰	Ground treated by grouting
▦	Fill
▼	Groundwater level
F T	Freeze-tube
O H	Observation hole
P R H	Pressure relief hole
T C	Thermocouple
G L	Ground level

Introduction

Natural freezing occurs seasonally in many areas of the globe, and can adversely affect the engineering performance of roads and pavements as ice lenses form and grow; on the other hand, advantage can be taken of the solid nature of frozen ground as French engineers found during the severe winter of 1852 while constructing a mineshaft in otherwise saturated unstable ground. Towards the polar regions the 'season' is year-round, and we experience permafrost.

Man-made freezing, on the other hand, is controllable and can often be used profitably by civil and mining engineers alike to temporarily stabilize the ground in the form of a cofferdam, that is to provide structural support and/or to exclude groundwater from an excavation until construction of the final lining provides permanent security. For want of a better term the process has become known as Artificial Ground Freezing (AGF) in Europe, and sometimes Construction Ground Freezing in North America.

Although patented by H. Poetsch in Germany in 1883, AGF was first practised in South Wales in 1862 and was mainly applied to vertical openings — shafts or pits. Later, as the ability to drill horizontally was achieved, horizontal applications, e.g. tunnels, were also treated. Apart from protecting excavations, ground freezing has also been used, *inter alia*, to stabilize slips, sample weak soils, construct temporary access roads, and maintain permafrost below overhead pipeline foundations and below heated buildings.

Being transient in nature the process does not affect the level of the water table or the quality of the groundwater, has a low environmental impact and a very high safety record. The prime cost may appear to be high, indeed often more expensive than other methods; equally the reverse is true and, when risk and final costs are included in the equation, the method frequently proves to be the most cost-effective.

A growing international awareness of the merits of AGF as a temporary expedient to aid construction in adverse ground conditions with minimal environmental impact, is closely related to progress in understanding the behaviour of naturally frozen soils — permafrost — and refrigerated enclosures. The construction of oil pipelines across the sub-arctic regions of North America, and storage of liquefied natural gas (LNG) below ground, are but two examples.

The study of problems of common interest, by practitioners, academics and researchers in related disciplines, has generated more scientific publications in the last 15 years than appeared during

INTRODUCTION

the preceding century. Particular examples are the collected contributions to the seven international symposia devoted to the method in the period from 1978 to 1994, with another due in 1997.

Many of the papers and other publications on the subject of ground freezing tend to emphasize or concentrate on the physics of frozen soils, rather than on the practical application of the method in mining or civil engineering. This book aims to provide an overview of the process, its relationship with other geotechnical methods, and its role (in civil engineering terminology) as 'temporary works'; in addition it includes summaries and details of AGF works, often derived from the literature, which inevitably will be subject to corrections and additions which the author will be pleased to receive and incorporate.

1. Geotechnical processes

Range of options

Excavation below the surface of the earth, however shallow, incurs consideration of the stability of the boundary — how best to counter the tendency to collapse, and how to control groundwater. At its simplest, the excavation will be above the water table and will only need to remain unsupported for a short period of time; cutting the side at a batter in cohesive soil, or at an angle which is less than the angle of repose of non-cohesive soil, may be sufficient.

Ground engineering is rarely that simple, and some form of barrier and support will be necessary if the works are to proceed safely and expeditiously without risk of collapse or flooding. The methods available vary in their suitability and complexity, in the time and cost to install and maintain them, and in their reliability.

Physical support

Physical types of support, which include trench sheets, H-beams with timber panels (lagging), driven sheet piling, bored or driven contiguous piles, bored large diameter secant piles or diaphragm walls, and caisson sinking (see Fig. 1.1) are perhaps the traditional

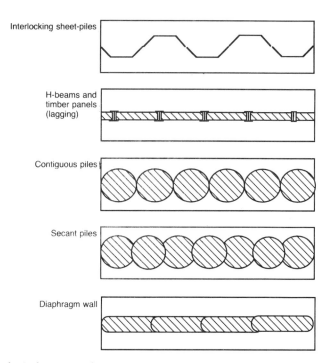

Fig. 1.1. Types of piling for physical support (plan views)

first choice of contractors. They all form a strong vertical boundary; if they are also to function as a cofferdam without the aid of dewatering the choice narrows to interlocking or secant piles and caisson sinking.

When the excavation is to be taken below the water table, the toe of the cofferdam must either terminate in an impermeable stratum or be taken far enough below the sump level of the excavation to preclude instability of the base due to buoyancy uplift. Depending on the depth and width of the excavation, these methods may require walings, struts and/or tie-backs to ensure that the cofferdam itself is sufficiently strong to withstand the active pressure of the soils being supported.

These methods can be used to depths of 25—30 m in most types of soils free of boulders. Recovery of trench sheets and sheet piles is often possible if they are not required to form part of the permanent works. Concrete piles and diaphragm walls will usually constitute the final structural wall, but their exposed face may need a cosmetic finish. Problems can arise during construction from noise or vibration in densely built areas, particularly with driven piles.

Caisson sinking requires the construction on the surface of the complete structure, ready for sinking into the ground with the aid of kentledge. The method relies on relatively uniform, weak soil through which the caisson can penetrate by cutting and displacement of the soil until the target depth is reached. A change of soil density can lead to frustration, while the random presence of hard obstructions (boulders) can deflect the caisson from the vertical; remobilizing a stood caisson or redirection of a deflected caisson is time consuming and costly, as occurred at the Humber Bridge, and at Ameria in Cairo.

Groundwater lowering

In a large open site where it is not necessary to limit the extent of the excavation during the construction stage, a low cost option is 'dewatering', i.e. depression of the water table, by pumping, to a level below that of the intended excavation. This may be achieved by a series of wellpoints inserted, usually by jetting techniques, at relatively close spacing (approximately 1 m) around the perimeter of the area. The method is effective within the suction range of surface pumps, i.e. for excavations no deeper than about 6 m, provided that the shallowest impermeable horizon is at least 1 m below the drawdown level. The method is most appropriate for large, shallow excavations in granular deposits.

Greater depths of drawdown can be achieved by installing a subordinate (multi-stage) wellpoint system within the base of the first (and subsequent) excavations, but each increment of depth will be limited to a maximum of about 6 m. Enlarged sites are needed to accommodate the side-slopes of each excavation and the level berm between each increment of depth for the wellpoints and their collector main (Fig. 1.2).

Submersible pumps are suitable for greater depths of drawdown; they can be purpose designed to suit a large range of depths and

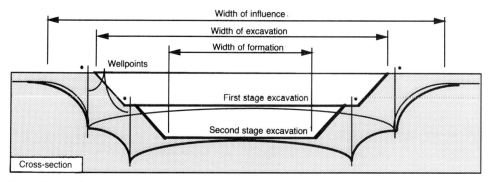

Fig. 1.2. *Principle of two stage wellpoint dewatering*

water yields. Each pump requires the drilling of a suitable borehole, but the number of wells is normally much fewer than for wellpoints — possibly as few as three or four.

Care must be taken to avoid the withdrawal of 'fines' — e.g. silt/fine sand particles — with the water yield, since this could lead to subsidence and/or loss of support to neighbouring property, depending on the size of the cone of depression. Well established techniques are available.

Note: Both methods are cost effective in suitable conditions. Ground freezing has been used to seal potentially serious leaks between parted sheet steel piles, and through the gaps between contiguous bored piles. See Chapters 4 and 7, Somerville (1986).

Electro-osmosis For very fine soils of low permeability, extraction of water by pumping may not be feasible, even with the assistance of vacuum. A little used method, appropriate only in the silt range, is electro-osmosis. If filter wells are treated as cathodes and nearby iron/steel rods as anodes, water migration towards the cathode is induced when a DC current is caused to flow. The few recorded instances of use suggest that currents of up to 100 A and power demands of up to 16 kW/m^3 have been needed to stabilize slopes.

Injection methods Injection of suitable materials (cement or bentonite grout, or chemical gel) under pressure into the soil voids reduces the permeability. By creating a continuous barrier around the desired excavation, the groundwater inflow can be significantly reduced to a manageable order. In practice it is found that a target thickness of less than 4 m rarely ensures the necessary continuity; the quantity of grout actually injected often greatly exceeds the theoretical minimum.

For coarse soils, and rocks with major fissures, cement based grouts, often mixed with clay or bentonite, are used. Silicates or resins may be used in fine/medium sands, particularly for underpinning works where strength and durability are crucial. Recent developments include jet and replacement grouting.

Pre-injection may be employed in association with ground freezing, particularly if the site to be treated includes a zone subject to groundwater movement. (See Chapters 6 and 7.) Injection methods are effective and popular, but the eventual time and cost often exceed the original estimates by large margins. See Raffle and Greenwood (1961), Cambefort (1977), Harris and Pollard (1985).

Compressed air   Groundwater can also be excluded from, or inflow reduced to, confined working spaces such as tunnels, by pressurizing the tunnel air supply to counterbalance the groundwater pressure. Conventionally, this is achieved by installing an air lock at the entrance to the tunnel, or within the access shaft, and a compressor system at the surface.

High establishment costs, not only of the equipment installation but of medical supervision and backup, mean that the method will mostly be used for relatively large scale projects when the unit cost is more reasonable. It is a valuable technique for dealing with saturated permeable soils that cannot readily be dewatered. The safe working limit is about 3·5 bar below the water table.

The hydrostatic pressure to be balanced increases with depth, and therefore varies over the height of the tunnel face. In some circumstances a single pressure that will cope with all variables of strata within the face, groundwater pressure and overburden cover may be impractical. Selection of the balancing pressure is therefore a compromise. Local injection or ground freezing in conjunction with, or even within, on-going compressed air works may be necessary. See for example, *Helsinki, Edinburgh* (Craigentinney) and *Three Valleys* in Chapter 7.

Developments include limitation of the pressurized volume to the face portion of a full-bore tunnelling machine, thus eliminating the need for an air-lock, and significantly reducing the time each operative spends in a pressurized atmosphere. Alternatively or additionally, bentonite slurry tunnelling machines can deal with many conditions. See Megaw and Bartlett (1981), Bartlett *et al.* (1973).

Artificial ground freezing (AGF)

Role of ground freezing
The impervious nature of ice, and the significant strength of frozen soil—water mixtures, are the principal properties that can be exploited for civil/mining engineering purposes. That the process is transient, i.e. thawing ensues once the source of chilling is removed, is a bonus, since it allows conditions to return to 'normal' without interference to or contamination of aquifers.

Thus creation of a vertical cylindrical ice-wall to act as a temporary cofferdam through wet, low-strength soils or weathered rocks, enabling dry shaft sinking to proceed, is probably the most widely known use of ground freezing. Small variations in the placement pattern of freeze-tubes allow excavations of varying shape and orientation, including tunnels and drifts, to be protected by this means.

The properties of frozen ground have also been usefully applied to stabilizing slips, underpinning structures, sampling weak or non-cohesive soils, providing temporary access/egress 'roads' over soft ground, the control of the stress/deflection relationship below sensitive equipment, the preservation of permafrost otherwise subject to man-induced thawing, and exploitation of submarine ore-bodies. Examples of some of these applications are recorded in greater detail in Chapter 7, e.g. *Burgos, São Paulo, Vienna, Green River Kentucky* and *Belmont*.

The first reported use of ground freezing was in 1862 for a mineshaft in Swansea, South Wales. Brine, chilled by a Siebe–Gorman ether engine, was circulated through a series of freeze-tubes sunk into the ground; the same basic principle, patented by Poetsch in 1883, still applies today — '. . . which consists in driving freezing pipes through said strata (quicksand or other waterbearing soils), next freezing a portion of said strata by circulating a refrigerating medium through said pipes, and then proceeding with the excavating operations, through or within the frozen strata . . . '. From humble beginnings, the method reached a record penetration from the surface of 900 m in *Saskatchewan*, Canada in the 1950s, and a greatest depth of 975 m at *Boulby Mine* in Yorkshire in the 1970s where the ring mains were housed in a chamber 550 m below surface.

In mining the method has proved to be reliable, safe and cost-effective, and therefore enjoys serious consideration throughout the feasibility and design stages of a project, often leading to its selection and specification by the designer at an early stage. The civil engineer, however, tends to regard the method as 'too expensive', overlook its well-known quality attributes, and treat it as a last resort — often to his embarrassment and the client's eventual cost. A reason may be that the onus for temporary works is vested in the Contractor under many forms of Conditions of Contract; the contractors have to tender for the Works against their competitors, and many will inevitably include in their bid the method offering the lowest *apparent* cost. See Chapter 9.

Principle of the ground freezing method
The primary objective is to remove heat from the ground until the temperature is below the freezing point of the groundwater system. This is achieved, for example, by providing a refrigeration plant to receive and (re)chill a heat transfer medium that is circulated through the target strata, and to dissipate the heat recovered.

The method is not limited by problems of scale. It can readily cope with very small situations, and has been used for excavations up to 45 m diameter and to depths of over 900 m. It is also versatile in that it can accommodate the full range of soil and rock types, the only limitations being (a) that each stratum must have an adequate moisture content and (b) that water flow through or beside the (intended) ice-body be nominal. Further discussion of these limitations, and other side-effects, will be found in Chapter 6.

Figure 1.3 offers a useful guide to the suitability of geotechnical methods according to soil types and permeab and summarizes their pros and cons. It will be noted th method unrestricted by soil grain size is ground freezing, which is particularly valuable when mixed strata have to b More precise criteria of ground acceptance of various typ are given by Littlejohn (1993).

The freeze-tube configuration will depend on the size and shape of the intended excavation. The simplest will be a circle of freeze-tubes around a cylindrical excavation being taken down vertically through unstable ground (Figs 1.4 and 1.7). As shafts and tunnels usually have circular cross-sections this is both convenient and economical. Application to other shapes is subject only to the ingenuity and ability of the drilling contractor to install freeze-tubes to correspond with the shape of the desired construction. There are many examples of simple and complicated geometries, some of which are illustrated in Figs 1.7–1.9.

The stages in ice-wall growth for a cylinder are illustrated in Fig. 1.4. The first stage (i) is the development of individual columns around each freeze-tube; after a time they merge (ii) to form a continuous hollow cylinder of nominal effective thickness. During the next stage (iii) the cusps of adjoining columns tend to smooth as the minimum design thickness is reached. A further short time may be needed for the inward ice-wall growth to reach the excavation line; as any unfrozen soil not removed as part of the excavation process will

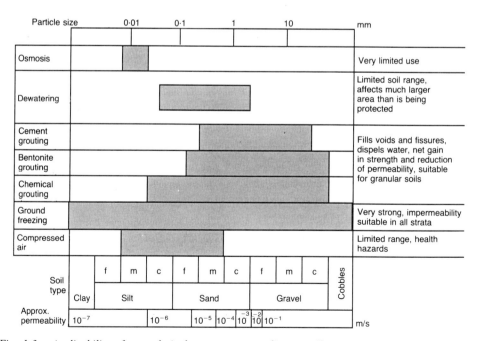

Fig. 1.3. *Applicability of geotechnical processes according to soil type*

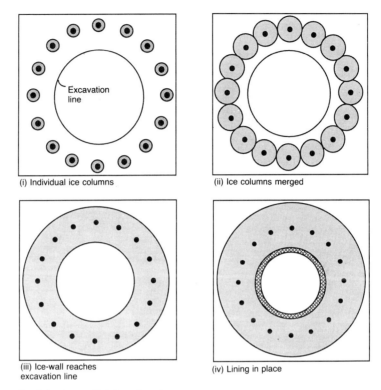

Fig. 1.4. *Stages in growth of the ice-wall (plan views)*

'slough' away from the frozen front, effectively creating an excessive overbreak which has to be filled later, most programmes will allow for this (short) time. The foregoing stages combined are termed the Primary (or Active) Freeze Period (PFP). Continued refrigeration during excavation and lining — the Secondary (or Passive) Freeze Period (SFP) — will lead to a solidly frozen core as the limited quantity of heat is removed, and to gradual further outward growth of the ice-wall, unless a significant reduction in refrigeration is applied. Security is achieved when the lining is in place (iv); natural or accelerated thawing can then commence. See Chapters 4 and 6.

The stages in the life of the ice-wall are typified by the successive temperature profiles illustrated in Fig. 1.5. Such profiles form part of the ice-wall monitoring process. See Chapter 6.

A standard freeze-tube is shown in Fig. 1.6(i). Flow of refrigerant will normally be out by the inner pipe and return by the annulus. This will generate a slightly 'pear' shaped column of frozen ground, which will be the approximate shape of the ice-wall in section. In profile the ice-wall may exhibit vertical cusps near the inlet which, if the groundwater is close to the ground surface, can lead to difficulty at the commencement of shaft excavation. This problem can be overcome either by reversing the flow direction for a period or by installing a small number of annular freeze-tubes laid in a shallow

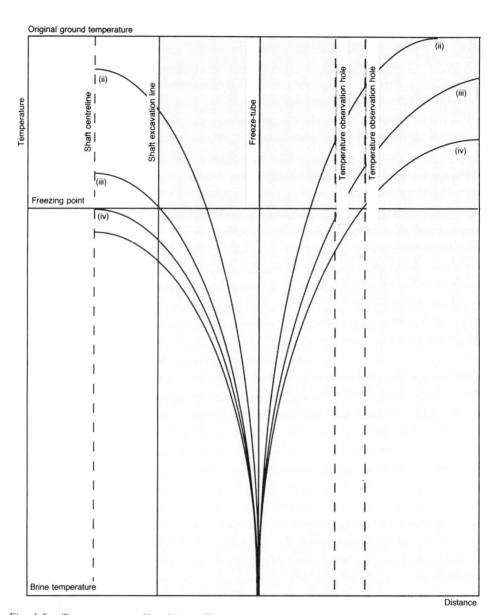

Fig. 1.5. *Temperature profile of ice-wall*

trench. The rate of growth varies between strata, as discussed elsewhere, leading to further modifications to the ice-wall profile as illustrated in Fig. 1.6(iii).

Freeze-tube configurations — shafts

A shaft or pit being taken sufficiently deep for the freeze-tubes to reach an impermeable cut-off stratum will require only a circle of freeze-tubes enclosing its plan area, as in Fig. 1.7(i). If no cut-off occurs within reasonable depth, central freeze-tubes will be needed

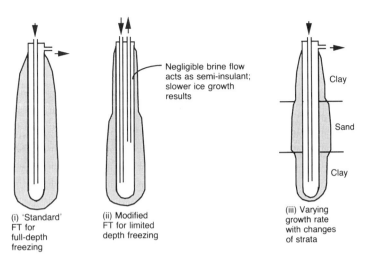

Fig. 1.6. *Freeze-tube assemblies and ice-wall profiles*

in addition to the basic circle in order to create a frozen seal across the base, as in Fig. 1.7(ii); unless steps are taken to minimize the extraction of heat from the shallower horizons by the central freeze-tubes, e.g. by insulation, the core of the excavation will also be solidly frozen and therefore difficult to remove. At some sites the insulation effect has been achieved by providing two inner pipes, the extra one recovering brine at the lower limit of the section not needing to be frozen rather than at the surface [see Fig. 1.6(ii)]. When the stratum to be treated occurs at depth, beneath sound strata that do not need treatment, economy of refrigeration can be achieved by drilling the freeze-tubes from a sub-surface drilling chamber, or by pre-drilling from the surface but equipping with freeze-tubes only over the desired depth range as illustrated in Fig. 1.7(iii) and (iv). It will be noted that method (iii) incurs a reduction in the diameter of the shaft on entering the frozen zone. Examples of (iv) at *Boulby* and *Asfordby* are described in Chapter 7.

Freeze-tube configurations — tunnels
Shallow tunnels may be dealt from the surface when space permits, but structures, buried services or natural features often preclude placement of freeze-tubes over all or part of the tunnel length. Three possible solutions using vertical or angled freeze-tubes, depending on the location of a cut-off and the presence of structures, are illustrated in Figs 1.8(i), (ii) and (iii). A scheme may combine elements of any of the typical freeze-tube patterns illustrated. Care must be taken, when adopting pattern 1.8(iii), to avoid an excavation sequence which results in flotation of the ice-trough through buoyancy.

For sections where surface drilling is impractical or the overburden very deep, horizontal freeze-tubes installed parallel with the tunnel may be chosen [see Fig. 1.9(i)]. Horizontal drilling is more expensive

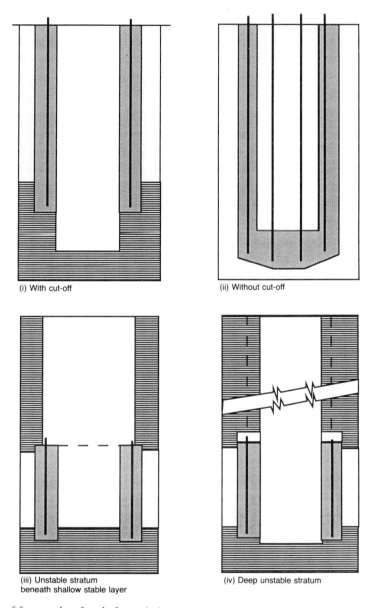

Fig. 1.7. *Configuration of freeze-tubes for shaft or pit (cross-sections)*

(i) With cut-off
(ii) Without cut-off
(iii) Unstable stratum beneath shallow stable layer
(iv) Deep unstable stratum

than surface drilling, and the necessary accuracy can only be relied on for penetrations of 30–40 m unless steerable drilling heads are employed. For tunnel lengths exceeding the accuracy limit, a solution practised on several sites is to install the freeze-tubes in a slightly fan-shaped array to permit the excavation of an enlargement, as the tunnel approaches the end of each frozen section; these can form drilling chambers from which further freeze-tube arrays can be installed [see Fig. 1.9(ii)].

GEOTECHNICAL PROCESSES

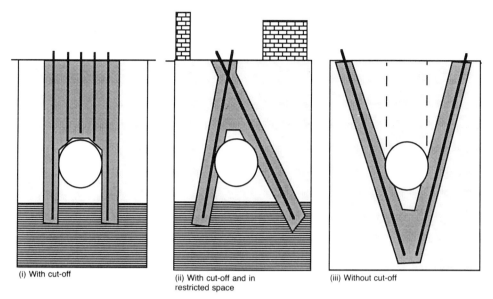

(i) With cut-off   (ii) With cut-off and in restricted space   (iii) Without cut-off

Fig. 1.8. Shallow tunnels frozen from the ground surface (cross-sections)

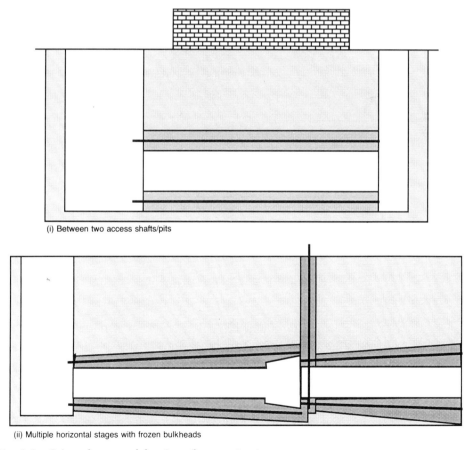

(i) Between two access shafts/pits

(ii) Multiple horizontal stages with frozen bulkheads

Fig. 1.9. Sub-surface tunnel freezings (long sections)

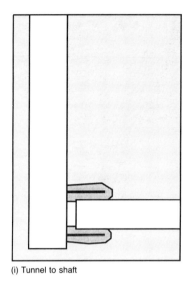 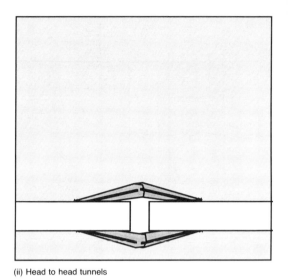

(i) Tunnel to shaft   (ii) Head to head tunnels

Fig. 1.10. Example freeze-tube configurations for junctioning

Freeze-tube configurations — junctioning
Ground freezing has proved to be a very convenient and effective method for facilitating the connection between a tunnel and a shaft, or between two tunnels, each of which have reached the junction using more conventional sinking or tunnelling methods. Generally the junction will be sub-surface, and probably sub-marine, requiring installation of freeze-tubes from within the converging constructions. Example configurations for tunnel to shaft, and head to head tunnels, are illustrated in Fig. 1.10.

Precautions against water/soil pressure/influx through the freeze-tube drilling ports (see *stuffing box arrangement* in Chapter 6), and careful monitoring, are the main prerequisites for successful junctioning. Numazawa *et al.* (1988) have described a project utilizing frozen head to head junctioning.

Freeze-tube configurations — underpinning
Two examples of freeze-tube patterns for underpinning applications are shown in Fig. 1.11. Figure 1.11(i) shows how temporary support for a structure, while excavations are made for the foundations of another building alongside, can be effected by a combination of vertical and angled freeze-tubes from alongside or within the neighbouring building. Any part of the ice-wall exposed to the atmosphere will need to be waterproofed and insulated against degradation by rain and heat. Fibreglass blankets between sheets of polythene are simple and effective.

Part (ii) shows how a chain of overlapping frozen pits is a proven means of providing both support and access while existing structural foundations are deepened (e.g. to create a basement car park). Each pit is treated as an entity, although the overlaps permit some of the freeze-tubes to serve two pits. Thus a programme of freezing,

GEOTECHNICAL PROCESSES

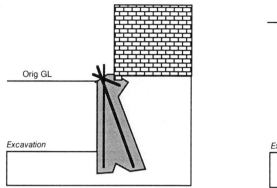
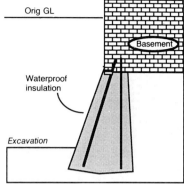

(i) Combined underpinning and retaining wall support (cross-sections)

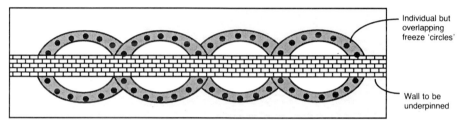

(ii) Chain of elliptical pits (plan view)

Fig. 1.11. Underpinning

excavation and construction sequences can be planned and executed such that no part of the structure being treated suffers concurrent loss of support at neighbouring panels. Examples of the use of most of the various configurations will be found in Chapter 7.

Other applications

Gap freezing It is sometimes necessary to undertake works with minimum interference to the existing groundwater flow regime. This technique was employed at *Duisberg*, as described in Chapter 7.

Slides An unstable slope can be protected by the temporary provision of an arch shaped frozen ground retaining wall. The design will follow the usual soil mechanics principles, but with the advantage that the only disturbance of the slope during creation of the retaining structure is that associated with the placement of freeze-tubes in boreholes. Such an exercise, at the *Grand Coulee Dam* (USA), was described by Mussche and Waddington (1946).

Preservation of permafrost In areas subject to marginal-permafrost conditions, it is essential that the frozen state of the foundation stratum is preserved below new buildings. Such conditions occur in North Asia, and the Russian literature records the adaptation of the heating and ventilation services of the building to contribute permanent supplementary refrigeration to the ground on which it stands.

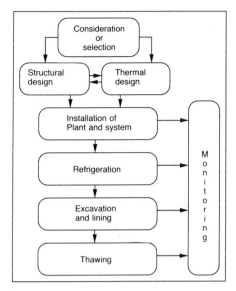

Fig. 1.12. Sequence of AGF events [after Jones (1982)]

Access roadways Transporting heavy plant across very soft alluvial valleys, and recovery of aircraft which have left the runway, are possible applications for temporarily frozen roadways, e.g. *Green River, Kentucky* in Chapter 7.

Procedure
Following recognition of the need for a geotechnical process to deal with ground conditions, and consideration or selection of ground freezing as an appropriate solution, a design will be prepared and costed. Selection may be supported by a risk assessment. See Chapter 9.

The structural and thermal elements of the design are fundamental and interrelated since, as will be seen in later chapters, the strength of frozen ground increases the more it is cooled below the freezing point temperature, and the ice-wall itself is subject to a characteristic temperature profile (Fig. 1.6). When these factors have been considered an ice-wall thickness and freeze-tube deposition are determined, permitting selection of plant capacity and freezing time.

The procedure is carried out in four stages, each of which is subject to detailed monitoring

- installation of the refrigeration plant and coolant distribution system
- refrigeration to construct the ice-wall
- refrigeration to maintain, and possibly increase, the thickness of the ice-wall while the excavation is made and the permanent lining installed
- allowing or controlling the thaw and the effect thereof.

These are shown in Fig. 1.12 and discussed in following chapters.

References

Anon 1937. Freezing arch across toe of east forebay slide, Grand Coulee dam. *Reclamation Era*, Jan, 12.

Bartlett J. V. et al. 1973. The bentonite tunnelling machine. *Proc. ICE*, **54**, Nov., 605−624.

Bell A. L. 1993 (ed.). *Grouting in the ground*. Thomas Telford.

Cambefort H. 1977. The principles and applications of grouting. *QJ Engng Geol.*, 10−2, 57−95.

Harris J. S. 1993a. Ground freezing. In Thorburn S. and Littlejohn G. S. (eds), *Underpinning and retention*, Blackie, 220−241.

Harris J. S. and Pollard C. A. 1985. Some aspects of groundwater control by the ground freezing and grouting methods. *Proc. NS on Groundwater in Engineering Geology*, Sheffield, Sept, 499−519.

Jones R. H. 1982. Closing remarks. *Proc. 3rd ISGF*, Hanover NH, **2***, 169−171.

Littlejohn G. S. 1993. Chemical grouting. In Thorburn, S. and Littlejohn, G.S. (eds), *Underpinning and retention*, Blackie, 242−273.

Megaw T. M. and Bartlett J. V. 1981. Compressed air working. In *Tunnels: planning, design and construction*, **1**, Ch 5, 125−156.

Mussche H. E. and Waddington J. C. 1946. Applications of the freezing process to civil engineering works. *Proc. ICE*, 14 May, 3−34.

Numazawa K. et al. 1988. Application of the freezing method to the undersea connection of a large diameter tunnel shield. *Proc. 5th ISGF*, Nottingham, **1**, 383−388.

Raffle J. F. and Greenwood, D. A. 1961. The relationship between rheological characteristics of grouts and their capacity to permeate soils. *Proc. 5th IC SM & FE*, **2**, 789−793.

Site Investigation Steering Group 1993. *Site investigation in Construction* (in 4 parts). Thomas Telford.

Somerville S. H. 1986. *Control of groundwater for temporary works. CIRIA Rpt 113*.

2. Description, sampling and testing frozen soils

For all works below ground level it is necessary to establish, by investigation, the nature of the strata and the groundwater regime; potential difficulties indicating the need for special techniques will be identified within this exercise. The investigative methods — boring, sampling and testing — in general use are appropriate.

Description of frozen soils

Standards for describing soil for engineering purposes apply in many countries and are generally alike or compatible. BS5930 is the principal document in the UK, ASTM D2487 and D2488 in the USA, and NRCC TM79 in Canada. They all apply to soil at ambient (unfrozen) temperatures, and do not cater for sub-freezing temperatures.

A visual–manual procedure for describing frozen soils was proposed in 1978 (principally for the permafrost establishment), and republished in 1982 as ASTM D4083; it is complementary to, and intended to be used in conjunction with, the aforementioned American standards. Definitions, a list of field equipment, and descriptions of the possible phases which can occur according to the degree of saturation and the freeze–thaw regime, are included.

Note: Guidelines for classification and laboratory testing of artificially frozen ground were presented by an ISGF Working Group in 1983 [Sayles *et al.* (1987)] and 1991.

Terms appropriate to AGF and in common use, which embody the definitions (modified if necessary) from ASTM D4083 (but not those relating specifically to permafrost), include the following.

Excess ice: ice in excess of that fraction that would be retained as water in the soil upon thawing,

Frost heave: the rise in level/increase in volume/increase in stress due to the formation of ice lenses within a soil stratum.

Frost susceptible: soils prone to frost heave are said to be frost susceptible.

Frozen ground or zone: a range of depth or thickness within which the soil is frozen due to the action of one or more freeze-tubes.

Ground ice: a body of more or less clear ice within frozen ground.

Ice lenses and *Ice segregation*: the growth of ice as distinct lenses, veins, and masses in soils, occurring essentially parallel to each other, generally normal to the direction of heat transfer, and often in repeated layers.

Ice-wall (or *Frost-wall* in USA): the monolithic body of frozen ground, created as a result of applying refrigeration through

DESCRIPTION, SAMPLING AND TESTING

a series of freeze-tubes, which provides cofferdam and/or structural protection to an excavation in soil or rock.

Permafrost: Ground in arctic or sub-arctic regions which becomes frozen permanently or cyclically on account of its geographical, low ambient temperature location.

Poorly-bonded: soil particles weakly held together by the ice so that the frozen soil has low resistance to chipping and breaking; poorly-bonded frozen soil often remains relatively permeable.

Thaw-stable: upon thawing, does not show loss of strength or produce significant settlement as a direct result of the melting of the ice in the soil, in comparison to normal long-term unfrozen values (converse *Thaw-unstable*).

Well-bonded: soil particles cemented together by the ice into a hard solid mass which has high resistance to chipping and breaking.

Sampling

The stages of recovery, transportation and storage of samples, and the preparation of test specimens, are discussed under this heading. As design precedes AGF, the usual tests will be made on samples recovered at normal temperatures and subsequently frozen in the laboratory, and standard sampling and descriptive procedures will apply.

Field frozen samples

There may be an occasional need to recover samples from the ice-wall itself, from a dedicated test site, from ground incidentally frozen by some other refrigerative operation, e.g. under cold-stores or ice-rinks, or from permafrost. When access can be gained, carefully cut block samples will most closely retain the important undisturbed characteristics of the frozen soil; the alternative is to recover continuous core samples with air-flush diamond/tungsten-bit rotary drilling techniques. As the cut surface suffers heat input, a core larger than the eventual test specimen size should be recovered to allow for subsequent trimming in the laboratory coldroom. Chilling and drying of the compressed air (flush) is particularly advantageous.

Protection of the frozen samples must (a) maintain their pre-existing temperature, as closely and constantly as possible, (b) preserve their moisture content. Temperature gradients lead to moisture migration within the sample, while changes in actual moisture content, either by loss or by sublimation, can significantly change the structural and strength properties. Each sample should be wrapped in insulating material before being placed in a stout polythene bag, from which all air is evacuated by vacuum pump before sealing (a very necessary procedure), prior to storage and transfer in a thermostatically controlled freezer. The addition of a small amount of snow or crushed ice in the bag before sealing is helpful in maintaining humidity.

Double walled wooden crates, the inner box insulated on the inside and the space between the two boxes packed with dry ice (solid CO_2), have proved successful and cheaper than refrigerated transport for the relatively short time period required for shipping from site to laboratory. Any space not occupied by samples should be filled with crushed newspapers or bagged crushed ice to provide further insulation and minimize the possibility of air movement.

Laboratory frozen samples

Uniaxial freezing in a controlled environment, simulating the in situ freezing regime, is considered to be essential if realistic values of the various behavioural properties of frozen soils are to be measured. This feature was recognized during the study of road foundations which had been subject to atmospheric freezing, and had failed by frost heave. A standard test apparatus and method was devised to measure the susceptibility of a soil to frost heave; it is the fundamental technique of uniaxial freezing between constant negative and positive temperatures at either end of the constrained specimen in the presence of a water supply, which is so important in preserving the integrity of the specimen during the freezing stage. The Frost Heave test is described later in this chapter.

Preparation of undisturbed specimens

All preparatory work should be carried out in a refrigerated workshop within the laboratory. The workshop will be equipped with turning, cutting, coring and filing facilities, i.e. lathe(s), mechanical and hand, diamond or tungsten saws, drill with diamond core-bits (set or impregnated as for concrete coring) and milling bits, and coarse wood rasps or files. All items will have been introduced to the coldroom at least 24 hours ahead of work on specimens. The method of preparation varies according to the soil type, the desired specimen shape, and the tolerances appropriate to the particular test.

All specimens should be handled only with insulated gloves. Machining is a difficult and painstaking operation. Hand saws and hot wires can only be used for rough cutting frozen clays; chain saws are suitable for cutting large blocks of frozen peat and silt; band saws are satisfactory for fine-grained through coarse sands; diamond saws cut true and clean for most soils including pebbly soils, but tend to clog in clay so need frequent cleaning.

Cylindrical specimens are best obtained by rough cutting to 10% oversize, followed by trimming in the lathe with mill or rasp, with wooden blocks between the frozen specimens and all metal faces, e.g. the lathe stocks, to minimize heat effects and equalize stresses.

Preparation of remoulded specimens

The object is to create specimens of specified density and degree of saturation at ambient temperature, and to freeze them in a manner which corresponds with the in situ regime. For saturated specimens, it has been shown [see Baker (1976a), Baker and Konrad (1985), Ladanyi (1983)] that compaction by rodding/drop hammer, and saturation under a vacuum, yield remoulded sand specimens with a more uniform degree of compaction and saturation than specimens prepared by vibration and saturation under a head of water. The preparation and freezing times should be kept short to reduce sublimation effects, and as similar as possible to ensure repeatability.

Two methods of sample preparation appear to be used in practice: (a) the mixing of oven-dried soil with de-aired, distilled water to the required moisture content, leaving in a sealed jar for 24 hours to equilibrate, then rod-compaction in 15 mm layers within a mould to produce a 38 mm diameter specimen 165 mm long, (b) compaction of air-dried soil in 25 mm layers, each scarified by 5 mm before starting the following layer, then saturating by connecting the base via porous stone/end cap to a water reservoir, and the top to a (low) vacuum; the differential pressure causes slow and steady water flow until the specimen is saturated — the water reaches the top of the specimen; it is then usual to allow reverse flow under 0·5 m head for one hour.

The specimen can then be frozen in a similar manner to that used for the frost heave test, i.e. from the top downwards while the base of the specimen is allowed free access to take up or dispel water. When the specified saturation level is below 90%, e.g. when a granular stratum above the water table is the subject, access to free water is inappropriate. The rate of freezing should be rapid: freezing too slowly can produce a reticulate ice structure, while freezing too quickly can cause air bubble entrapment. Experience suggests that sandy soils react best to a temperature of $-8°C$ to $-12°C$, while clay soils need $-25°C$ to $-30°C$. Following unidirectional freezing the specimen is trimmed in a cold room, to produce the 76 mm long test specimen.

Testing

Laboratory testing

Strength and creep behaviour

In principle, the strength and creep properties of a frozen soil, summarised by Jones (1989), are easily determined. A specimen is obtained either by sampling frozen ground or by freezing an undisturbed or remoulded sample. It is then inserted into a test cell which is either self-refrigerated or operated in a cold room.

In practice, the accuracy of the result obtained at a given temperature is influenced by

- the method of specimen preparation
- the salinity of the pore water
- the size and shape of the specimen
- the system stiffness
- the end conditions of the specimen
- the deformation rate, etc.

Classification and laboratory testing of artificially frozen ground have been discussed by Sayles *et al.* (1987) [on behalf of the International Organising Committee (IOC) for Ground Freezing Symposia], following an earlier initiative by Jessberger *et al.* (1980), who recommended that

- reference testing be undertaken on specimens with a height : diameter ratio of not less than 2
- deformation should be at rates of 0·1% and 1%/min
- end caps should be lubricated where possible
- the loading system should have a stiffness at least five times that of the specimen
- (common) test temperatures should be -2, -5 and $-10°C$.

The ISGF IOC Working Party on testing of frozen soils favours procedures that freeze laboratory specimens from all surfaces concurrently, i.e. sides, bottoms and top. The authors have experienced problems with this technique, e.g. thermal cracking and, as advocated earlier, prefer rapid, one-dimensional freezing in an open system similar to that used for frost heave tests.

Test rigs may consist of any of the following.

- Conventional loading frames and cells placed in a cold room, or enclosed by a refrigerated chamber. A Hoek cell can be conveniently adapted for the latter case for instantaneous strength tests.
- Purpose-built self-contained loading frame with an integral self-refrigerated cell for creep test at low pressure as illustrated in Fig. 2.1. See Gardner *et al.* (1982).

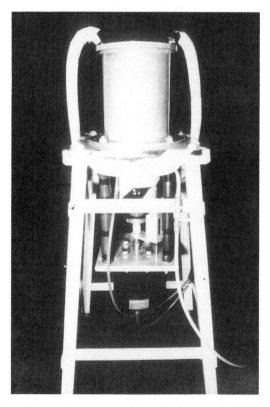

Fig. 2.1. Low-pressure test cell for refrigerated creep tests [after Gardner *et al.* (1982)]

- Conventional loading frames with self-refrigerated high pressure cells using either cooling coils for creep test at larger pressures.

Regulation of the temperature in a cold room, serviced by a heat exchanger and forced cold air, can be difficult — variations lead to evaporation of pore ice and its redisposition as frost. Such sublimation problems can be minimized by wrapping specimens during storage, and installing a protective sheath during testing. Temperature stabilization techniques include supplementary enclosed chambers within the cold room, and localized backheating. A progression, therefore, is to conduct the tests in small temperature controlled test chambers enclosing the apparatus, and independent of the storage and preparation cold room.

Short term strength of frozen soil is usually measured as total stress at a constant, rapid deformation rate, standardized for comparative purposes at $0\cdot1\%$/min and $1\cdot0\%$/min. This parameter, quoted as unconfined compressive strength K, or the Mohr–Coulomb values c and ϕ, is utilized in the equations given in Chapter 4. (See Chapter 3 for discussion of stress, strain and creep deformation.)

Long term strength of frozen soil is a measure of its time dependent creep behaviour, and the test is conducted for an extended period of time at a constant deformation rate of $1\cdot0\%$/min at loading stresses σ_1 of 30%, 40%, 50% and 70% of the uniaxial compressive strength determined by the short term strength test. It is usual to assume that the specimen deforms as a right cylinder at constant volume, and that the cross-sectional area remains constant over the full length. A programmable servo-mechanism is advantageous in maintaining constant axial stress, by adjusting the load as the specimen deforms.

Also for comparative purposes between the work from different establishments, standard temperatures of $-2°C$ and $-10°C$ are recommended; this does not, of course, preclude tests at other temperatures to suit particular projects.

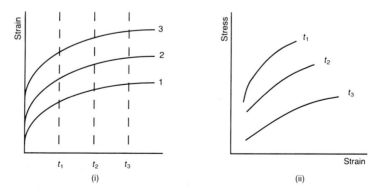

Fig. 2.2. Generalized creep relationships

Creep deformation is non-linear and time dependent, as shown in Fig. 2.2. Figure 2.2(i) shows the non-linear time/strain behaviour for various applied loads, while Fig. 2.2(ii) shows the isochronous stress/strain behaviour for various times. See also Chapter 3.

The creep equation (see Chapters 3 and 4, equation [4.11]) takes the basic form

$$\epsilon_c = A\sigma^B t^C$$

where ϵ_c is the creep strain, σ is the applied stress, t is time, A is a temperature dependent modulus with units stress^{-B} time^{-C}, B is a dimensionless exponent >1, and C is a dimensionless exponent <1.

The creep modulus A and exponents B and C can be obtained graphically from the test results as follows.

- Plot strain against time for each constant stress on log−log graph paper to produce straight parallel lines: their gradient strain/time is the power C.
- For any time on the above strain against time curves, plot strain against stress on log−log graph paper to produce a straight line: the gradient strain/stress gives the power B, and the intercept on the strain axis gives the parameter A.

Frost heave

The British frost heave test [Roe and Webster (1954), Pike *et al.* (1990)] was developed to check whether an aggregate from a particular source is likely to prove frost susceptible when used as a sub-base material in road construction. It is conducted in a self-refrigerated unit. Three sets of three cylindrical specimens 102 mm diameter and 152 mm high, mounted on porous discs and with their sides wrapped in waxed paper, are stood in a tray with the discs in contact with water. The intervening space between the specimens is filled with loose dry sand. Push rods, bearing on the top caps, enable the heave to be measured. The water temperature is maintained at $4 \pm 0.5°C$, and the air temperature above the specimens at $-17 \pm 1°C$.

Remoulded specimens are compacted to their maximum dry density at optimum moisture content to BS 5835:1980 or BS 1377:1990. After compaction and extrusion the specimens are left to imbibe water for five days before commencing refrigeration.

A test is accepted as valid if the mean heave is more than 2 mm and less than 18 mm providing the range of individual results is not more than 6 mm. Repeat tests are otherwise required. The criteria for frost susceptibility [DTp (1986)] were

(i) <9 mm non- and >15 mm frost susceptible
(ii) if 9.1 mm-14.99 mm, repeat tests elsewhere, when <12 mm classified non-frost susceptible.

The procedure was amended [DTp (1991a,b)] to include a reference specimen which generated a heave of 13·6 ± 4 mm, when the materials being tested are classified non-frost susceptible if the heave is 15 mm or less.

In situ testing Design and construction of the ice-wall generally precedes the opportunity to test the ground in its frozen state, unless a specific small scale test is commissioned. Even then, the majority of tests are undertaken in the laboratory; it has been the norm for development of in situ testing techniques to measure the properties of frozen ground (or ice) to rest with those concerned with permafrost. Much of this work has been regularly reported by Ladanyi (and co-authors) since 1972, and relates to use of the pressuremeter and the cone penetrometer. Seismic acoustic properties have been investigated by Kurfurst and Pullan (1985, 1988).

Pressuremeter
The Ménard pressuremeter and the CSM (Colorado School of Mines dilatometer) cell have been used in boreholes in frozen ground, and interpretive methods developed to estimate the short term stress—strain relationships, the creep parameters, and prediction of the long term strength. It is considered that conventional short term pressuremeter tests, a multistage test of about 15 min per stage, and a series of long term tests at successive pressure increments, are needed for a full analysis which confirms the linearity of creep lines for long time intervals. A method of evaluation is given by Ladanyi and Johnston (1973).

Later work suggests that the Ménard is most suited to stress controlled tests, and the CSM cell for strain controlled tests. In both cases the results are most appropriate when the direction of the test matches that of the anticipated loading.

Cone penetrometer
Ladanyi (1982) has demonstrated that the cone penetrometer test can yield data from which foundation designs (in permafrost) can be performed with reasonable confidence. See Ladanyi and Johnston (1973b) and ISGF WG (1991).

Seismic refraction
Studies of the acoustic wave velocities of frozen sediments, in situ (seismic) and on core samples in the laboratory, showed close agreement within the range 2200—3800 m/s. Compressional wave velocities for sea ice samples varied from 3300—3600 m/s. Variations were observed to be temperature dependent, i.e. the wave velocities are functions of, and more dependent on, the ice content and unfrozen pore water in the specimens than on the soil type. The principal acoustic elastic constants, Young's modulus, shear modulus and bulk modulus increase with increased ice content

and decrease temperature, although Poisson's ratio remains relatively constant. The contrast in these characteristics between frozen and unfrozen states is useful for verifying the thickness of the ice-wall during excavation. See Kurfurst and Pullan (1985, 1988).

References

ANDERSLAND O. B. and ANDERSON D. M. (eds) 1978. *Geotechnical engineering for cold regions*. McGraw-Hill, 566 pp.

BAKER T. H. W. 1976a. Preparation of artificially frozen sand specimens. *NRCC Div. of Building Research*, Paper 682, 15 pp.

BAKER T. H. W. 1976b. Transportation, preparation and storage of frozen soil samples for laboratory testing. *ASTM Publication 599*, 88–112.

BAKER T. H. W. and KONRAD J. M. K. 1985. Effect of sample preparation on the strength of artificially frozen sand. *Proc. 4th ISGF*, Sapporo, Japan, **2**, 171–176.

BRITISH STANDARDS INSTITUTION 1972. CP 2004. *Code of practice for foundations*. BSI.

BRITISH STANDARDS INSTITUTION 1980. BS 5835. *Recommendations for testing aggregates*. Part 1 Compactibility test for graded aggregates. BSI.

BRITISH STANDARDS INSTITUTION 1981. BS 5930. *Code of practice for site investigations*. BSI.

BRITISH STANDARDS INSTITUTION 1990a: BS 812. *Testing aggregates.* Part 124 Method for determination of frost heave. BSI.

BRITISH STANDARDS INSTITUTION 1990b. BS 1377. *Methods of testing soils for civil engineering purposes*. Part 4 Compaction — related tests. BSI.

DEPARTMENT OF TRANSPORT 1986. *Specification for highway works*, 6th edition, HMSO.

DEPARTMENT OF TRANSPORT 1991a. *Specification for highway works*, 7th edition, HMSO.

DEPARTMENT OF TRANSPORT 1991b. *Notes for guidance on the specification for highway works*, 7th edition, HMSO.

GARDNER A. R. et al. 1982. Strength and creep testing of frozen soils. *Proc. 3rd ISGF*, Hanover NH, USA, **1**, 53–60.

HAMPTON C. N. et al. 1988. The time dependent response of frozen soils subject to triaxial stress. *Proc. 5th ISGF*, Nottingham, **2**, 559–560.

ISGF Working Group 1. 1991. Testing methods for frozen soils. *Proc. 6th ISGF*, Beijing, **2**, 493–502.

JESSBERGER H. L. and EBEL W. 1980. Proposed method for reference tests on frozen soil. *Engng Geol.*, **18**–1, 31–34.

JONES R. H. 1989. Basic properties of freezing, frozen and thawing soils. *Proc. 5th NSGF*, BGFS, Nottingham, 14–29.

KURFURST P. J. and PULLAN S. 1985. Field and laboratory measurements of seismic and mechanical properties of frozen ground. *Proc. 4th ISGF*, Sapporo, **1**, 255–264.

KURFURST P. J. and PULLAN S. E. 1988. Acoustic properties of frozen near-shore sediments, southern Beaufort Sea. *Proc. 5th ISGF*, Nottingham, **1**, 197–204.

LADANYI B. 1980. Stress and strain rate controlled borehole dilatometer tests in permafrost. *Proc. Wksp on Permafrost Engineering*, Quebec, NRCC Tech memo ACGR 130, 57–69.

LADANYI B. 1982a. Determination of geotechnical parameters of frozen soils by means of the cone penetration test. *Proc. 2nd EurS on Penetration Testing*, Amsterdam, May, 671–678.

LADANYI B. 1993. Remoulded test specimen preparation (private communication).

LADANYI B. and JOHNSTON G. H. 1973. Evaluation of in situ creep properties of frozen soils with the pressuremeter. *Proc. 2nd IC on Permafrost*, Yakutsk, 310–318.

PIKE D. C. *et al*. 1990. The BS frost heave test: development of the standard and suggestion for further improvement. *Quarry Management*, Feb., 25–30.

ROE P. G. and WEBSTER D. C. 1984. Specification for the TRRL frost heave test. *Supp Rpt 829*, 39 pp.

SAYLES F. H. *et al*. 1987. Classification and laboratory testing of AGF. *J. Cold Regions Engng*, **1**–1, 22–48.

SHOCKLEY W. G. and THORBURN T. H. 1978. Suggested practice for description of frozen soils (visual-manual procedure). *Geotech. Testing J. (ASTM)*, **1**–4, 228–233.

SITE INVESTIGATION STEERING GROUP. 1993. *Site investigation in construction*. Thomas Telford, London. In four parts of 40, 32, 128 and 40 pp.

UFF J. F. and CLAYTON C. R. I. 1991. Role and responsibility in site investigation. *CIRIA Spl Publication 73*.

WELTMAN A. J. and HEAD J. M. 1983. Site investigation manual. *CIRIA Special Rpt 25*.

3. Properties of freezing, frozen and thawed soils

by R. H. Jones

The design of a successful 'ice-wall' for an AGF scheme requires thermal and structural calculations. In addition, precautions may be necessary to ensure that frost heaving, thaw weakening and thaw settlement do not have any deleterious effects on either the freeze pipe system, associated works or nearby structures. This chapter considers the thermal properties of soils, the mechanism and modelling of freezing and thawing, the residual effects of freeze—thaw and finally the mechanical properties of frozen ground. Both the freezing process and the mechanical properties of frozen soil at a given temperature are modified by dissolved salts in the pore water. This aspect is treated in the appropriate sections. Throughout, a basic knowledge of soil mechanics as it applies to unfrozen soils is assumed.

Formulae have been transposed into SI units. An attempt has been made to use a consistent notation (see p.xiii) but where more appropriate the original author's notation has been retained to assist readers should they wish to refer to source papers.

Thermal considerations

Thermal conductivity

The thermal conductivity K (W/mK) is the amount of heat flowing through a unit area in a unit time under a unit temperature gradient. Thus steady state heat flow is governed by the equation

$$q = -K \frac{dT}{dx} \qquad [3.1]$$

where q is the heat flow per unit area (W/m^2), T is the temperature (K), and x is the distance in the direction of the heat flow (m). The thermal conductivity of a soil depends on

- its volume fractions, that is, its porosity n (or dry density ρ_d and particle density ρ_s) and degree of saturation
- the thermal conductivities of the constituents (mineral particles, air and water/ice)
- the micro-geometry
- the temperature.

It can be measured in the laboratory, either under steady state conditions (e.g. guarded hot plate) or transient conditions (thermal probe). The various methods, reviewed by Farouki (1981), require the use of equipment and techniques which are not commonly available in geotechnical laboratories.

The alternative is to use empirical equations based on previous measurements. Kersten (1949) developed equations for thermal conductivity based on the results of measurements. These were presented in chart form by Harlan and Nixon (1978).

Johansen and Frivik (1980) tested Kersten's data against his equations and noted that the greatest discrepancies occurred when the quartz content was high. Quartz has a much higher thermal conductivity than other minerals commonly found in soils (see Table 3.1). They put forward an improved model as follows:

$$K = K_d + (K_s - K_d) \cdot K_e \quad [3.2]$$

where K_d is dry conductivity $= 0.034 n^{-2.1}$, n is porosity, K_s is the saturated conductivity $= K_m^n K_p^{(1-n)}$, K_m is the conductivity of moisture $= 0.57$ W/mK (water) or 2.3 W/mK (ice), K_p is the particle conductivity $= K_q q K_r^{(1-q)}$, q is the quartz content, K_q is the conductivity of quartz $= 7.7$ W/mK, K_r is the conductivity of the rest of the constituents ($= 2$ W/mK for granitic rocks, 3 W/mK for others), K_e is the Kersten number $= S_r$ (frozen) or $a \log S_r + 1$ (unfrozen), a is 0.68 (coarse) or 0.94 (fine), and S_r is the degree of saturation as a ratio. This model gives a considerable improvement in the accuracy of prediction.

It should be noted that the thermal conductivity of a coarse soil can be 1·5 or more times that of a clay soil when both are in the unfrozen state. The factor for frozen soils is somewhat lower.

At temperatures around $-20°C$, such as occur when soil is frozen with re-circulating brine, the thermal conductivities of the particles do not change significantly with temperature. However, around the freezing point, the thermal conductivity of the soil increases as the ice content increases. Although there is a sharp change in ice content just below 0°C, some water will remain unfrozen until lower temperatures are achieved (see section on frost heave). The thermal conductivity of partially frozen soils, K, may be estimated by

$$K = K_F + (K_U - K_F) V \quad [3.3]$$

where K_F is the conductivity at $-4°C$, K_U is the conductivity at $+4°C$, V is w_u/w_t, w_u is the unfrozen water content, and w_t is the total water content.

For many soils the unfrozen water content can be expressed by an equation in the form:

$$w_u = \alpha' (\Delta T)^\beta \quad [3.4]$$

where ΔT is the temperature depression below zero (expressed as a positive number in °C) and α' and β are characteristics of the particular soil related to its specific surface area [Anderson and Tice (1972)]. However, dissolved salts depress the freezing point and thus increase the amount of water unfrozen at a given

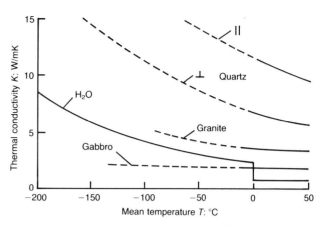

Fig. 3.1. Thermal conductivity against mean temperature [after Frivik (1980)]

temperature (see later and Chapter 6). At temperatures associated with LN freezing, the thermal conductivity of the mineral grains may be considerably higher than that around 0°C (Fig. 3.1).

Heat capacity

The volumetric heat capacity C (J/m³K) is the amount of heat required to raise the temperature of a unit volume of substance by 1 K (i.e. one degree kelvin). The specific heat capacity c (J/kgK) is the amount of heat required to raise the temperature of a unit mass of substance by 1 K. Thus

$$C = \rho c \text{ J/m}^3\text{K} \qquad [3.5]$$

where ρ = density (kg/m³).

If there is a change of phase, then latent heat must be considered. The latent heat of fusion of water is the quantity of heat released when a unit mass of water is converted into ice. Its value is 334 kJ/kg at temperatures close to 0°C. The heat released on freezing a unit volume (1 m³) of soil is thus 334 $w\rho_d$ kJ, where w is the water content (expressed as a ratio).

The apparent heat capacity can be calculated from a weighted average of the values for the components plus an additional term for the latent heat released. Thus

$$c_a = c_s x_s + c_i x_i + c_u x_u + c_a x_a + \frac{1}{\Delta T} \int_{T_1}^{T_2} L \frac{\partial w_u}{\partial T} \, dT \qquad [3.6]$$

where L = latent heat of fusion, T = temperature, w_u = unfrozen water content as a ratio of dry weight, x = weight fraction and the subscripts s, i, u, a refer to soil particles, ice, unfrozen water and air phases respectively.

The variation of specific heat capacities with temperature is shown in Fig. 3.2 while data on volumetric heat capacities are shown in Table 3.1.

Assuming all the water to be frozen, which is a reasonable assumption for coarse soils, the volumetric heat capacity, is given by

PROPERTIES OF SOILS

$$C_u = \rho_d/\rho_w(0 \cdot 18 + w/100)C_w \qquad [3.7a]$$

$$C_f = \rho_d/\rho_w(0 \cdot 18 + 0 \cdot 5w/100)C_w \qquad [3.7b]$$

where $w\%$ is the water content and the subscripts, u, f and w refer to unfrozen soil, frozen soil and water respectively.

Thermal diffusivity

Transient heat problems are defined by the equation

$$\frac{\partial T}{\partial t} = \alpha \frac{\partial^2 T}{\partial x^2} \qquad [3.8]$$

where α (m^2/s) is the thermal diffusivity $= K/C$.

High values of α imply a capability for rapid and considerable changes in temperature. Ice has a thermal diffusivity about eight times that of liquid water — see Table 3.1. In the practice of AGF, this accounts, in part, for the increased rate of temperature drop during 'pull-down' once the freezing point has been passed (subject to refrigerative capacity being available) and also for the gradient of the temperature profile of the ice-wall being steeper within the

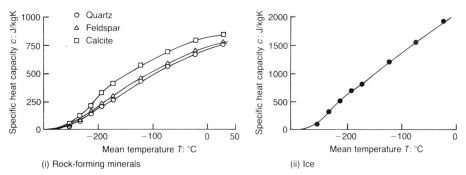

Fig. 3.3. Schematic of rhythmic ice lens formation [after Konrad and Morgenstern (1980)]

Table 3.1 Typical values of thermal parameters

Material	Thermal conductivity: W/mK	Volumetric heat capacity: MJ/m³K	Thermal diffusivity: m²/s × 10⁻⁷
Water	0·602	4·18	1·45
Ice	2·22	1·93	11·5
Air	0·024	0·001 26	190
Quartz	8·4	1·9	44
Many soils minerals	2·9		
Organic soil	0·25	2·5	1
Limestone	1·7–2·9	2·4–4·2	4–12
Dolomite	5·02	2·51	2
Sandstone	1·8–4·2	2·51	7–16·7
Shale	1·5	1·84	8·2
Quartzite	4·5–7·1		
Concrete	1·3–1·7		

frozen section than in the chilled section just beyond (see Chapter 4).

Frost heave

When frost penetrates into the ground, the surface may heave. If movement is restrained, for example by a building foundation, a significant heaving pressure may be developed. Our understanding of the physical processes underlying these phenomena commenced with the classical studies of Taber (1918, 1929, 1930) and Beskow (1935) [see also Black and Hardenberg (1991)]. They showed that frost heave was mainly due to the freezing of transported water, with the 9% increase in volume of the water as it freezes making only a minor contribution. The ice was segregated into a series of lenses (rhythmic ice banding) as shown in Fig. 3.3. Application of a surcharge pressure reduced the heave.

Suctions are set up when soils are frozen and the freezing point is depressed, essentially in accordance with the equation:

$$u_i - u_w = \frac{\Delta T L}{T_0 V_w} \qquad [3.9]$$

where u_i is ice pressure, u_w is pore water pressure, ΔT is the

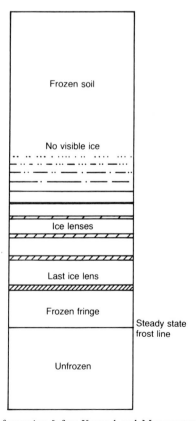

Fig. 3.3. Schematic of rhythmic ice lens formation [after Konrad and Morgenstern (1980)]

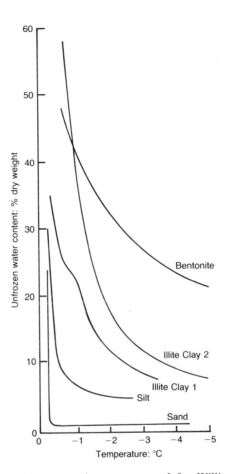

Fig. 3.4. *Typical curves of unfrozen water content against temperature* [after Williams (1988)]

freezing point depression, T_0 is the normal freezing point of water (K) and V_w is the specific volume of water. Not all the water will freeze at the same temperature and significant quantities of unfrozen water may be present (equation [3.4] and Fig. 3.4).

It should be noted that equation [3.9] is a particular case of the Clausius–Clapeyron equation

$$(V_w \Delta u_w - V_i \Delta u_i) T_0 = \Delta TL \qquad [3.10]$$

where the subscripts w and i refer to water and ice, Δu is the pressure change and T is the temperature. Equation [3.9] is for the situation where the ice pressure remains constant, $(\Delta u_i = 0)$ and all the pressure change takes place in the water. An alternative form in which all the pressure change is accommodated in the ice phase may be relevant to heaving pressure calculations and is obtained by setting $\Delta u_w = 0$.

Capillary model In the capillary model [Penner (1963), Everitt and Haynes (1965), Williams (1967)], the suction was related to the radius of curvature of ice trying to penetrate through a pore neck (Fig. 3.5) by the

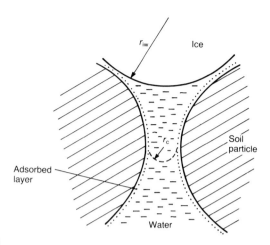

Fig. 3.5. *Idealized capillary pore*

Kelvin equation, giving

$$u_i - u_w = \frac{2\sigma_{iw}}{r_{iw}} \quad [3.11]$$

where σ_{iw} = interfacial energy ice/water ($\approx 0\cdot 0331$ Nm^{-1}) and r_{iw} = radius of curvature.

With low ice pressures, u_w will be negative and water will be sucked continuously to the freezing front where it will freeze and release its latent heat. If the rate of heat extraction is sufficiently high, the temperature will drop resulting in an increase in suction and a decrease in the radius of the ice–water interface; ice can then penetrate through the pore neck. This theory explains why clean gravels are non-frost susceptible (low suction) but that as soils become finer, they become more frost susceptible until the lack of permeability of the unfrozen soil restricts the frost heave. It also suggests that rapid freezing will give rise to less frost heave than slow freezing because there is less time for water to be transported.

The capillary theory enjoyed some success in predicting both the suction at the freezing front and also the frost heave in non-colloidal soils. However, evidence began to emerge that in restrained heave tests, the heaving pressures predicted by the theory were too low [Penner (1967), Hoekstra (1969b), Sutherland and Gaskin (1973)].

The frozen fringe and Miller's model

The observed heaving pressures corresponded to a lower temperature (larger ΔT in equation [3.9]), implying that ice lenses formed some way behind the freezing front. The concept of a frozen fringe in front of the lowest ice lens [Fig. 3.6(ii)] was introduced into a model by Miller (1972). The existence of the frozen fringe was subsequently confirmed by Loch and Miller (1975), Loch and Kay (1978) and Penner and Goodrich (1980).

It is now considered that the original capillary model is concerned

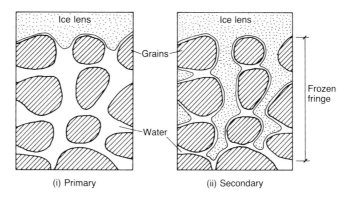

Fig. 3.6. *Primary and secondary heaving*

with 'primary' heaving [Fig. 3.6(i)] which only applies to ice formed adjacent to a cooling surface prior to frost penetration in the soil or in special circumstances with terminal lenses when the fringe disappears [Piper et al. (1988)].

With a frozen fringe model, the problems are to determine the location of the base of the ice lens and to describe the flow through the frozen fringe in the region where ice and water co-exist. Miller (1978) modified an equation by Bishop (1955) for partitioning of pressures to give

$$u = \chi u_w + (1 - \chi) u_i \qquad [3.12]$$

where u is neutral stress, u_w and u_i are pore pressures in water and ice respectively, and $0 \leq \chi \leq 1$ is the stress partition factor. Knowing the total stress σ, the effective stress, σ' can be found from the well known Terzaghi effective stress equation

$$\sigma' = \sigma - u \qquad [3.13]$$

A new lens is assumed to form when σ' falls to zero.

The sequence of events is depicted in Fig. 3.7. In Fig 3.7(i) just before a new lens is initiated, the distribution of water pressure u_w is shown by the solid line. u_i, u and σ' are then controlled by equations [3.9], [3.11], [3.12] and [3.13]. At the level where $\sigma' = 0$, a new lens forms and u_i falls immediately to σ as required by vertical equilibrium; the other stresses adjust as shown in Fig. 3.7(ii). The new lens then grows [Fig. 3.7(iii)] until σ' falls to zero on some plane and the cycle is repeated. Note that at the initiation of the lens u_i is greater than the total stress σ.

The other feature of Miller's model is that moisture is transmitted through both the liquid and solid (ice) phases; this is referred to as series-parallel transport. Transport through the ice phase is accomplished by a process of microscopic regelation.

Mathematical modelling of heave

Figure 3.8 shows the components of heat flow in a section of soil subject to freezing from the top. It is assumed that the bottom of the section is completely unfrozen and that no further phase change

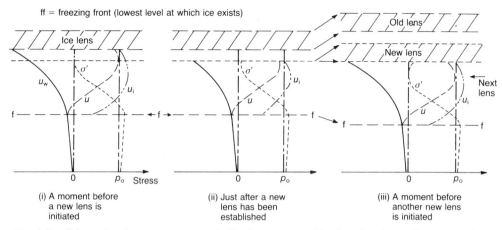

Fig. 3.7. Schematic of pore pressure and effective stress profiles in a heaving soil column [after Miller (1978)]

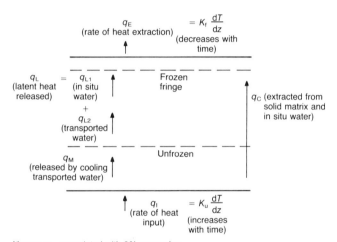

Fig. 3.8. Heat flow rates through a freezing front

occurs beyond the top of the section. Of the various components of heat flow defined in Fig. 3.8 q_M can be shown to be negligible, thus

$$q_E - q_I = q_C + q_{L1} + q_{L2} \qquad [3.14]$$

which can be written as:

$$\frac{\partial}{\partial z}\left(K\frac{\partial T}{\partial z}\right) = C_s\frac{\partial T}{\partial t} + \left(\rho_w L\frac{\partial \theta_i}{\partial T}\right)\frac{\partial T}{\partial t}$$

$$+ \rho_w L\left(k\frac{\partial h_w}{\partial z}\right) \qquad [3.15]$$

where z is depth, C_s is the volumetric heat capacity of the soil, t is time, ρ_w is the density of water, θ_i is volumetric ice content, and h_w is suction, expressed as metres of water $= u_w/\gamma_w$. Equation [3.15] is essentially the hydrodynamic model proposed by Harlan (1973). It is valid at every level, that is, for all values of z. Over the section as a whole:

$$\frac{\partial}{\partial z}\left(K\frac{\partial T}{\partial z}\right) = K_f\left(\frac{\partial T}{\partial z}\right)_f - K_u\left(\frac{\partial T}{\partial z}\right)_u \qquad [3.16]$$

where the subscripts f and u refer to frozen and unfrozen soils respectively.

The last term of equation [3.15] which incorporates Darcy's law, represents the coupling between the flows of heat and water. Additional information is required to solve the equation. Incorporation of Miller's lensing criterion is one way of providing this information.

The resulting model, referred to as a rigid ice model (because the ice in the frozen fringe is rigidly attached to the growing lens) considers the water flow through the frozen fringe where suction and therefore unfrozen water contents and permeability are varying rapidly with depth (Fig. 3.9). The original formulation was very demanding of computing power [O'Neill and Miller (1980, 1985)] and although simplified quasi-static solutions have been developed (Piper et al. 1988) they still demand input parameters which are difficult to measure or assess. Fowler and Noon (1993) have presented an asymptotic reduction of the O'Neill−Miller model which not only renders the model tractable but also permits its extension, for example to saline soils and to three-dimensional situations.

In particular, the permeability is extremely sensitive to variation in suction (or ice content). Use has been made of relations such as

$$k = k_o\left(1 - \frac{\theta_i}{\theta_o}\right)^m \qquad [3.17]$$

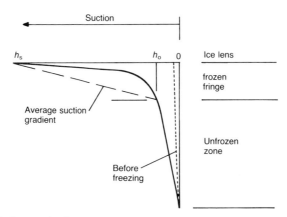

Fig. 3.9. Variation of suction below an ice lens
Note: h_s = suction at lensing front, h_0 = suction at freezing front

where k is the coefficient of permeability at a given ice content, k_0 is the coefficient of permeability of the unfrozen soil, θ_i is the volume fraction occupied by ice, θ_0 is the porosity, and m is a parameter (often taken in the range 7 to 9). Another uncertainty is the value of χ in equation [3.12]. Piper *et al.* (1988) used

$$\chi = \left(1 - \frac{\theta_i}{\theta}\right)^{1 \cdot 5} \qquad [3.18]$$

Ladanyi and Shen (1989) summarize other approaches to the variation of χ.

Adsorbed film models

Capillary models do not apply to colloidal soils in which all water is adsorbed. However, suctions are again set up on freezing, essentially in accordance with equation [3.9]. Transport to the lensing front is through adsorbed films (Fig. 3.10) but mathematical versions of this type of theory have yet to be developed.

Segregation potential

One way of overcoming these difficulties is to use the concept of Segregation Potential (SP) [Konrad and Morgenstern (1981)] such that

$$V = SP \times \text{temperature gradient} \qquad [3.19]$$

where V is water intake velocity (i.e. velocity in unfrozen zone).

$$SP = \frac{h_s - h_0}{T_s} \bar{k}_{ff} \qquad [3.20]$$

where T_s is the negative temperature at the base of the ice lens (°C) and $\bar{k}_{ff}$ is the average coefficient of permeability in the

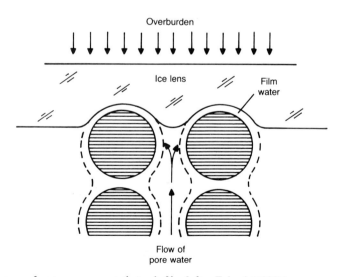

Fig. 3.10. *Schematic diagram of water movement through film* [after Takagi (1980)]

frozen fringe. Both h_s and h_0 are expressed as an equivalent height of water (see Fig. 3.9).

Note that the temperature gradient is essentially that in the frozen zone. Lenses are assumed to form when the temperature reaches T_s. The heave rate $\dot{h}$ is then $1 \cdot 09v$. The experimental determination of segregation potential requires fairly sophisticated apparatus and careful measurements [Konrad (1987), Jessberger et al. (1988)].

SP is dependent on the rate of cooling, the suction at the freezing front and the surcharge. The effects of surcharge can be expressed by

$$SP = SP_0 \exp(-a\sigma) \qquad [3.21]$$

where SP is the segregation potential under surcharge σ, SP_0 is the value obtained at zero pressure, and a is a constant for a given soil [Konrad and Morgenstern (1984)]. However, results obtained by Jessberger et al. (1988) with both a sandy gravel and a slightly plastic clay, showed a linear reduction on a log–log plot of SP against σ which does not accord with equation [3.21] thus highlighting the need for specific tests over the appropriate range.

The segregation potential method has been criticized for its semi-empirical nature and because it fails to forecast some of the features observed in laboratory freezing tests, such as the tendency for water to be expelled in the early part of the freezing period. Nevertheless it is a useful index test and it has also been used successfully to predict heave in connection both with chilled pipelines [Konrad and Morgenstern (1984), Nixon (1987)] and artificial ground freezing [Jessberger et al. (1988)]. To apply the method, it is necessary to determine grad T by a detailed thermal analysis.

A review of observed measurements of heave and settlement where AGF has been used as a construction expedient is given in Chapter 6.

Discrete ice lens theory

Nixon (1991) has re-examined critically the segregation potential theory and suggested a number of modifications to remedy its shortcomings. He proposed that the lenses form when the ice pressure exceeds the total stress by some value, σ_{sep}. The concept of a separation pressure was first advanced by Gilpin (1980) and may be similar to the 'internal' pressure measured by Williams and Wood (1985). This criterion is simpler to apply than Miller's $\sigma' = 0$ criterion discussed earlier. It has already been noted that u_i exceeds σ at the point of lens formation (Fig. 3.7).

The variation of permeability is determined by

$$k = k_0/(-T)^\alpha \qquad [3.22]$$

where k_0 is hydraulic conductivity at $-1°C$, and T is temperature (°C) below freezing, which is understood to be divided by a reference temperature of 1°C to maintain the correct dimension. Equation [3.22] appears to be less sensitive to uncertainty than equation [3.17].

Nixon (1991) proposes that the segregation potential should be redefined as

$$SP = V_{ff}/G_{ff} \qquad [3.23]$$

where V_{ff} is the average velocity in the frozen fringe and G_{ff} is the temperature gradient in the frozen fringe. Nixon also proposes a method by which k_0 and α in equation [3.22] can be inferred from the results of laboratory heave tests.

The discrete ice lens approach has been calibrated against published frost heave results and encouraging agreement obtained between prediction and observation. Certainly the alternative definition of SP avoids the problem caused by additional water arriving at the lensing front due to desiccation of the underlying soil [Baba (1993)].

Postscript on hydrodynamic models

The above models lack consideration of the mechanical properties of the frozen ground although Shen and Ladanyi (1988) have proposed a method for considering stress effects. Williams and Smith (1990) have summarized evidence that ice can grow at higher levels than the bottom of the lowest ice lens as the overlying ground becomes colder. Even so, although the models help in the understanding of the processes of frost heave, many uncertainties remain [Black and Hardenberg (1991)].

Thermo-mechanical theory

A completely different approach, based on continuum mechanics and macroscopic thermodynamics has been advanced by Frémond and Mikkola (1991). The model accounts for mechanical effects as well as thermal and hydraulic aspects but does not require a lensing criterion. The model has yet to be verified by comparisons between predictions and actual results.

Frost susceptibility of soils

Soils which are particularly prone to frost heaving are said to be frost susceptible. Generally, for significant frost heave, three conditions are necessary

(*a*) prolonged freezing temperatures
(*b*) a frost susceptible material
(*c*) a supply of water from the unfrozen region (open system).

Several index tests for frost susceptibility are in use in various parts of the world [Chamberlain *et al.* (1984), Jones (1987)]. Table 3.2 gives a summary of a number of methods of determining frost susceptibility [ISSMFE (1989)] but some care should be exercised in transferring frost susceptibility criteria developed for short term seasonal freezing of roads to AGF projects involving longer freeze periods and lower temperatures. The British Frost Heave Test [BSI (1989), DoT (1991)] although developed essentially for aggregates, can also give useful guidance on the frost susceptibility of soils.

Table 3.2 Frost susceptibility of various soil types [after ISSMFE (1989)]

Frost susceptibility class	Soil type (USCS)	Plasticity chart I_p^1	Plasticity chart w_L^1	Capillary rise[2]: m H_C	Liquidity index[3] I_L	Fines factor[4] R_f	Segregation potential[5]: mm²/hK SP_c	Frost heave rate[6]: mm/d
Negligible	GW, GP SW, SP	<1		<1	≤0	<2·5	<0·5	<0·5
Low	CH	≥7	>50	1·0–1·5	<0·25	2·5–5	0·5–1·5	0·5–2
Medium	CI	≥7	35–50	1·5–2·0	0·25–0·50	5–10	1·5–3·0	2–4
	OH[7] MH	≥7	>50					
Strong	CL	≥7	<35	>2·0	>0·50	>10	>3·0	>4
	ML	<4	≤50					
	OL[7]	≥7	35–50					

Note: Soils above the A-line with I_p between 4 and 7 are borderline cases requiring the use of several methods.
1 Merkblatt für die Vertnutung von Frostschaden in Strassen. Entwurf 03/87 (unpublished). 2 Beskow (1949). 3 SNiP II — 15–74.

4 Fines factor, $R_f = \dfrac{(\% \text{ fines}) \times (\% \text{ clay sizes in fine fraction})}{\text{liquid limit of fine fraction}}$

where fine fraction is below 74 μm [Rieke *et al.* (1983), Vinson *et al.* (1987)].
5 Konrad and Morgenstern (1981), Konrad (1987). 6 CRREL test — Chamberlain (1981). 7 Under A-line.

Table 3.3 Frost susceptibility criteria according to thaw CBR [after Chamberlain (1987)]

Frost susceptibility	Thaw CBR: %
Negligible	>20
Very low	20–15
Low	15–10
Medium	10–5
High	5–2
Very high	<2

However, since thaw weakening can be a problem (see later) it may be advisable to use a frost susceptibility test which incorporates a thaw CBR [Chamberlain (1987)] — see also Table 3.3.

Care is needed in assessing the effects of frost in soils consisting of alternate clay and silt layers, such as varved clays. An experimental study showed that lenses formed at the face of the clay layer when it encountered the freezing front but that heave in the silt was inhibited [Penner (1986)].

Effects of salinity on freezing

When the pore water contains significant quantities of solutes, usually sodium chloride (NaCl), the freezing point is depressed and the salts are concentrated in the unfrozen water. This further depresses the freezing point and increases the local concentration of solutes. The processes are limited by the solubility as shown

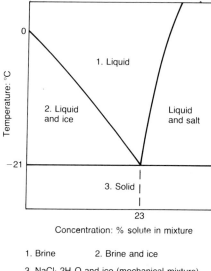

1. Brine 2. Brine and ice
3. NaCl· 2H₂O and ice (mechanical mixture)

Fig. 3.11. Salt-water system-phase composition at atmospheric pressure

in the salt-water system phase diagram at atmospheric pressure (Fig. 3.11) (see also Chapter 6).

The amount of unfrozen water in a soil can be determined by Nuclear Magnetic Resonance (NMR) [Tice *et al.* (1978, 1984)] or Time Domain Reflectometry (TDR) [Patterson and Smith (1981), (1985)]. Zhang Lixin (1991) using NMR techniques on three soils showed that there was a step change in the unfrozen water content of frozen saline soils at around $-21\,°C$ (Fig. 3.12). The step appears to correspond with the eutectic point of the brine solution (Fig. 3.11) (which varies somewhat with pressure/suction). On the warm side of the step, ice is being formed adjacent to brine. On freezing, crystals of $NaCl \cdot 2H_2O$ are formed. Only some ion-enriched film water remains unfrozen or uncombined. When the solubility limit is exceeded, precipitation occurs. Chen

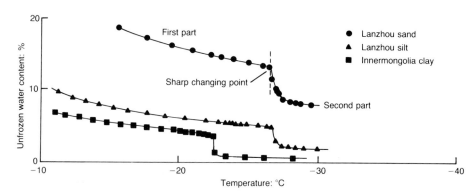

Fig. 3.12. Unfrozen water content against temperature of saline frozen soil [after Zhang Lixin (1991)]

et al. (1988) describes how 'salt heave' arises as the result of precipitation of sodium sulphate crystals.

Solutes are transported with the water towards the freezing front, but excluded by the ice on freezing. Changes in concentration gradients not only give rise to diffusion but also cause a gradient in the temperature depression on the warm side of the freezing front. The decrease in freezing point may prevent freezing adjacent to the freezing front. A little further away, although it is warmer, the freezing point depression may be significantly less, thus favouring ice nucleation. The freezing front jumps over these solute enriched bands entrapping them in a frozen matrix [Kay and Perfect (1988)]. This mechanism appears to be responsible for the ice-banding in saline soils observed by Chamberlain (1983). Kay and Groenevelt (1983) also produce theoretical and practical evidence to support this phenomenon which in their study occurred at micro-scale with no net effect at macro-scale.

The importance of the direction of drainage was shown by Baker and Osterkamp (1988) by laboratory experiments in which columns of uniform medium sand, saturated with 35 parts per thousand (ppt) of sodium chloride solution, were frozen both from the top down and the bottom up. For downward freezing, significant salt rejection occurred at a rate which increased with decreasing freezing rate, but none was observed with upward freezing. It appeared that gravity drainage accounted for the discrepancy. Many ground freezing applications, for example a frozen arch over a tunnel, involve freezing in an upward direction.

The effects of thawing

On thawing, water is released by frozen ground. If the ground has heaved, there will be excess water present which must drain away. The rate at which water is released is controlled by thermal considerations whilst the Terzaghi consolidation theory can be applied to the drainage process. Major studies in this area, reviewed by Morgenstern (1981) have established the basic mechanics involved.

Thaw settlement According to ISSMFE (1989) thaw settlement is the generally uneven downward movement of the ground surface due to thaw consolidation; it has three possible components, namely phase change, self weight of the soils and applied loading. Thaw consolidation tests were first performed in a standard oedometer with a heated top cap. The procedure for this type of test is reviewed by Nixon and Ladanyi (1978).

However, the modern 'permode' (= permafrost oedometer) [Morgenstern and Smith (1973)] is a specially designed oedometer in which both the top and bottom temperatures are closely controlled by thermoelectric devices. Radial heat flow is minimized by good insulation. One way drainage is provided through the top of the specimen. The rate of thaw, from the top down, is monitored

GROUND FREEZING IN PRACTICE

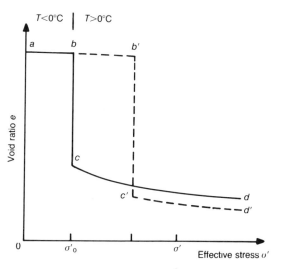

Fig. 3.13. *Typical thaw-settlement behaviour of frozen soils* [after Tsytovich et al. (1965)]

by thermocouples or thermistors located at two intermediate heights on the sides of the specimen. As soon as the thaw plane reaches the base of the specimen, the pore pressure is measured by a transducer. Settlement is monitored throughout the process.

Typical thaw settlement curves are shown schematically in Fig. 3.13. A small settlement of the frozen soil occurs under increasing stress between a and b, then a large settlement, (from b to c), when the thaw takes place followed by consolidation again under increasing stress of the completely thawed soil between c and d. Thawing at higher stress, indicated by the dashed lines may produce different post-thaw behaviour. Since most of the settlement occurs actually on thawing, its magnitude may be estimated if the ice content is known. Crude estimates may be possible from measuring the visible ice lenses in a core, but a more satisfactory solution is to measure the frozen bulk density [Watson et al. (1973)].

Theory of thaw consolidation

Imposition of a constant positive temperature to the upper surface of a semi-infinite frozen soil will cause the thaw plane to descend through the soil a distance X, in time t where

$$X = \alpha t^{1/2} \qquad [3.24]$$

where α is a constant which depends on the rate of heat extraction.

In fine grained soils, the drainage of excess water will be impeded by the low permeability, resulting in excess pore pressures, u, which will dissipate according to the Terzaghi consolidation theory. The solution to this problem (Fig. 3.14) [Morgenstern and Nixon (1971)] depends on the thaw consolidation ratio, R, where

$$R = \frac{\alpha}{2\sqrt{c_v}} \qquad [3.25]$$

and c_v is the coefficient of consolidation.

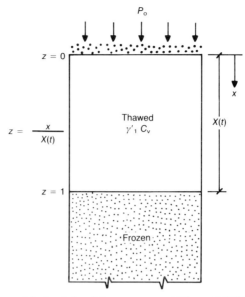

Fig. 3.14. *One-dimensional thaw consolidation* [after Morgenstern and Nixon (1971)]

For a soil consolidating under its own weight γ' it was found that

$$\frac{u}{\gamma' X} = \frac{1}{1 + 1/(2R^2)} \qquad [3.26]$$

This is shown in chart form in Fig. 3.15(i).

A solution for a weightless soil subject to a surcharge p_0 is given in Fig. 3.15(ii). In both cases, it is found that the dimensionless pore pressures at a given depth increase with increasing thaw consolidation ratio R but are independent of time.

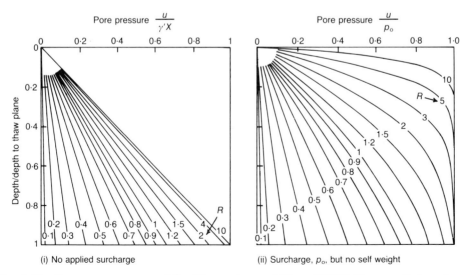

Fig. 3.15. *Excess pore pressures* [after Morgenstern and Nixon (1971)] *(i) no applied surcharge (ii) surcharge, p_0, but no self weight*

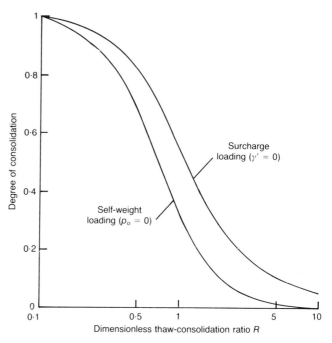

Fig. 3.16. Variation of degree of consolidation with thaw consolidation ratio [after Morgenstern and Nixon (1971)]

The relation between degree of consolidation and thaw consolidation ratio, determined by integrating the pore pressure distribution is shown in Fig. 3.16.

A series of tests on the thawing of undisturbed permafrost samples accorded well with the above theory [Nixon and Morgenstern (1974)]. The extension of the theory to non-linear behaviour and layered systems is discussed in Nixon and Ladanyi (1978). One difficulty in applying thaw consolidation theory is that parameters such as c_v are likely to change as freeze-thaw cycles modify the structure of the soil (see later).

Residual stress

The effective stress obtained by thawing under undrained conditions, σ'_0, termed the residual stress [Nixon and Morgenstern (1973)], is an important boundary condition in theories of thaw consolidation. Its meaning is illustrated by the schematic settlement curve for an undrained freeze−thaw test shown in Fig. 3.17. A specimen of unfrozen soil is normally consolidated under an effective stress of p_0 (point A). On freezing, the volume expansion involved in the phase change causes an increase in thickness (and *average* void ratio) at the same effective stress, B.

Thawing under undrained conditions reduces the *average* void ratio back to the initial value. However, there has been a change in stress conditions. The suctions set up during freezing cause an increase in effective stress in the soil and consolidation to point

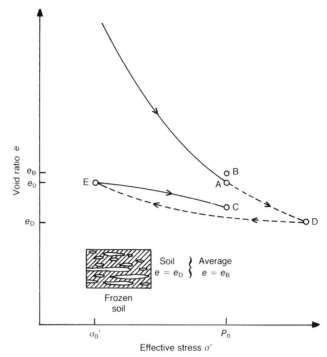

Fig. 3.17. Changes in void-ratio in a closed system freeze–thaw cycle [after Nixon and Morgenstern (1973)]

D. The overconsolidated elements in the soil absorb water from macropores or lenses and swell almost instantaneously. If the soil elements are capable of taking up all the water, point E will be reached, and the corresponding effective stress is the residual stress. If drainage is then permitted the soil will consolidate to point C. Note, however, that if the soil elements are incapable of absorbing all the water, the residual stress is zero and excess pore water remains in the soil. Residual stresses can be measured in special apparatus in which drainage and lateral yield can be prevented.

It should be noted that there is a lower limit of moisture content below which clays cannot be consolidated by freezing and thawing. Theoretically this is the shrinkage limit but Chamberlain (1980) considered that in practice, the plastic limit w_p is appropriate. Chamberlain (1980) found that thaw strain increases linearly with the w/w_p ratio. Leroueil et al. (1991) found, in tests on a sensitive clay, which had not been subject to previous freezing in its geological history, that there was a linear relation between thaw consolidation (expressed as a change of liquidity index) and increase of initial liquidity index. Yamomoto et al. (1988) found that the overconsolidation ratio, OCR, was an important factor in freeze–thaw consolidation. In their particular tests with a clay of medium plasticity net thaw settlements were obtained only when the OCR was less than 4.

Thaw weakening

Under undrained conditions at the end of a freeze−thaw cycle (point E in Fig. 3.17) the effective stress may be reduced to a very low value. According to the Mohr−Coulomb theory, the shear strength, τ, of the unfrozen soil is given by

$$\tau = c' + \sigma_n' \tan \phi' \qquad [3.27]$$

where c' is the cohesion with respect to effective stresses, ϕ' is the angle of shearing resistance, and σ_n' is the effective stress acting normally to the shear plane. Thus the soil may exhibit a very low strength immediately after thawing [Broms and Yao (1964)]. The soil will recover its strength as it consolidates. This problem is well known and causes the imposition of temporary load restrictions on roads after seasonal thaw in, for example, North America, until the road foundation has recovered its strength.

Implications for artificial ground freezing

Thawing of artificially frozen ground may be achieved either naturally or by circulating warm fluid through the freezing tubes. Thaw settlement and/or thaw weakening are thus potential problems. However, there are some important differences between naturally frozen ground and ground artificially frozen as a construction expedient, for which temperatures are generally lower. Furthermore, the block of frozen ground is finite and the heat flow and/or drainage may not be remotely one-dimensional. In particular, horizontal drainage, controlled by the coefficient of consolidation, c_h, which is normally greater than c_v, may occur. Thawing of the inner zone around a shaft or tunnel will initially be under undrained conditions since drainage is restricted by the lining on one side and frozen ground on the other (see Fig. 3.18).

When field observations are made it is usually not possible to identify the components of settlement due to freeze−thaw and those due to stress relaxation and volume change due to the excavation. However, in principle these can be separated for design purposes. In practice, a design which limits the frost heave should also limit settlement solely due to thaw. The greatest problem is likely to be thaw weakening when frost susceptible ground is frozen from the surface, for example in shaft construction. An estimate of the

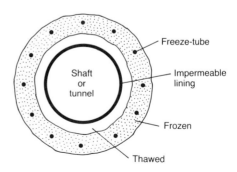

Fig. 3.18. Thawing of an artificially frozen excavation

Changes in properties of soils after freezing and thawing

time required to regain a specified strength can be obtained using the principles outlined above and appropriate boundary conditions.

The processes of frost heave, thaw consolidation and thaw weakening have been discussed in the previous section. Freeze−thaw action may result in the soil shrinking and possibly suffering from microcracks. Clays become fissured but coarser soils with little or no clay fraction will be unaffected. Fissuring may be exacerbated by thermal contraction and differential strains between layers. A scheme to store liquefied gas in unlined cavities excavated in frozen ground at Canvey Island had to be abandoned due to leakage through shrinkage cracks [Graham et al. (1983)]. See Chapter 7.

Structural changes in clay soils due to freeze−thaw

Clay soils are made up of individual fabric units or peds. The peds contain micropores and are separated by macropores. In soils which have never been subject to freezing, the form of the peds will be determined by the mineralogy, depositional conditions and bonding due either to particle surface charge interaction or to cementation. This is shown as Stage 1 in Fig. 3.19. When the temperature drops, freezing begins in the macropores with water migration from adjacent micropores or films. The expanding ice will compress the peds and break some of the bonds between them. Eventually new peds can be formed (Stage 3). These processes are likely to cause the greatest changes in sensitive clays, where cementitious bonds are broken irreversibly, and in silty clays, which are likely to be more affected than clayey silts.

Shrinkage cracking, both vertical and horizontal, has been observed by many investigators. Chamberlain and Gow (1978) presented photographs of thin sections which showed polygonal cracks 2 to 5 mm across, in the horizontal plane. This cracking may originate in the unfrozen clay which is desiccated in advance of freezing [Skarzynska (1985)]. Ice subsequently forms in the cracks. Van Vliet-Lanoë and Dupas (1991) found that changes in structure tended to stabilize after four to five cycles. The induced fabric tended to remain stable after thaw especially if the colloid content was high, the freezing temperature was low and the initial water content corresponded to pF 2·5 (the water potential at field capacity).

The size of the aggregates resulting from freeze−thaw depends on many factors including the initial (pre-freezing) structure, the moisture content, the rate of freezing and the lowest temperature. Although the combined effects tend to be soil specific, and thus require individual testing [Chamberlain (1989)] the published information enables an initial appraisal to be made. It should be noted that AGF is often applied to soils which have previously been frozen in their geological history, for example, in the Pleistocene, or as a result of seasonal freezing of shallow strata.

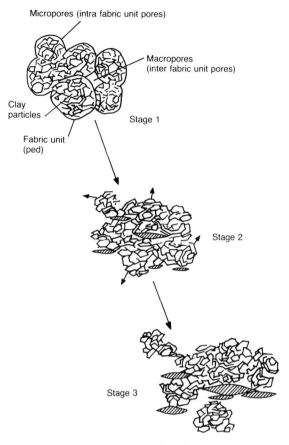

Fig. 3.19. Ice crystal growth and its effects on soil structure [after Yong et al. (1985)]

Permeability and consolidation characteristics

The fissuring of clays due to freeze–thaw results in an increase to their mass permeability. Nagasawa and Umeda (1985) showed that this increase was related to liquid limit (Fig. 3.20). Chamberlain (1980) showed that the permeability in both the vertical and horizontal directions of the Ellsworth Clay (Fig. 3.21) increased, often by two orders of magnitude, although the void ratio decreased. The ratio of permeability was stress dependent as shown in Fig. 3.22. When specimens were subjected to several freeze–thaw cycles, the greatest change in permeability was due to the first cycle. The observation that the permeabilities both normal and parallel to the direction of freezing are increased has important implications for thaw consolidation associated with artificial freezings which are often three-dimensional situations.

The coefficient of consolidation c_v is directly related to the coefficient of permeability k, so that freeze–thaw would be expected to increase its value. This has been confirmed by several experimental studies — for example Chamberlain and Blouin (1978), whose results on a plastic clay are summarized in Fig. 3.23, Ryden (1985) and Leroueil et al. (1991).

PROPERTIES OF SOILS

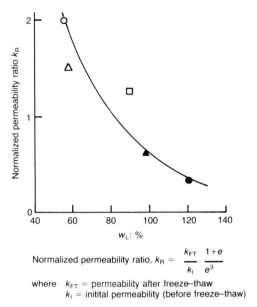

Normalized permeability ratio, $k_R = \dfrac{k_{FT}}{k_I} \dfrac{1+e}{e^3}$

where k_{FT} = permeability after freeze–thaw
k_I = initital permeability (before freeze–thaw)

Fig. 3.20. *Increase in permeability ratio after freeze–thaw as a function of liquid limit* [after Nagasawa and Umeda (1985)]

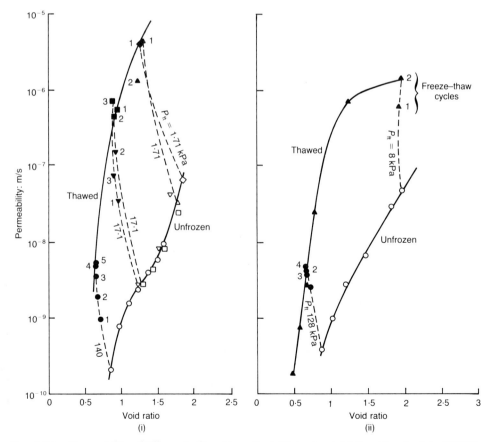

Fig. 3.21. *Permeability of Ellsworth Clay (i) vertical (ii) horizontal* [after Chamberlain (1980)]

49

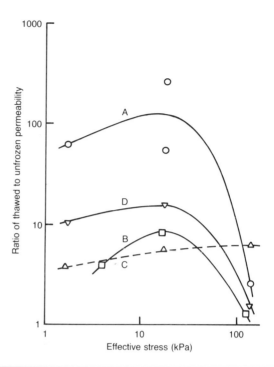

Curve	Material	w_L	w_P	w_s	I_p	G_s	O	Class	Sand (%)	Silt (%)	Clay (%)
A	Ellsworth Clay	45	25	14	20	2.66	3.2	CL	20	40	40
B	Morin Clay	26	19	18	7	2.76	0.9	CL	4	63	33
C	CRREL Clay	28	23	23	5	2.77	0.4	CL–ML	6	84	10
D	Hanover Silt	25	NP	24	NP	2.79	1.2	ML	25	70	5

w_s = shrinkage limit, O = organic content (%)
Class = Unified soil classification

Fig. 3.22. Ratio of thawed to unfrozen permeability for a range of silty and clayey soils [after Chamberlain and Gow (1978)]

The preconsolidation pressures resulting from freeze−thaw can be high. Chamberlain (1980) estimated values of up to 660 kPa. It is possible that a new structure, constructed after completion of both AGF and thaw consolidation may impose a total loading less than the preconsolidation pressure and consequently experience much lower settlements than if the clay had been normally consolidated.

Strength

Freeze−thaw may either increase or decrease the undrained shear strength of soils. The initial effects of thaw weakening have been discussed in a previous section. After thaw consolidation is complete the overconsolidation effect will tend to increase the strength. On the other hand, clays gain a pseudo-preconsolidation pressure due to ageing and this is presumably lost by freeze−thaw. Also the effects of cementation may be lost or very much reduced. As an example, Ogata *et al.* (1985) reported a decrease in strength

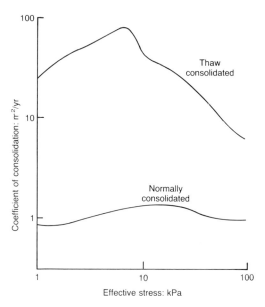

Fig. 3.23. Variation of the coefficient of consolidation with effective stress for normally consolidated and thaw consolidated clay [after Chamberlain and Blouin (1978)]

for a highly plastic clay when the confining stress was less than the preconsolidation pressure but an increase in strength for a kaolin clay for an OCR ≤ 8. They also noted an increase in the angle of shearing resistance, ϕ and a decrease in cohesion. Aoyama et al. (1985) also observed a decrease in cohesion for a highly plastic volcanic ash clay, but found almost no effect on ϕ. Chuvilin and Yazynin (1988) observed an increase of an order of magnitude in the strength of clay soils. Chamberlain (1989) concluded that increases in strength can be expected when consolidation increases the density. The special cases where reductions occur include cemented clays and overconsolidated clays.

Frost heave

The changes to the soil structure affect its future frost susceptibility. Both Chamberlain (1986) and Czurda and Schababerle (1988) noted increases in frost heave during a second cycle of freezing.

Liquid limits and suction characteristics

Several authors [Aoyama et al. (1985), Yong et al. (1985), Vahaaho (1988)] have noted reductions in the liquid limits of some highly plastic clay due to freeze−thaw which are associated with changes in particle aggregation. Little change in either aggregation or liquid limit was observed with the primary clay minerals of kaolin and bentonite [Yong et al. (1985)].

Nagasawa and Umeda (1985) noted a change in the suction characteristics of a clay, often with increases in the amount of water retained at low suction and decreases at higher suctions, depending on the plasticity. The suction characteristics reflect the pore size distribution and its modification by freeze−thaw behaviour.

Mechanical properties of frozen soils

Artificial ground freezing applications rely on the increased strength and stiffness of soils, when they are frozen, to provide temporary support to excavations. Frozen soil is a complex material which may contain four components, namely the soil grains, structural unfrozen water, ice and air. Applied stresses are shared between the ice and the soil skeleton. In very simple terms the cohesion of the ice matrix is responsible for the often high instantaneous strengths of frozen ground. However, ice deforms under very low stresses and transfers the stress to the soil skeleton up to the limit of its frictional resistance. If this limit is exceeded failure will occur.

The transfer of stress from the ice to the soil skeleton at micro-scale is aided by local pressure melting at highly stressed grain contacts, followed by water flow to a lesser stressed region and refreezing. As indicated in earlier sections, the unfrozen water content is influenced by the mineralogy, temperature and salinity of the pore water. The structural unfrozen water, in equilibrium with the ice, can resist compression and tension but has very little resistance to shear so that strength and stiffness decrease with increase in unfrozen water content. The behaviour of frozen soil is therefore strongly time, temperature and stress dependent.

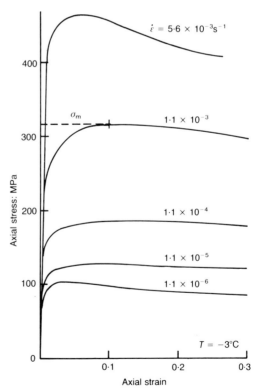

Fig. 3.24. Stress—strain curves for uniaxial compression of a remoulded silt [after Zhu and Carbee (1984)]

PROPERTIES OF SOILS

Stress–strain relations

Typical stress–strain curves for uniaxial compression tests on a frozen silt ($I_p=4$) are shown in Fig. 3.24. As the strain rate increases the strength increases and the failure mode changes from ductile to brittle. There is evidence to suggest that the sharp bend in the curves which occurs at less than 1% strain is due to initial cracking of the ice matrix. For the ductile failure, the relatively constant post peak stress is attributed to friction. The strain rate at which the transition to brittle behaviour occurs is higher for clays than gravels, probably due to their greater unfrozen water contents. Once in the brittle range, the strain rate has little effect on the strength [Sayles (1988)].

Figure 3.25 shows the effect of confining pressure on the stress–strain behaviour of frozen Ottawa sand. At high confining pressures a second yield occurs at around 10% strain. Mohr's circles for the two yield points are shown in Fig. 3.26. For first yield, the envelope is curved at low stresses then reaches an almost constant value. For second yield, the failure envelope shows an angle of shearing resistance close to that obtained from unfrozen specimens. This suggests that the first yield is related to the ice matrix and the second yield represents the frictional resistance and the residual strength. For frozen clays the effect of confining pressure may be less significant.

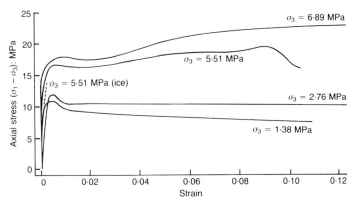

Fig. 3.25. Axial stress–strain curves for Ottawa sand and ice at various confining stresses, σ_3

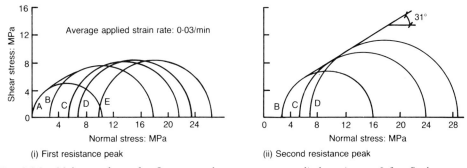

Fig. 3.26. Mohr envelopes for Ottawa sand at a constant applied strain rate [after Sayles (1973)]

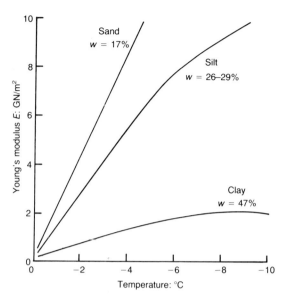

Fig. 3.27. Average Young's modulus from cyclic loading tests [after Tsytovich (1975)] — see Table 3.4 for soil gradations

Table 3.4 *Poisson's ratio for frozen soils* [after Tsytovich (1975)]

Material	% Passing 250 μm	% Passing 50 μm	% Passing 5 μm	Water content: %	Temperature: °C	Stress: kPa	Poisson's ratio
Sand	7	9·4		19·0	−0·2	196	0·41
				19·0	−0·8	588	0·13
Silt		64·4	9·2	28·0	−0·3	147	0·35
				28·0	−0·8	196	0·18
				25·3	−1·5	196	0·14
				28·7	−4·0	588	0·13
Clay			>50	50·1	−0·5	196	0·45
				53·4	−1·7	392	0·35
				54·8	−5·0	1176	0·26

Values of Young's modulus obtained from cyclic loading tests and of Poisson's ratio [Tsytovich (1975)] are shown in Fig. 3.27 and Table 3.4. Such values give useful bench marks, but care is needed as many materials exhibit non-linear behaviour.

Creep

'Creep' in this context is the irrecoverable time dependent deformation which occurs at constant stress. In studies of creep behaviour it is usual to work in terms of true strain ϵ rather than conventional or engineering strain e. The meaning of and the relation between these terms, assuming compressive strains to be positive, are

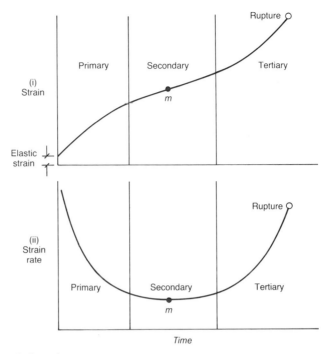

Fig. 3.28. Conventional description of creep curves.

$$e = \frac{\Delta L}{L_0} = \frac{L_0 - L}{L_0} \qquad [3.28a]$$

$$\epsilon = \int_{L_0}^{L} \frac{\delta L}{L} = \ln \frac{L_0}{L} \qquad [3.28b]$$

so that

$$\epsilon = \ln \frac{1}{1 - e} \qquad [3.28c]$$

where L is length and L_0 is initial length. Conventionally, creep curves (Fig. 3.28) have been considered to consist of three stages: primary (strain-hardening), secondary (linear) and tertiary (strain-softening). Accurate, precise measurements suggest that the secondary stage is often absent in frozen soils and is replaced by a point of inflection, m, which corresponds to a minimum strain rate. For the purpose of engineering design, the inflexion point is effectively 'failure'. Behaviour of this type has been observed in many tests, for example that shown from a uniaxial test under a stress of 7 MPa in Fig. 3.29. At low stresses the strain–time relation became almost linear and the inflexion point was not reached. An interesting feature of the results is that the product of the minimum strain rate $\dot{\epsilon}_m$ and the time to reach the inflexion point (t_m) was constant, which agreed with observations made on ice [Mellor and Cole (1982)].

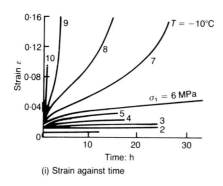 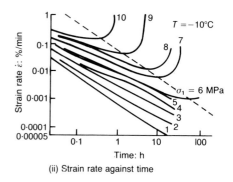

Fig. 3.29. *Uniaxial creep tests on saturated medium sand at $-10°C$ (i) strain, ϵ_f and (ii) strain rate $\dot{\epsilon}_f$ against time* [after Orth (1988)]

Models of creep behaviour

There are two approaches to modelling creep behaviour

(a) theoretical, based on an attempt to understand and quantify the physical processes involved
(b) empirical, based on curve fitting — these tend to be simpler and to be more immediately useful in engineering design.

Of many models which have been proposed, the empirical approach based on the work of Vyalov et al. (1962) is widely used. The total strain ϵ is made up of initial and delayed elastic strains (ϵ_0 and ϵ_d respectively) and irrecoverable creep strain (ϵ_c) such that

$$\epsilon = \epsilon_0 + \epsilon_d + \epsilon_c \qquad [3.29]$$

Generally ϵ_c is very much greater than $(\epsilon_0 + \epsilon_d)$.

For primary creep, their equation, modified as suggested by Assur (1963) becomes

$$\epsilon_c = \left[\frac{\sigma t^\lambda}{w\left(\dfrac{T}{T_\theta} + 1\right)^k} \right]^{1/m}$$

where σ is applied stress, t is time, T is the temperature (°C) below freezing, T_θ is the reference temperature, and w, λ, m and k are constants which are characteristic of the material. In the ground freezing industry, this is usually simplified [Klein (1979)] to

$$\epsilon_c = A\sigma^B t^C \qquad [3.31]$$

Some values of A, B and C are shown in Table 3.5.

An expression of similar form to equation [3.31] was used by Thimus et al. (1991) who undertook tests on Boom Clay (I_p = 25 to 48) down to $-20°C$ (for a brine freezing) and the Ypresian Clay (I_p = 66 to 108) down to $-120°C$ for liquefied natural gas storage. Specimens of the Ypresian Clay were obtained both parallel and perpendicular to the stratification.

The influence of stratification is seen in the results of the tests

Table 3.5 Creep parameters of frozen soils [after Jessberger (1987)]

Material	T: °C	A: $\text{mPa}^{-B} \times h^{-C}$	B	C
Ottawa sand	−9.4	3.50×10^{-4}	1.28	0.44
Manchester fine sand	−9.4	1.90×10^{-4}	2.63	0.63
Clayey fine sand	−10	8.20×10^{-3}	2.25	0.24
Sand	−10	1.67×10^{-3}	2.80	0.42
Bat-Baioss clay	−10	1.60×10^{-3}	2.50	0.45
Callovian sandy loam	−10	5.50×10^{-4}	3.70	0.37
Emscher marl	−10	7.60×10^{-5}	4.00	0.10
Silt	−10	7.90×10^{-6}	5.60	0.88
Silty clay	−10	5.99×10^{-3}	2.63	0.38
Oil sand	−10	1.18×10^{-2}	1.60	0.44
	−20	2.11×10^{-3}		

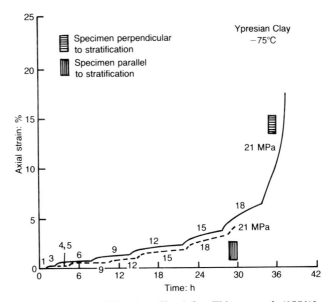

Fig. 3.30. *Influence of stratification on creep of Ypresian Clay* [after Thimus et al. (1991)]

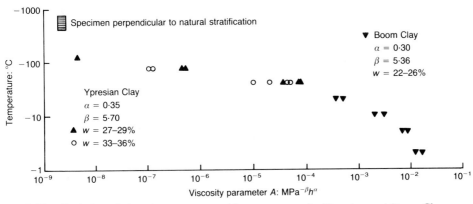

Fig. 3.31. *Variation of viscosity parameter with temperature for Ypresian and Boom Clays* [after Thimus et al. (1991)]

at $-75°C$ in Fig. 3.30. The influence of temperature on the parameter A is shown for specimens loaded perpendicular to their natural stratification in Fig. 3.31.

Several attempts have been made to model the entire creep curve. For example, Fish (1982, 1983) used the rate process theory, supplemented by some empirical elements to derive the equation

$$\dot{\epsilon} = \dot{\epsilon}_0 \, t^{-\delta} \exp(t\delta/t_m) \qquad [3.32]$$

where $\dot{\epsilon}$ is the strain rate after time t, $\dot{\epsilon}_0$ a parameter calculated at $t=1$ min with dimensions of $t^{(\delta-1)} t_m$ is the time to the inflection point and δ is a dimensionless parameter.

Equation [3.32] is consistent with the unified constitutive equation (Fish 1982)

$$\dot{\epsilon} = \frac{C}{t_m} \left(\frac{\sigma}{\sigma_c}\right)^n \left(\frac{t_m}{t}\right)^\delta \exp\left(\frac{t\delta}{t_m}\right) \qquad [3.33]$$

where σ_c = unconfined compressive strength obtained in a loading time t_0, $t_m = t_0 (\sigma/\sigma_c)^{-m}$, and C, m and n are dimensionless parameters. A family of creep curves can be obtained using equation [3.33].

In a parallel study, Ting (1983a) developed a tertiary creep model for frozen sands based on

$$\dot{\epsilon} = At^{-m} \exp(\beta t) \qquad [3.34]$$

where A, m and β are experimentally determined constants.

Integration of strain rate equations gives a series expansion which can be used to produce strain–time creep curves. These tend to be too complex for use in design and give rise to other problems. For example, Gardner et al. (1984) found that the approximate integrated version of equation [3.32] did not satisfy the criterion that the strain rate was a minimum at $t=t_m$. This led to the development of a new equation

$$\frac{\epsilon_c}{\epsilon_{cm}} = \left(\frac{t}{t_m}\right)^c \exp[(c^{1/2} - c)(t/t_m - 1)] \qquad [3.35]$$

where $\epsilon_{cm} = \epsilon_m - \epsilon_0$ = creep strain at the inflection point and c is a dimensionless parameter.

Hampton et al. (1985) compared the predictions developed from equations [3.31], [3.32], [3.34] and [3.35]. If the power law equation [3.31] is applied to the inflection point then

$$C = \frac{\dot{\epsilon} t_m}{\epsilon_{cm}} \qquad [3.36]$$

in which case c, in the Gardner equation $= C^2$ in the Klein equation.

In the comparison, the Fish approach was modified so that it was forced to fit the strain and strain rate data at $t=t_m$. In a sample comparison, all the models were within 10% for $t/t_m >$

0·3 and all except the Ting model gave very close agreement for $t/t_m > 0·7$. The poorest fit was at low values of t/t_m.

Of particular interest was the performance of the power law equation [3.31] which gave a maximum error of under 10% (albeit upon the unsafe side) for $t/t_m > 0·05$. It was concluded that although the other approaches are more soundly based on physical and thermodynamic principles, the simple power law had many attractions for engineering design.

The power law approach was extended to triaxial conditions [Hampton et al. (1988)] who developed the equation

$$\epsilon_c = A(\Delta\sigma)^B t^N (\text{ASR})^H \qquad [3.37]$$

where ϵ_c is the true irrecoverable creep strain, t is time, and A, B, N and H are empirically determined constants (A and B are the same as in equation [3.31]). ASR is the applied stress ratio ($\Delta\sigma/\Delta\sigma_f \times 100\%$) where $\Delta\sigma$ is the applied deviator stress and $\Delta\sigma_f$ is the deviator stress at failure. Note that the effect of confining pressure is automatically taken into account in the ASR term.

The agreement between observed results (Table 3.6) and equation [3.37] is shown in Fig. 3.32. The normalized form of these plots is recommended as a standard format for comparison purposes.

Table 3.6 Values for the constants obtained from tests at $-10°C$

Material	A	B	N	H
Silty sand	0·75	2·56	0·015	0·80
Medium sand (uniform)	0·54	0·60	0·042	0·62
Keuper marl	0·06	5·60	0·019	0·82

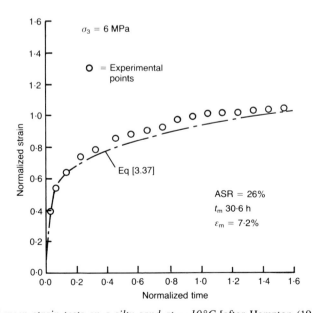

Fig. 3.32. Typical results of creep strain tests on a silty sand at $-10°C$ [after Hampton (1986)]

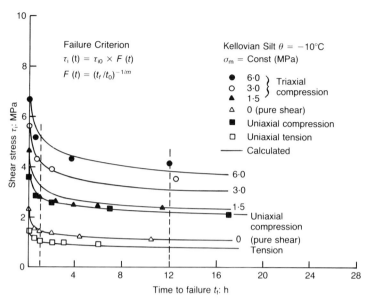

Fig. 3.33.   Long term strength of frozen Kellovian silt under combined stresses [after Fish (1991)]

Time dependent creep strength

The creep strength of a frozen soil may be defined as the stress level that can be resisted up to a finite time at which instability occurs. Variations of creep strength for a variety of tests are shown in Fig. 3.33.

An expression for long term strength σ_f was given by Vyalov (1959)

$$\sigma_f = \frac{\beta}{\ln\left[\dfrac{t_f}{B}\right]} \qquad [3.38]$$

where β and B are temperature dependent parameters relating to the particular soil. For long term strength, it is appropriate to set $t_f = t_m$ (the time to reach the inflection point).

Fish (1991) has developed new yield and creep strength criteria for frozen soils under combined stress states. Local melting of the ice causes the shear strength to reach a maximum at a certain level of mean normal stress σ_{max}, referred to as the ice melting pressure. In a plot of principal stress space this can be represented by a parabolic yield surface given by

$$\tau_i = c + b\,\sigma_m - \frac{b}{2\sigma_{max}} \cdot \sigma_m^2 \qquad [3.39]$$

where

$$\tau_i = [(\sigma_1 - \sigma_2)^2 - (\sigma_2 - \sigma_3)^2 - (\sigma_3 - \sigma_1)^2]^{1/2}/\sqrt{6} \quad [3.40]$$

$$\sigma_m = (\sigma_1 + \sigma_2 + \sigma_3)/3 \qquad [3.41]$$

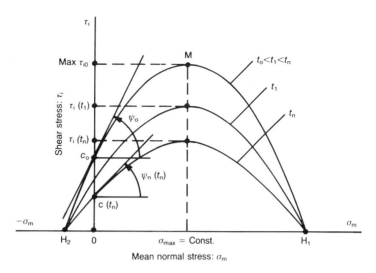

Fig. 3.34. *Parabolic criterion for creep strength* [after Fish (1991)]

c is essentially the cohesion mobilized on the octahedral shear plane, b is $\tan \Psi$ where Ψ is the angle of internal friction mobilized on the octahedral shear plane, σ_{max} is the stress corresponding to τ_{imax}. It is considered a fundamental characteristic which depends on soil structure, density, unfrozen water and ice contents, temperature etc. τ_i, c and b are thus all time dependent but the ratio c/b is assumed constant. A series of yield curves at different times is given in Fig. 3.34. Curves for all times are assumed to go through points H_1 and H_2 (Fig. 3.34) with the same σ_{max}.

Equation [3.39] can be considered a generalized extended von Mises–Drucker–Prager criterion. With $b = 0$ it reduces to the von Mises criterion for frictionless materials. Equation [3.39] is also fully applicable to the Mohr–Coulomb rupture criteria.

The time to failure t_f is taken as

$$t_f = t_0 \left[\frac{\tau_{i0}}{\tau_i} \right]^m \qquad [3.42]$$

where τ_{i0} is the instantaneous yield criterion at $t = t_0$ and $m \geq 1$ is a dimensionless parameter.

It is found that at any time t

$$\frac{\tau_i}{\tau_{i0}} = \frac{b}{b_0} = \frac{c}{c_0} = F(t) \qquad [3.43]$$

Fish refers to $F(t)$ as a dimensionless time function. Although the ratios in equation [3.43] are clearly functions of time, their purpose is to describe yield (or strength) and so the present author will use 'normalized shear stress' for τ_i/τ_{i0}. Use of this model reduces all the data shown in Fig. 3.33 to a single curve (Fig. 3.35).

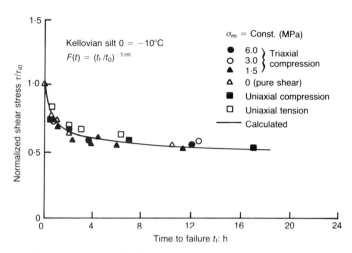

Fig. 3.35. Normalized curve of long term strength [after Fish (1991)]

Temperature effect on strain rate and strength	In general terms, lowering the temperature increases both the long term and the instantaneous strengths although for some soils there is a limiting temperature beyond which little or no increase in strength occurs [Fig. 3.36(i)] probably due to thermally induced damage [Bourbonnais and Ladanyi (1985a)].
An extensive study on an overconsolidated clay by Bourbonnais and Ladanyi (1985b) showed a continuous increase in strength [Fig. 3.36(ii)] with decreasing temperature. However, the behaviour was split into three zones. Zone 1 down to $-60°C$, Zone 2 from $-60°C$ to $-110°C$ and Zone 3 below $-110°C$. Zone 1 is characterized by significant unfrozen water and thermal expansion on freezing. In Zone 2, the very small quantity of unfrozen water is thought to be in a semi-solid state which evolves considerably as the temperature drops to $-110°C$. At still lower temperatures the protons become immobile and the behaviour changes from ductile to brittle. Further evidence is provided by Thimus *et al.* (1991) who noted that the Ypresian Clay (Fig. 3.31) changes from ductile to brittle behaviour between $-75°C$ and $-120°C$.
Postscript on models of creep behaviour for AGF	In relation to AGF tenders and contracts, it is highly beneficial to obtain the maximum amount of information from the least number of tests since both samples and time are usually in short supply. The constants A, B, N and H required by equation [3.37] can, in theory, be determined by one short term strength test and three long term creep tests, all at the same confining pressure. Two further short term tests are needed to define the strength envelope so that the equation can be used at other confining pressures. However, sample variability can be a problem since it can lead to significant errors in determining the ASR (equation [3.37]). The multi-stage (step loading) technique employed by Thimus *et al.* (1991) appears to have considerable advantages. Not

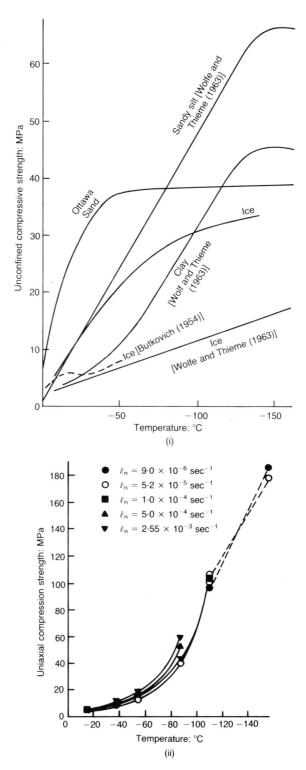

Fig. 3.36. *Influence of temperature on strength (i) various soils and ice* [after Sayles (1986)] *(ii) saturated clay* [after Bourbonnais and Ladanyi (1985b)]

only are the tests short (six to eight hours for the Boom Clay, 26–36 hours for the Ypresian Clay) but a maximum amount of data is extracted from a single specimen.

Effects of salinity on mechanical properties

The strength of frozen saline soils, at least at temperatures down to $-20°C$, is less than that of salt-free soils at the same temperature because of the higher unfrozen water contents [Leonards and Andersland (1960)]. The strength in uniaxial compression of saline sands with salt concentrations up to 100 ppt increased with increasing applied strain rates [Sego et al. (1982)]. This effect increased as the salt concentration increased. Ogata et al. (1983) confirmed these general trends for both sand and clay. They suggest a relationship with unfrozen film thickness be used to describe the decrease in compressive strength with increasing concentration.

Chamberlain (1985) showed that frozen soils are anisotropic with respect to shear strength although the effect is less marked in saline than in freshwater soils. He found that below $-1°C$ frozen clay was stronger than frozen sand at the same salinity, probably due to salt exclusion on freezing having a disproportionate effect on the sand. Creep strain also increases with increase in salt concentrations [Ogata et al. (1983)], Nixon and Pharr (1984)]. Aas (1980) and Nixon and Lem (1984) give data on strength and creep tests on specific frozen saline clays. Further information is given in Chapter 6.

References

AAS G. 1980. Laboratory determination of strength properties of frozen salty marine clay. *Proc. 2nd ISGF*, Trondheim, **1**, 144–156 and *Engng Geol.*, **18**, 67–78.

ANDERSON D. M. and TICE A. R. 1972. Predicting unfrozen water contents in frozen soils from surface area measurements. *US Highway Research Record*, 393, 12–18.

AOYAMA K. et al. 1985. Temperature dependencies of mechanical properties of soils subjected to freezing and thawing. *Proc. 4th ISGF*, Sapporo, Japan, **1**, 217–222.

ASSUR A. 1963. Discussion on creep of frozen soils. *Proc. 1st IC on Permafrost*, Indiana, 339–340.

BABA H. U. 1993. *Factors influencing the frost heave of soils*, University of Nottingham *PhD Thesis*, 453 pp.

BAKER G. C. and OSTERKAMP T. E. 1988. Salt redistribution during laboratory freezing of saline sand columns. *Proc. 5th ISGF*, Nottingham, **1**, 29–33.

BESKOW G. 1935. Tjälbildningen och tjällyftiningen med särskild hänsyn til vägar och järnväger. *Statens Väginstitut, Stockholm*, Meddelande 48, 242 pp. Also published as Sveriges geologiska undersökning. *Avh och Uppsats*, Ser C 375, and transl. into English: *Tech Inst Northwestern Univ Evanston, ILL*, Nov. 1947.

BESKOW G. 1949. Amerikansk och svensk jordklassifikation; Speciellt för vägar och flyfält. *Statens Väginstitut, Stockholm*, Meddelande 76.

BISHOP A. W. 1955. The principle of effective stress. (Lecture delivered in Oslo.) *Technisk Ubeblad*, **39**, 1959, 859–863.

BLACK P. B. and HARDENBERG M. J. 1991. Historical perspectives in frost heave research. The early works of S. Taber and G. Beskow. *CRREL Spl Rpt 91-23*, 169pp. [Includes reprints of Taber 1929, 1930, and Beskow 1935. (English transl.).]

BOURBONNAIS J. and LADANYI B. 1985a. The mechanical behaviour of frozen sand down to cryogenic temperatures. *Proc. 4th ISGF*, Sapporo, **1**, 235–244.

BOURBONNAIS J. and LADANYI B. 1985b. The mechanical behaviour of a frozen clay down to cryogenic temperatures. *Proc. 4th ISGF*, Sapporo, **2**, 237–244.

BRITISH STANDARDS INSTITUTION 1989. BS 812 *Testing aggregates*. Part 12f Method of determination of frost heave.

BROMS B. B. and YAO L. Y. C. 1964. Shear strength of a soil after freezing and thawing. *ASCE SM & FE Div.*, SM4, 1–25.

BUTKOVICH T. R. 1954. Ultimate strength of ice. *US Army Research Rpt 11*.

CHAMBERLAIN E. J. 1980. Overconsolidation effects of ground freezing. *Proc. 2nd ISGF*, Trondheim, **1**, 325–337, and in *Engng Geol.*, **18**, 97–110.

CHAMBERLAIN E. J. 1981. Frost susceptibility of soil: review of index tests. *CRREL Monograph 81–2*, 110pp.

CHAMBERLAIN E.J. 1983. Frost heave of saline soils. *Proc. 4th IC on Permafrost*, Fairbanks, 121–126.

CHAMBERLAIN E. J. 1985. Shear strength anisotropy in frozen saline and freshwater soils. *Proc. 4th ISGF*, Sapporo, **2**, 189–194.

CHAMBERLAIN E. J. 1986. Evaluation of selected frost susceptibility test methods. *CRREL Rpt 86–14*, 51pp.

CHAMBERLAIN E. J. 1987. A freeze–thaw test to determine the frost susceptibility of soils. *CRREL Rpt 87–1*, 90pp.

CHAMBERLAIN E. J. 1989. Physical changes in clays due to frost action and their effect on engineering structures. In Rathmeyer H. (ed.), *Frost in Engineering*, IS VTT95, Espoo, **2**, 863–893.

CHAMBERLAIN E. J. and BLOUIN S. E. 1978. Densification by freezing and thawing of fine material dredged from waterways. *Proc. 3rd IC on Permafrost*, Edmonton, 623–628.

CHAMBERLAIN E. J. et al. 1984. Survey of methods for classifying frost susceptibility in Frost Action and its control. *ASCE Tech. Comm. Cold Regions Engng Monograph*, 102–142.

CHAMBERLAIN E. J. and GOW A. 1978. Effect of freezing and thawing on the permeability and structure of soils. *Proc. 1st ISGF*, Bochum, **1**, 31–44, and in *Engng Geol.*, **13**, (1979) 73–92.

CHEN X. B. et al. 1988. On salt heave of saline soil. *Proc. 5th ISGF*, Nottingham, **1**, 35–39.

CHUVILIN YE M. and YAZYNIN O. M. 1988. Frozen soil macro- and microtexture formation. *Proc. 5th IC on Permafrost*, Trondheim, **1**, 320–328.

CZURDA K. A. and SCHABABERLE R. 1988. Influence of freezing and thawing on the physical and chemical properties of swelling clays. *Proc. 5th ISGF*, Nottingham, **1**, 51–58.

DEPARTMENT OF TRANSPORT 1991. *Specification for highway works*. 7th edition. HMSO, London, Clause 705.

EVERITT D. H. and HAYNES J. M. 1965. Capillary properties of some model pore systems with special reference to frost damage. *Bulletin RILEM*, June, NS **27**, 31–38.

FAROUKI O. T. 1981. Thermal properties of soils. *CRREL Monograph 81–1*, 136pp, and in *Trans. Tech. Publications*, 1986.

FISH A. M. 1982. Comparative analysis of USSR construction codes and the US Army technical manual for designs of foundations on permafrost. *CRREL Rpt 82–14*.

FISH A. M. 1983. Thermodynamic model of creep at constant stresses and constant strain rates. *CRREL Rpt 83–33*, 18pp.

FISH A. M. 1991. Strength of frozen soil under a combined stress state. *Proc. 6th ISGF*, Beijing, **1**, 135–145.
FOWLER A. C. and NOON C. G. 1993. A simplified numerical solution of the Miller model of secondary frost heave. *Cold Regions Sci. and Tech*, **21**, 327–336.
FRÉMOND M. and MIKKOLA M. 1991. Thermomechanical modelling of freezing soil. *Proc. 6th ISGF*, Beijing, **1**, 17–24.
FRIVIK P. E. 1980. State of the Art report. Ground freezing: thermal properties, modelling of processes and thermal design. *Proc. 2nd ISGF*, Trondheim, **1**, 354–373, and in *Engng Geol.*, **18** (1981), 115–133.
GARDNER A. R. 1985. *The creep behaviour of frozen ground in relation to artificial ground freezing*. University of Nottingham PhD Thesis, 331pp.
GARDNER A. R. et al. 1984. A new creep equation for frozen soils and ice. *Cold Regions Sci and Tech.*, **9**, 271–275.
GILPIN R. R. 1980. A model for the prediction of ice lensing and frost heave in soils. *Water Resources Research*, **16**–5, 918–930.
GRAHAM E. B. et al. 1983. The British Gas Canvey Island LNG Terminal: a review of developments. *Proc. IC LNG7*, Jakarta, May, 17pp.
HAMPTON C. N. 1986. *Strength and creep testing for artificial ground freezing*. Univ of Nottingham PhD thesis, 221pp.
HAMPTON C. N. et al. 1985. Modelling the creep behaviour of frozen sands. *Proc. 3rd NSGF*, BGFS Nottingham, 27–33.
HAMPTON C. N. et al. 1988. The time dependent response of frozen soils subject to triaxial stress. *Proc. 5th ISGF*, Nottingham, **2**, 559–560.
HARLAN R. L. 1973. Analysis of coupled heat-fluid transport in partially frozen soil. *Water Resources Research*, **9**–5, 1314–1323.
HARLAN R. L. and NIXON J. F. 1978. Ground thermal regime. In Andersland, O.B. and Anderson, D.M. (eds) *Geotechnical engineering for cold regions*. McGraw-Hill, 103–163.
HOEKSTRA P. 1969b. Water movement and freezing pressures. *Soil Science Soc. of America, J.*, **33**–4, 512–518.
ISSMFE 1989. Work report 1985–1989 of the Technical Committee on Frost, TC8. *Frost in Engineering* (ed. Rathmeyer, H.), IS VTT94, Espoo Finland, **1**, 15–70.
JESSBERGER H. L. 1987. Artificial freezing of the ground for construction purposes. In Bell, F.G. (ed.), *Ground Engineer's Reference Book*, Ch 31, 1–17.
JESSBERGER H. L. et al. 1988. Thermal design of a frozen soil structure for stabilisation of the soil on top of two parallel metro tunnels. *Proc. 5th ISGF*, Nottingham, **1**, 349–356.
JOHANSEN Ø. and FRIVIK P. E. 1980. Thermal properties of soils and rock minerals. *Proc. 2nd ISGF*, Trondheim, **1**, 427–453.
JONES R. H. 1987. Developments in the British approach to prevention of frost heave in pavements. *Transportation Research Record 1146*, 33–40.
KAY B. D. and GROENEVELT P. H. 1983. The redistribution of solutes in freezing soil: exclusion of solutes. *Proc. 4th IC on Permafrost*, Fairbanks, 584–588.
KAY B. D. and PERFECT E. 1988. State of the Art: Heat and mass transfer in freezing soils. *Proc. 5th ISGF*, Nottingham, **1**, 3–21.
KERSTEN M. S. 1949. Laboratory research for the determination of the thermal properties of soils. *ACFEL Tech Rpt 23*, and in *Univ of Minn Engineering Stn Bulletin 28*.
KLEIN J. 1979. The application of finite elements to creep problems in ground freezing. *Proc. 3rd IC on Numerical Methods in Geomechanics*, Aachen, April, 493–502.
KONRAD J. M. 1987. Procedure for determining the segregation potential of freezing soils. *Geotech. Testing*, June, **10**–2, 51–58.
KONRAD J. M. and MORGENSTERN N. R. 1981. The segregation potential of a freezing soil. *Cndn Geotech. J.*, **18**, 482–491.

KONRAD J. M. and MORGENSTERN N. R. 1984. Frost heave prediction of chilled pipelines in unfrozen soils. *Cndn Geotech. J.*, **21**, 100–115.

LADANYI B. and SHEN M. 1989. Mechanics of freezing and thawing in soils. In Rathmeyer, H. (ed.), *Frost in geotechnical engineering*, IS VTT94, Espoo, **1**, 73–103.

LEONARDS G. A. and ANDERSLAND O. B. 1960. The clay water system and the shearing resistance of clays. *C. on Shear Strength of Cohesive Soils*, ASCE, Boulder, 793–818.

LEROUEIL S. et al. 1991. Effects of frost on the mechanical behaviour of Champlain Sea clays. *Cndn Geotech*, **28**, 690–697.

LOCH J. P. G. and KAY B. D. 1978. Water distribution in partially frozen saturated silt under several temperature gradients and overburden loads. *Soil Sci. Soc. of America J.*, **42**–3, 400–406.

LOCH J. P. G. and MILLER R. D. 1975. Tests of the concept of secondary frost heaving. *Soil Sci. Soc. of America J.*, **39**–6, 1036–1041.

MELLOR M. and COLE D. M. 1982. Deformation and failure of ice under constant stress or constant strain rate. *Cold Regions Sci. and Tech.*, **5**–3, 201–219.

MILLER R. D. 1972. Freezing and heaving of saturated and unsaturated soils. *Highway Research Record*, **393**, 1–11.

MILLER R. D. 1978. Frost heaving in non-colloidal soils. *Proc. 3rd IC on Permafrost*, Edmonton, **1**, 708–713.

MORGENSTERN N. R. 1981. Geotechnical engineering and frontier resource development (29th Rankine Lecture). *Géotechnique*, **31**–3, 305–365.

MORGENSTERN N. R. and NIXON J. F. 1971. One-Dimensional consolidation of soils. *Cndn Geotech. J.*, **8**, 558–565.

MORGENSTERN N. R. and SMITH L. B. 1973. Thaw consolidation tests on remoulded clays. *Cndn Geotech. J.*, **10**–1, 25–40.

NAGASAWA T. and UMEDA Y. 1985. Effects of the freeze–thaw process on soil structure. *Proc. 4th ISGF*, Sapporo, **2**, 219–224.

NIXON J. F. 1987. Thermally induced heave beneath chilled pipelines in frozen ground. *Cndn Geotech. J*, May, **24**–2, 260–266.

NIXON J. F. 1991. Discrete ice lens theory for frost in soils. *Cndn Geotech. J.*, **28**, 843–859.

NIXON J. F. and LADANYI B. 1978. Thaw consolidation. In Andersland, O. B. and Anderson, D. M. (eds), *Geotechnical engineering for cold regions*. McGraw-Hill, 164–215.

NIXON J. F. and LEM G. 1984. Creep and strength testing of frozen saline fine-grained soils. *Cndn Geotech. J.*, **21**, 518–529.

NIXON J. F. and MORGENSTERN N. R. 1973. The residual stress in thawing soils. *Cndn Geotech. J.*, **10**–4, 571–580.

NIXON J. F. and MORGENSTERN N. R. 1974. Thaw consolidation tests on undisturbed fine-grained permafrost. *Cndn Geotech. J.*, **11**–1, 202–214.

NIXON M. S. and PHARR G. M. 1984. The effects of temperature, stress and salinity on the creep of frozen saline soil. *J. Energy Resources Tech. (ASME)*, **106**, 344–348.

OGATA N. et al. 1983. Effects of salt concentration on strength and creep behaviour of artificially frozen soils. *Cold Regions Sci. and Tech.*, **8**, 139–153.

OGATA N. et al. 1985. Effect of freezing-thawing on the mechanical properties of soil. *Proc. 4th ISGF*, Sapporo, **1**, 201–207.

O'NEILL K. and MILLER R. D. 1980. Numerical solutions for rigid-ice model of secondary frost heave. *Proc. 2nd ISGF*, Trondheim, **1**, 656–669, and in *CRREL Rpt 82–13*.

O'NEILL K. and MILLER R. D. 1985. Explorations of a rigid ice model of frost heave. *Water Resources Research*, **21**, 281–296.

ORTH W. 1988. A creep formula for practical application based on crystal mechanics. *Proc. 5th ISGF*, Nottingham, **1**, 205–211.

PATTERSON D. E. and SMITH M. W. 1981. The measurement of unfrozen

water content by time domain reflectometry: results from laboratory tests. *Cndn Geotech. J.*, **18**, 131–144.
PATTERSON D. E. and SMITH M. W. 1985. Unfrozen water content in saline soils: results using time domain reflectometry. *Cndn Geotech. J.*, **22**, 95–101.
PENNER E. 1963. Frost heaving in soils. *Proc. 1st IC on Permafrost*, Indiana (Building Res Advistory Bd NRC), 197–202.
PENNER, E. 1967. Heaving pressure in soils during unidirectional freezing. *Cndn Geotech. J.*, **4**, 398–408.
PENNER E. 1986. Ice lensing in layered soils. *Cndn Geotech. J.*, **23**–3, 334–340.
PENNER E. and GOODRICH L. E. 1980. Location of segregated ice in frost susceptible soil. *Proc. 2nd ISGF*, Trondheim, **1**, 626–639.
PIPER D. et al. 1988. A mathematical model of frost heave in granular materials. *Proc. 5th IC on Permafrost*, Trondheim, **1**, 370–376, and abbreviated in *Proc. 5th ISGF*, Nottingham, **2**, 569–70.
RIEKE R. D. et al. 1983. The role of specific surface area and related index properties in the frost heave susceptibility of soils. *Proc. 4th IC on Permafrost*, Fairbanks, 1066–1071.
RYDEN C. G. 1985. Pore pressure in thawing soil. *Proc. 4th ISGF*, Sapporo, **1**, 223–226.
SAYLES F. H. 1966. Low temperature soil mechanics. *CRREL Tech. Note*.
SAYLES F. H. 1973. Triaxial and creep tests on frozen Ottawa sands. *Proc. 2nd IC on Permafrost*, Yakutsk, 384–392.
SAYLES F. H. 1988. State of the Art: Mechanical properties of frozen soils. *Proc. 5th ISGF*, Nottingham, **1**, 143–165.
SEGO D. C. et al. 1982. Strength and deformation behaviour of frozen saline sand. *Proc. 3rd ISGF*, Hanover NH, **1**, 11–18.
SHEN M. and LADANYI B. 1988. Calculation of the stress field in soils during freezing. *Proc. 5th ISGF*, Nottingham, **1**, 121–127.
SKARZYNSKA K. M. 1985. Formation of soil structure under repeated freezing-thawing conditions. *Proc. 4th ISGF*, Sapporo, **2**, 213–218.
SNiP II-15-74 1975. *Bases and foundations of buildings and structures*. Moscow.
SUTHERLAND H. B. and GASKIN P. N. 1973. Pore water and heaving pressures developed in partially frozen soils. *Proc. 2nd IC on Permafrost*, Yakutsk, 409–419.
TABER S. 1918. Ice forming in clay soils will lift surface weights. *Engineering News Record*, **80**–6, 262–263.
TABER S. 1929. Frost heaving. *J. Geol.*, **37**–5, 428–461.
TABER S. 1930. The mechanics of frost heaving. *J. Geol.*, **38**–4, 303–317.
TAKAGI S. 1980. The adsorption force theory of frost heaving. *Cold Regions Sci. and Tech.*, **3**, 57–81.
THIMUS J. F. et al. 1991. Rheological behaviour of overconsolidated clay measured by creep test — application to cryogenic storage. *Proc. 6th ISGF*, Beijing, **1**, 181–188.
TICE A. R. et al. 1978. Determination of unfrozen water in frozen soil by pulsed nuclear magnetic resonance. *Proc. 3rd IC on Permafrost*, Edmonton, **1**, 149–155.
TICE A. R. et al. 1984. The effects of soluble salts on the unfrozen water content of the Lanzhou PRC silt. *CRREL Rpt 84–16*.
TING J. M. 1983a. Tertiary creep model for frozen sands. *J. Geotech. Engng*, **109**–7, 932–945.
TSYTOVICH N. A. 1975. *The mechanics of frozen ground*. McGraw-Hill, 426pp.
TSYTOVICH N. A. et al. 1965. Consolidation of thawing soils. *Proc. 6th ICSMFE*, Montreal, **1**, 390–394.
VAHAAHO I. T. 1988. Soil freezing and thaw consolidation results for a major project in Helsinki. *Proc. 5th ISGF*, Nottingham, **1**, 219–223.

VAN VLIET-LANOË and DUPAS A. 1991. Development of soil fabric by freeze/thaw cycles — its effect on frost heave. *Proc. 6th ISGF*, Beijing, **1**, 189–195.

VINSON T. S. *et al.* 1987. Factors important to the development of frost heave susceptibility criterion for coarse-grained soils. *Transportation Research Record*, **1089**, 124–131.

VYALOV S. S. 1959. Rheological properties and bearing capacity of frozen soils. *CRREL Trans 74*, 120pp.

VYALOV S. S. *et al.* 1962. The strength and creep of frozen soils and calculations for ice-retaining structures. *CRREL Trans 76*.

WATSON G. H. *et al.* 1973. Determination of some frozen and thawed properties of permafrost soils. *Cndn Geotech. J.*, **10**–4, 592–606.

WILLIAMS P. J. 1967. Properties and behaviour of freezing soils. *Norwegian Geotechnical Institute*, **72**, 120pp.

WILLIAMS P. J. 1988. Thermodynamic and mechanical conditions within frozen soils and their effects. *Proc. 5th IC on Permafrost*, Trondheim, **1**, 493–498.

WILLIAMS P. J. and SMITH M. W. 1990. *The frozen earth: fundamentals of geocryology.* Cambridge University Press, 306pp.

WILLIAMS P. J. and WOOD J. A. 1985. Internal stresses in frozen ground. *Cndn Geotech. J.*, **22**–3, 413–416.

WOLFE L. H. and THIEME J. O. 1963. Physical and thermal properties of frozen soil and ice. *J. Soc. Petroleum Engineering*, **4**–1, 67–72.

YAMAMOTO H. *et al.* 1988. Effect of overconsolidation ratio of saturated soil on frost heave and thaw subsidence. *Proc. 5th IC on Permafrost*, Trondheim, **1**, 522–527.

YONG R. N. *et al.* 1985. Alteration of soil behaviour after cyclic freezing and thawing. *Proc. 4th ISGF*, Sapporo, **1**, 187–195.

ZHANG L. 1991. The law of unfrozen water content change in frozen saline (NaCl) soils. *Proc. 6th ISGF*, Beijing, **1**, 113–119.

ZHU Y. and CARBEE L. 1984. Uniaxial compressive strength of frozen silt under constant deformation rates. *Cold Regions Sci. and Tech.*, **9**, 3–15.

4. Engineering design of frozen ground works

by Dr F. Alan Auld and John S. Harris

Ground information

Constructional work below ground level is normally preceded by a ground investigation to establish the strata sequence, soil/rock properties and hydrogeology of the site; such exploration must be sufficiently deep and comprehensive to establish the long term stability of the intended structure, and of any temporary works needed to effect the excavation within which the sub-surface permanent works can safely be constructed. Procedures for ground investigations, which will generate this data, are now well documented. See Chapter 2.

The results of such investigations will identify the location and extent of weak and/or saturated strata, alert the designer to potential problems or difficulties, and indicate the need for preventive measures. The same data will enable a ground freezing engineer to prepare a scheme, but supplementary information, as indicated below, may be needed before the design can be finalized.

A feature of ground freezing is that a total cofferdam is created. This is achieved, as we have seen, either by terminating the freeze-tubes in a cut-off stratum or by solidly freezing the whole zone. Frequently site investigation borings have only been taken to a predetermined depth, related to the perceived deepest foundation level, and terminated there even if they are then within a permeable stratum. Later, when the need to use a geotechnical process is recognized, the actual depth to a possible cut-off stratum is not available. The cheapest freezing solution could well incur freeze-tubes 20 m or more deeper than the intended foundation in order to reach a cut-off, in preference to freezing solid.

Groundwater flow is one of the greatest hazards to effective freezing. It is particularly important, therefore, that the existence of flow conditions is recognized and evaluated at investigation stage. Flow may be due to natural sub-surface drainage, proximity to a river (hydraulic gradient) or a subterranean water-course, (sub)artesian flow between aquifers, or artificial extraction by pumping or recharge. Similarly the groundwater level may fluctuate due to tidal effects at sites close to the coast or an estuary.

The freezing point may be depressed below zero if salts are present in the groundwater, or in any stratum. Chemical analyses will be useful in this context, but actual measurement of the freezing point is preferable. The concentration of salt (NaCl) in seawater is about 28 g/l, and freezing occurs at around −2·5°C. The presence of salts in soils reduces the strength of the frozen material compared with that for the same stratum free of salt. See Chapter 3.

Refrigeration load, and the length of the primary freeze-period, are affected by the range of temperatures to be dealt with, i.e. the ground temperature will determine the quantity of sensible heat to be removed before reaching the freezing point, while the ambient temperature and humidity will influence plant performance. Although the ground temperature in the UK is generally around 15°C, surface temperatures vary with the season and with location; in some parts of the world the temperature gradient with depth below surface can be as high as 1°C per 30 m.

The purpose of ground freezing is to improve the strength of unstable soils (sands, silts and clays) in addition to the prevention of water ingress. Where the strata is competent (self-supporting when excavated) rock, only water exclusion is required.

As frozen soils behave elasto-plastically, the time-dependent deformation characteristics of the frozen soil must be considered if the ice-wall is to be subject to high stresses (e.g. for deep mineshafts) and/or will be unsupported for long periods. In this case the laboratory testing programme will need to include suitable tests to establish the strength, Young's modulus, Poissons ratio and creep parameters for a range of sub-zero temperatures. Guidelines on transportation, preparation and storage of frozen samples, description of frozen soils, and classification and laboratory testing of frozen ground have been published, and are discussed in Chapter 2.

The property of volume increase as water freezes is well known, while the highways engineer is all too familiar with frost heave when ice lenses form in the sub-base or sub-grade. The two phenomena are separate, and their effect is markedly different. The most significant soil properties are water permeability and frost susceptibility; these parameters should be measured when the soil is cohesive (silts and clays) if ground heave could lead to distress of nearby structures. See Chapter 3.

Structural design

The most common ice-wall configuration is based on the circle since, in the ideal case, this is the only shape which offers uniform compressive stress and minimal tensile stress in the ice-wall during its working life. Thus circular constructions are the simplest to design and construct.

Rectangular excavations are best enclosed by a circular ice-wall, or one made up of a series of arcs; only in exceptional circumstances will straight sided ice-wall cofferdams be engineered, and their design will almost certainly call for supplementary support or reinforcement (unless the dimensions are small) if the excavation is to be reliant on the ice-wall for its structural security. See *Milwaukee*, Chapter 7.

GROUND FREEZING IN PRACTICE

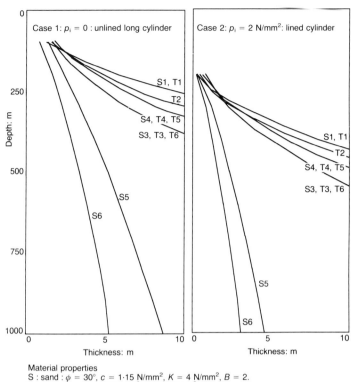

Fig. 4.1. Freeze-wall thickness against depth for unstable soils (clays and sands) [after Klein (1980a)]

Shafts and pits   *Determination of ice-wall thickness (unstable ground)*

For many years the structural design of an unlined frozen ground barrier has been based on empirical formulae named after their authors. The earliest and simplest (equations [4.1] and [4.2] below) yield thicknesses which invariably are much larger than those obtained from equations developed more recently, as illustrated in Fig. 4.1. Many of the variable properties of frozen ground, which are not considered in the simpler formulae, include the temperature gradient across the ice-wall, the variation in frozen strength with reduction in temperature, and time related temperature dependent deformation under constant load (creep). Experience has generally proven that designs based on the later formulae are adequate, implying that the earlier ones embody a higher factor of safety (ignorance).

The characteristic load and stress patterns are illustrated in Figs 4.2, 4.3 and 4.4, and defined by formulae as discussed. The various formulae have assumed fully elastic, partly elastic/partly plastic, fully plastic, and fully viscous behaviour of the ice-wall.

The Lamé and Clapeyron proposal (equation [4.1]) in 1833 assumed fully elastic behaviour, and is known to be over conservative

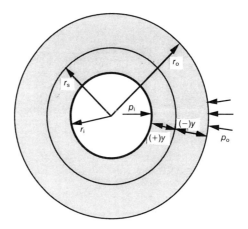

Fig. 4.2. Shaft nomenclature

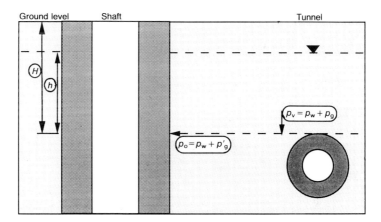

Fig. 4.3. Load nomenclature (cross-section)

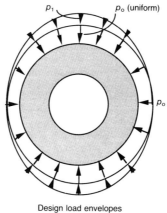

Design load envelopes

Fig. 4.4. Shaft load nomenclature (plan view); where $p = p_o + p_1$, $p_1 = p_z/2(1 + \cos 2\phi)$ and $p_z = 0 \cdot 0013H$, i.e. $0 \cdot 13 \times$ hydrostatic pressure from the surface [Link et al. (1976)]. This modification to 'uniform pressure' allows for differential pressures caused by freezing and thawing irregularities in anisotropic soil.

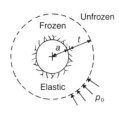

$$t = a\left[\sqrt{\left(\frac{q}{q - 2p_o}\right)} - 1\right] \quad [4.1]$$

where q is unconfined compressive strength and p_o is overburden pressure. (Note: in equations [4.1–4.5] and [4.10], $a = r_i$ and $b = r_o$.)

Equation [4.2] was developed by Domke (1915) who, after consideration of each mode, treated the ice-wall behaviour outside the freeze-tube circle as elastic and inside as plastic, it relates to thick-walled cylinders of infinite height and does not recognize time related deformation

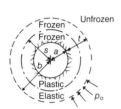

$$t = a\left[0\cdot 29\left(\frac{p_o}{q}\right) + 2\cdot 30\left(\frac{p_o}{q}\right)^2\right] \quad [4.2]$$

for $s = \sqrt{ab}$.

Vyalov (1962) investigated material behaviour according to Mohr interpretations yielding straight- and non-straight-line envelopes, studied creep of a frozen cylinder of infinite height, and the interaction of the latter with the surrounding soil mass. Their main proposal (equation [4.3]) assumed a fully plastic ice-wall

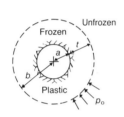

$$t = a\left[\left(\frac{p_o}{q}\frac{2\sin\phi}{(1-\sin\phi)} + 1\right)^{(1-\sin\phi)/2\sin\phi} - 1\right] \quad [4.3]$$

where ϕ = angle of internal friction.

Hall and Scott (1968) studied the influence of plastic–elastic phenomena and concluded that equation [4.3] was a good general approach for the unlined case.

Klein (1981) proposed a modified form of the Domke formula (equation [4.4]) which introduced the internal friction parameter ϕ

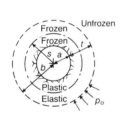

$$t = a\left[(0\cdot 29 + 1\cdot 42\sin\phi)\left(\frac{p_o}{q}\right) + (2\cdot 30 - 4\cdot 60\sin\phi)\left(\frac{p_o}{q}\right)^2\right] \quad [4.4]$$

with $s = \sqrt{ab}$.

At Panji no 3 colliery, Pang (1991) reported having used an alternative modification of Domke (equation [4.5]), proposed by Shangdong Mining College and others, to take account *inter alia* of plastic deformation of the ice-wall

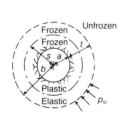

$$t = a\left[0\cdot 56\frac{p_o}{q} + 1\cdot 33\left(\frac{p_o}{q}\right)^2\right] \quad [4.5]$$

However, serious creep deformation and basal heave is reported to have occurred at 257 m depth, together with associated freeze-tube breakages, suggesting that in this case the design gave a factor of safety approximating to 1 at that depth.

With mud-flush drilling practice it is unreasonable to attempt

to install freeze-tubes at spacings of less than about 0·8 m. This spacing easily generates an ice-wall thickness of at least 0·7 m. A spacing of 1 m, yielding 0·9 m minimum thickness, is more common: for most shallow excavations even the most conservative formulae — e.g. Domke (equation [4.2]) — will indicate a thickness of these orders or more, in which case there is little point in undertaking more sophisticated designs. As depth, and therefore load, increases so does the necessary ice-wall quality (thickness—temperature relationship).

Frozen thickness in competent ground
The conversion of intergranular water in permeable rocks (e.g. sandstones), or fissure water in other massive rock formations, is often needed to exclude water from the excavation. The formulae given in the preceding section yield thicknesses which are both adequate and generally less than that which is generated by the usual range of spacing between freeze-tubes.

Determination of ice-wall stresses for shallow shafts (unstable ground)
Ice-walls for *shafts of shallow depth* in soft ground may be designed on elastic criteria, when the relationships can be expressed in the following equations.

Axial stresses [after Link et al. (1976)]

$$\sigma_1 = \frac{p_o r_o}{t \times 1}\left(1 + \frac{y}{r_s}\right) \qquad [4.6a]$$

$$\sigma_2^* = \frac{p_z r_o}{2 \times t \times 1}\left(1 + \frac{y}{r_s}\right) \qquad [4.6b]$$

where σ_1 = axial stress due to uniform load p_o (Fig. 4.4), σ_2^* = axial stress due to non-uniform load p_1, r_o = external radius of ice-walls, r_s = radius to centroid of ice-wall, y as in Fig. 4.2, $p_z = 0\cdot 13 \times$ hydrostatic pressure.

Bending stresses [after Link et al. (1976)]

$$\sigma_2^{**} = \pm \frac{p_z r_o}{6 \times t \times 1}\left(1 + \frac{6r_s}{t}\right)\frac{X}{(X-1)} \quad \theta=0 \text{ and } \theta=90° \qquad [4.7]$$

with $X = \dfrac{3E_f I}{p_o r_o r_s^2}$

where E_f = Young's modulus (frozen ground), $I = t^3/12$ and σ_2^{**} = bending stress due to non-uniform load.

Combined stresses

$$\sigma = \sigma_1 + \sigma_2^* \pm \sigma_2^{**} \qquad [4.8]$$

Note 1 $\sigma_1 + \sigma_2^*$ must always be greater than σ_2^{**} otherwise tensile stresses will develop.

Note 2   Allowable stress:

$$\sigma_a = \frac{\sigma}{\text{Factor of Safety}} = \frac{q_f(t)}{\text{Factor of Safety}}$$

where $q_f(t)$ = time dependent unconfined compressive strength (as defined by ISGF Working Group 2, 1991).

Deformation due to bending

$$\delta_b = \pm \frac{\sigma_2^{**} \, z \, r_s^2}{3 E_f I} \qquad [4.9a]$$

where z = section modulus.

Deformation due to axial stresses

$$\delta_a = \frac{(2 p_o + p_z) r_i \, r_o^2}{E_f(r_o^2 - r_i^2)} \qquad [4.9b]$$

where r_i = internal radius of ice-wall.

Maximum deformation

$$\delta = \delta_a + \delta_b$$

Deformation controlled design for deep shafts (unstable ground)
For *deep lined shafts* to be sunk through cohesive soils which, when frozen, exhibit significant creep deformation characteristics, an elastic design approach will be inappropriate. Advancement of any form of structural lining, as the shaft is excavated, introduces reaction against deformation pressures to within a short distance of the sump; it will be seen that when $p_i = 0$ many of the following equations simplify to one of the earlier equations.

Five formulae for determining the thickness t of a long thick cylindrical ice-wall with an inner lining (internal pressure), attributed to Tresca, Mises, Mohr–Coulomb and Drucker–Prager (plastic) and Klein (viscous) have been quoted by Klein (1980a) and Auld (1985 and 1988b); of these, equation [4.10] (Klein, viscous) gives the minimal solution

$$t = a \left\{ \left[1 - \frac{3}{B} \frac{(p - p_o)}{q_f(t)} \right]^{-B/2} - 1 \right\} \qquad [4.10]$$

where B = non-linear stress–time related parameter.

Four formulae, attributed to Vyalov (1962), Sanger and Sayles (1979), and Liberman (1960), have been quoted for short thick unlined cylinders. This condition is not often relevant as the unsupported excavation length is rarely less than the excavation diameter.

The foregoing are all strength related. For deep shafts control of deformation becomes the critical element, for which time dependent theories for long thick lined cylinders subject to internal pressure are the most appropriate. A suitable form of stress–strain relationship which is representative of the time and temperature dependent creep behaviour of cohesive, frictionless soils (clays) is given by Klein (1978) (equation [4.11])

$$\epsilon_c = A\sigma^B t^C \qquad [4.11]$$

where ϵ_c is the creep strain, σ is the applied stress, t is time, A is a temperature dependent modulus with units stress^{-B} time^{-C}, B is a dimensionless exponent >1, and C is a dimensionless exponent <1.

Refinements of equation [4.11] have been the subject of research by Gardner (1985) and Hampton (1986) who have proposed versions which apparently yield closer curve fits with results of experiments on remoulded specimens.

Of three formulae, attributable to Ladanyi (1982), Jessberger, Klein and Ebel (1979), and Klein (1980) derived using this stress relationship, equation [4.12] (Klein), is considered to be an appropriate and practical deformation (of the inside face of the ice-wall) formula for clays

$$\delta_i = -\left(\frac{\sqrt{3}}{2}\right)^{B+1} r_i \left[\frac{(p_o - p_i)(2/B)}{1-(b/a)^{-2/B}}\right]^B A t^C \qquad [4.12]$$

where A, B and C are parameters defined by equation [4.11].

Klein (1981, 1985) developed the above approach further to include the behaviour of soils containing different friction angles (e.g. sandy soils). The temperature dependent viscosity parameter A modifies to A^* when dependent additionally on the angle of friction. Closed form solutions, based on the analysis of thick cylinders, are given (by Klein) for the cases where $\phi>0$ and $B=2$, 1 and ∞. This set of the standard results encompasses the full range of friction angle soil conditions and permits a judgement to be made on soils which do not specifically fit the standard form. The creep deformation at the inside face of the freeze-wall when the friction angle $\phi>0$ and $B=2$ is

$$\delta_i = \left(\frac{\sqrt{3}}{2}\right)^{B+1} r_i \left[\frac{2\lambda(1-\mu)}{(r_i/r_o)-\mu}\right] A^* t^C \qquad [4.13]$$

where λ = coefficient of soil pressure, $\mu = [(p_o+\lambda)/(p_i+\lambda)]^{1/2}$ and $1/\lambda = \tan\phi/c$.

The friction angle ϕ is constant with time at the particular temperature under consideration, and c is the cohesion which is both temperature and time dependent. Alternatively λ = Uniaxial Long-term Strength (ULS) $(1 - \sin\phi)/2\sin\phi$, where the ULS is also temperature and time dependent.

The various conditions and methods or stages of construction, which pose differing critical elements, must be considered individually. For example shaft sinking practice in the UK and North America, which is normally in competent rock, is to freeze to the centre of the shaft space, and sink and line permanently in stages as the excavation proceeds (the 'hanging lining' method). The practice in Germany is to freeze to the centre, sink and line temporarily providing support to control deformation, followed by installation of the permanent lining continuously from sump to surface once the excavation has reached its total depth. Chinese practice is to control the inward growth of the ice-wall to coincide, more or less, with the excavation line, thus leaving a 'soft' core for ease of excavation. Inevitably this means that the average temperature of the inner ice-wall is warmer, the consequence of which is a lesser gain in the strength and more creep on freezing; the freeze-circle diameter/number of freeze-tubes/total ice-wall thickness are usually larger than in western practice.

The design procedure will firstly estimate an ice-wall thickness, as illustrated in Fig. 4.1, then ensure acceptability by imposing constraints on the rate of advance and the distance of the lining behind the sump as illustrated in Fig. 4.5. The aims must be to preclude overstressing the lining, and avoiding damage to the freeze-tubes. Confidence can be enhanced by use of the Finite Element Method (FEM).

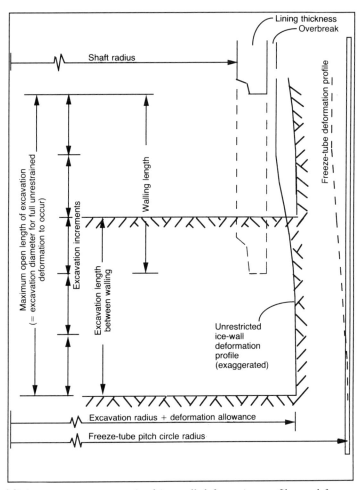

Fig. 4.5. Excavation and lining sequence, unrestrained ice-wall deformation profiles and freeze-tube deformation profile (cross-section)

The FEM, which may be applied to problems concerned with statics, dynamics, heat transfer, linear and non-linear behaviour, is a useful numerical technique for analysing complex continuum structures which are not easily treated using simple formulae. With the FEM a structure is discretized into small regularly shaped regions (elements); the stress—strain and displacement behaviour of each element is defined, then combined with that of adjoining elements so that the equilibrium and compatibility of the family of elements (the mesh) is preserved. Using the principle of minimum potential energy for displacement of the corners of the elements (nodes), subject to general equations containing the material parameters, e.g. elastic modulus, Poissons ratio etc, the deflections and stresses of the whole can be deduced in a reasonably short time with the aid of a suitable computer program. Complexity of shape, loading and material parameters is dealt with by varying the number of elements according to the sensitivity of each portion of the structure. Since we are dealing with a stress—strain regime of a body subject to a temperature gradient, composed of a material whose properties vary with temperature, it is first necessary to perform a thermal analysis (see later in this chapter) to determine the temperature distribution; only then can appropriate material parameters be attributed to each group of elements.

A sequence of operations might reasonably be

- discretize the structure, or the smallest (axi-)symmetric part of it, into a number of finite elements of rectangular and/or triangular shape, connected at their nodes by frictionless pins
- express the unknowns as functions, e.g. displacements
- define any material relationships, e.g. stress—strain
- define the properties of individual elements using the principle of minimum potential energy, e.g. load—deformation characteristics
- assemble individual characteristics to form the stiffness matrix for the whole body and introduce boundary conditions
- solve for primary unknowns, e.g. displacements
- using material relationships find derived quantities
- interpret the results in tabular, graphical and/or contour form.

Reliable behavioural parameters and appropriate boundary conditions are essential in such analyses, a requirement which imposes stringent demands on the sampling and low-temperature testing operations which are the subject of continuing research and study in many countries. See Chapters 2 and 3.

The design problems that must be addressed include

- short term stability of the exposed/unsupported ice-wall from the excavated sump to the bottom of the permanent lining, and of the sump itself, combined with the strength gain against induced pressure on the deepest section of the in situ concrete lining (British/American and Chinese practice)

- long term stability of the ice-wall protected by the temporary lining support, which is eventually incorporated into the full lining thickness (German practice) — theories based on long thick cylinders with internal pressure are appropriate
- restriction of creep deformation by adjustment of ice-wall thickness and/or temperature (usual German/Chinese practice), or control by introduction of a (sacrificial) stress-absorbent primary lining prior to placing the permanent structural lining (usual Chinese practice)
- short or longer term radial stability of a thin disc — annular section — of frozen unstable strata sandwiched between layers of competent rock, and thereby restrained above and below by shear forces developed in conjunction with the overburden pressure; this is a special case for which short cylinder theory may be appropriate.

Summarizing, the design procedure is

- to estimate the necessary ice-wall thickness using an appropriate method of analysis
- to estimate the likely deformation and ensure that it remains within acceptable limits, i.e. that overstressing of the lining and of the freeze-tubes is avoided — this is achieved by imposing limits on the height of the unlined section immediately above the sump, which in turn infers control of the rate of advance of the excavation (see Fig. 4.5).

Tunnels and drifts

As with shafts, many tunnels will approximate in profile to a circle, and much of the design approach will be similar if the freezing is to be undertaken with freeze-tubes parallel to the tunnel axis. This requires horizontal drilling which is slower and more expensive than vertical drilling. If the desired tunnel is at a relatively shallow depth and is without surface encumbrances (e.g. buildings, railways, certain watercourses), a design based on vertical or sub-vertical freeze-tubes will be more attractive.

The Muir Wood design method is regarded as the standard for thin tunnel linings taking into account the interaction of the lining with the ground. It has tentatively been suggested as a method for designing circular ice-walls around tunnels, but further investigations are necessary to confirm the validity of the method in relation to the relatively thick ice-walls involved. The characteristic load, stress and temperature patterns are illustrated in Figs 4.6, 4.7 and 4.8, and are defined by formulae for elastic analysis of shallow tunnels in soft ground.

Axial stress [after Link et al. (1976)]

$$\sigma_1 = \frac{N}{A} = \frac{p_v r_o}{t \times 1}\left(1 + \frac{y}{r_s}\right) \qquad [4.14]$$

where N = normal force, A = cross-sectional area, p_v = vertical pressure

ENGINEERING DESIGN

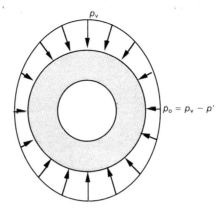

Envelope of initial load on undeformed tunnel

Fig. 4.6. Tunnel load nomenclature (cross-section, FTs parallel to tunnel)

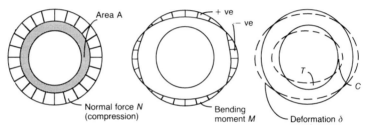

Fig. 4.7. Axial force, bending moments and deformation

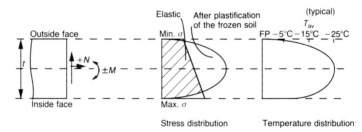

Fig. 4.8. Stress and temperature distributions

Bending stresses [after Muir Wood (1975)]

$$\sigma_2 = \pm \frac{M_{max}}{z} \qquad \theta = 0 \text{ and } \theta = 90° \qquad [4.15]$$

with

$$M_{max} = \pm \frac{p' \, r_s^2 \, E_f I (1+\nu)(5-6\nu)}{6 E_f I (1+\nu)(5-6\nu) + 2 \, r_s^3 \, E_u}$$

where M = bending moment, $p' = p_v - p_o$, ν = Poisson's ratio, E_u = Young's modulus unfrozen, $I = t^3/12$, $z = t^2/6$ and E_f is replaced by $E_f/(1-\nu_f^2)$ for a continuous length of frozen tunnel.

Combined stresses

$$\sigma = \sigma_1 \pm \sigma_2 (\theta = 0 \text{ and } \theta = 90°) \qquad [4.16]$$

Note 1 σ_1 must always be greater than σ_2 to avoid tensile stresses
Note 2 Allowable stress

$$\sigma_a = \frac{\sigma_1}{\text{Factor of Safety}} = \frac{q_f(t)}{\text{Factor of Safety}}$$

Deformation
Due to axial stress

$$\delta_a = \frac{2p_o r_i r_o^2}{E_f(r_o^2 - r_i^2)} \qquad [4.17a]$$

Due to bending [after Muir Wood (1975)]

$$\delta_{\text{maxb}} = \pm \frac{p' r_s^4 (1+\nu)(5-6\nu)}{18 E_f I (1+\nu)(5-6\nu) + 6 r_s^3 E_g} = \pm \frac{M_{\text{max}} r_s^2}{3 E_f I} \qquad [4.17b]$$

giving maximum deformation of

$$\delta = \delta_a + \delta_{\text{maxb}} \qquad [4.17c]$$

Underpinning

Apart from the cofferdam effect, the strength property of frozen ground may be utilized to safeguard existing structures during new works by underpinning or soil retention. This is particularly suitable in confined spaces as illustrated in Fig. 1.9 and described in examples in Chapter 7 at *Blackpool*, *Bromham Bridge*, *Brussels* (Belgium), *Burgos* (Spain), *Gibraltar* and *São Paulo* (Brazil).

The bearing capacity of frozen ground is estimated using the standard soil mechanics approach embodied in the Terzaghi equation [4.18], as given by Andersland and Anderson (1978). However, as the strength is temperature and time dependent, the parameter values selected must be carefully chosen to match the warmest thermal conditions which will apply during the working life of the ice-body.

$$q = cN_c S_c + pN_q S_q + \tfrac{1}{2} \gamma B N_\gamma S_\gamma \qquad [4.18]$$

Thermal design

Cooling proceeds in three distinct and successive stages, i.e. from ambient temperature to the freezing point, at constant temperature through the latent heat transition, and below the freezing point. Thus sensible heat has to be extracted from unfrozen and from frozen soil, and the latent heat during the change of state from liquid (water) to solid (ice). This activity takes place simultaneously along every freeze-tube, the 'field' from each being affected by its neighbours as refrigeration progresses, yielding a temperature field similar to that of overlapping magnetic fields, and temperature distribution curves as characterized in Fig. 1.5.

The theoretical quantity of heat to be removed can be readily calculated, examples of simple calculation methods having been given by Jumikis (1966), Collins and Deacon (1972), Shuster (1972), Sanger and Sayles (1979) Muzás (1980), Frivik and Thorbergsen (1980) and Jessberger et al. (1985). But the three-dimensional estimate of the thermal transfer-rate and efficiency in a medium whose thermal conductivity and heat capacity parameters vary — from stratum to stratum, with direction of heat flow, with variation of water content, and between the frozen and unfrozen states — is complex. Refined computations are possible with FE techniques, the 'proven' programs for which are not always readily available; they incorporate the experience of the specialist practitioners and the operational characteristics of their equipment; users can take into account all available information about the site conditions.

The selection of number and spacing of freeze-tubes must take account of many factors. The refrigeration capacity needed is directly proportional to the total length of freeze-tubes; the closer the freeze-tubes are placed the greater their number and total length, and the greater the cost of installing them; however ice-wall closure is achieved more quickly when the freeze-tubes are closely spaced. When the necessary ice-wall thickness is small, usually when the depth is small, the advantage lies with closely spaced freeze-tubes.

When the scale of the (freezing) project is small, as with many in the civil engineering field, the refrigeration plant can conveniently be sized to cool the brine to 'working' temperature — the pulldown — within a few days (two to five), and to generate the ice-wall in a few weeks; as the scale increases so this becomes less practical and several days (up to 20) may be needed for the pulldown, and the PFP may be up to four months. See Chapter 5.

Other design considerations

Water movement In general groundwater flow-rates of up to 2 m/day at brine temperatures, and 20 m/day at Liquid Nitrogen (LN) temperatures, are not significant. To minimize the possibility of a weak or incomplete ice-wall on the upstream side, greater flow-rates can be accommodated by installing more closely spaced, or multiple rows, of freeze-tubes, particularly on the upstream side. Lowering the freezing temperature, or pre-grouting the voids with a water−cement−bentonite mix to reduce the in situ permeability (taking care not to reduce the moisture content below 10%), or a combination of these, are further positive steps which can be effective.

If water is being extracted nearby by pumping it may be necessary to curtail the rate of extraction, or stop pumping altogether, to reduce the flow-rate at the ice-wall boundary to an acceptable value during the currency of refrigeration. The radius of influence, determined by rate of pumping and extent of drawdown, rarely

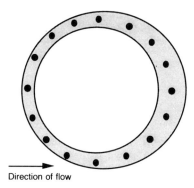

Fig. 4.9. Effect of groundwater flow on ice-wall growth

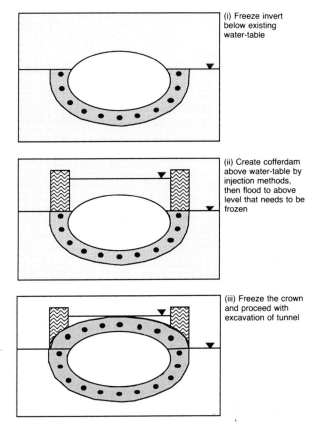

(i) Freeze invert below existing water-table

(ii) Create cofferdam above water-table by injection methods, then flood to above level that needs to be frozen

(iii) Freeze the crown and proceed with excavation of tunnel

Fig. 4.10. A method for local increase in water content (cross-section)

exceeds 50 m. The effect of cross flow or extraction to one side of the area being frozen is illustrated in Fig. 4.9.

With mechanical refrigeration a change of secondary refrigerant to (e.g.) lithium chloride will be required if the circulation temperature is to be reduced below $-40°C$; alternatively LN or liquid carbon dioxide can be considered (see Chapter 5). It is recorded that flow-rates of up to 50 m/day have been successfully frozen using LN, albeit with significant increase in consumption [see Shuster (1972)].

An analytical method, verified against a full-scale flow of 4 m/day, has been suggested by Weiler and Vagt (1980).

Low moisture content

In order to achieve both strong and impermeable characteristics, it is essential that the moisture content be sufficient to bond the soil grains and complete the monolithic matrix. A minimum value of 10% is usually required. When the in situ moisture content is below this value water must be introduced in a controlled manner. This may be achieved by an inground 'sprinkler' irrigation system, i.e. via strategically placed perforated tubes in horizontal borings above the target volume [Gonze *et al.* (1985), Maishman (1988)], or by flooding a contained volume, e.g. within a boundary of grouted ground (Fig. 4.10). Further details are given in Jessberger (1991).

High voids ratio

A high voids ratio is conducive to groundwater flow, with a consequent adverse effect on freezing, or to rapid drainage leading to a reduction in moisture content below the critical value. Both effects have been discussed above.

References

ANDERSLAND O. B. and ANDERSON D. M. (eds) 1978. *Geotechnical engineering for cold regions.* McGraw-Hill, 566pp.

AULD F. A. 1983. Design of temporary support for frozen shaft construction. *Proc. 1st NSGF Progress in AGF.* BGFS, Nottingham, 1–8.

AULD F. A. 1985a. Freeze-wall strength and stability design problems in deep shaft sinking — is current theory realistic? *Proc. 4th ISGF, Sapporo,* **1**, 343–350.

AULD F. A. 1988a. Design and installation of deep shaft linings in ground temporarily stabilised by freezing: Part 1: shaft lining deformation characteristics. *Proc. 5th ISGF,* Nottingham, **1**, 255–262.

AULD F. A. 1988b. Design and installation of deep shaft linings in ground temporarily stabilised by freezing: Part 2: shaft lining and freeze-wall deformation compatibility. *Proc. 5th ISGF,* Nottingham, **1**, 263–277.

COLLINS S. P. and DEACON W. G. 1972. Shaft sinking by ground freezing in Ely-Ouse Essex scheme. *Proc. ICE,* 7506S, May, 129–156, 319–336.

DOMKE O. 1915. On the stresses in a frozen cylinder of ground used for shaft sinking (in German). *Glückauf,* **51–47**, 1129–1135.

EINCK H. B. and WEILER A. 1982. Experiences and investigations using gap freezing to control ground water flow. *Proc. 3rd ISGF,* Hanover NH, **1**, 193–204.

FRIVIK P. E. and THORBERGSEN E. 1980. Thermal design of artificial soil freezing systems. *Proc. 2nd ISFG,* Trondheim, **1**, 556–567.

GARDNER A. R. 1985. *The creep behaviour of frozen ground in relation to artificial ground freezing.* University of Nottingham, *PhD thesis,* 331pp.

GONZE P. *et al.* 1985. Sand ground freezing for the construction of a subway station in Brussels. *Proc. 4th ISGF,* Sapporo, **1**, 277–284.

HALL E. W. and SCOTT R. A. 1968. Strength of mineshaft walls in shaft sinking by freezing. *Tech Mem TM52, Cementation Research Ltd.*

HAMPTON C. N. 1986. *Strength and creep testing for artificial ground freezing,* University of Nottingham, *PhD thesis,* 221pp.

ISGF Working Group 2 1991. Frozen ground structures — basic principles of design. *Proc. 6th ISGF*, Beijing, **2**, 503–516.

JESSBERGER H. L. 1991. Opening address. *Proc. 6th ISGF*, Beijing, **1**, 399–403.

JESSBERGER H. L., BÄSSLER K. H. and JORDAN P. 1985. Thermal calculation for ground freezing with liquid nitrogen. *Proc. 4th ISGF*, Sapporo, **2**, 95–101.

JESSBERGER H. L., KLEIN J. and EEBEL W. 1979. Shaft sinking in oil sand formations. *Proc. 7th EC SM & FE*, Brighton, Sept., **1**, 189–194.

JUMIKIS A. R. 1966. Thermal soil mechanics. Rutgers University Press.

KLEIN J. 1978. Time and temperature dependent stress–strain behaviour for frozen Emscher-marl (in German). *Rpt Ruhr University*, Dec., Series G.

KLEIN J. 1980a. Calculating the strength of frost walls in freezing shaft construction (in German). *Shaft construction study group, Bochum, 14pp.*

KLEIN J. 1980c. Structural design of freeze shafts in frictionless clay formations taking account of the time factor (in German). *Glückauf*, **41**, 51–56.

KLEIN J. 1981a. Dimensioning of ice-walls around deep-freeze shafts sunk through sand formations of type $B=2$, under consideration of time (in German). *Glückauf Forschungshefte*, **42**–3, June, 112–120.

KLEIN J. 1985b. Influence of friction angle on stress distribution and deformational behaviour of freeze shafts in nonlinear creeping strata. *Proc. 4th ISGF*, Sapporo, **2**, 307–315.

LADANYI B. 1982. Ground pressure development on artificially frozen soil cylinder in shaft sinking. *Special Vol: Prof De Beer: Inst Geotech Brussels*, 21pp.

LADANYI B. 1992. Design of shaft linings in frozen ground. In Bandopadhyay and Nelson (eds), *Mining in the Arctic*. Balkema, 51–60.

LIBERMAN Y. M. 1960. Method of calculation of the thickness of the ice-soil cylinder wall. *Mining Inst. USSR Academy of Science*.

LINK H., LÜTGENDORF H. O. and STOSS K. 1976. Instructions for the design of shaft linings in unstable ground (in German). *Glückauf*, 43pp.

MAISHMAN D. 1988. A short tunnel in Seattle frozen using liquid nitrogen cascades. *Proc. 5th ISGF*, Nottingham, **2**, 561–562.

MUIR WOOD A. M. 1975. The circular tunnel in elastic ground. *Géotechnique*, **25**–1, 115–127.

MURAYAMA S. et al. 1988. Ground freezing for the construction of a drain pump chamber in gravel between the twin tunnels in Kyoto. *Proc. 5th ISGF*, Nottingham, **1**, 377–382.

MUZÁS F. 1980. Thermal calculations in the design of frozen soil structures. *Proc. 2nd ISGF*, Trondheim, **1**, 545–555.

PANG J. 1991. The construction of East air shaft, Panji #3 colliery by freezing method 415 m depth. *Proc. 6th ISGF*, Beijing, **1**, 345–350.

SANGER F. J. and SAYLES F. H. 1979. Thermal and rheological computations for artificially frozen ground construction. *Proc. 1st ISGF*, Bochum, **2**, 95–118.

SHOCKLEY W. G. and THORBURN T. H. 1978. Suggested procedure for description of frozen soils. *Geotech. Testing J.*, **1**–4, 228–233.

SHUSTER J. A. 1972. Controlled freezing for temporary ground support. *Proc. 1st RETC*, Chicago, 863–894.

VYALOV S. S. 1962. The strength and creep of frozen soils and calculations for ice–soil retaining structures. *CRREL Transl 76*, 300pp.

WEILER A. and VAGT J. 1980. The Duisberg method of metro construction — a successful application of the gap freezing method. *Proc. 2nd ISGF*, Trondheim, **1**, 916–927; **2**, 126–128.

5. Refrigeration systems

Refrigeration methods

The 'conventional' and long-proven method, often referred to as the *Brine Method*, utilizes industrial mechanical refrigeration plant to cool calcium chloride brine, a heat transfer coolant — sometimes termed secondary refrigerant — which is then circulated through a closed circuit pipework system before returning to the plant for re-cooling. A variant of this system dispenses with the coolant, the primary refrigerant itself being circulated through a single closed circuit system.

The principal alternative method is to pass a *cryogenic* liquid refrigerant, e.g. liquid nitrogen, directly to the circuit, controlled to evaporate within the freeze-tubes, and then allow the resultant gas to exhaust to the atmosphere. It is impractical and uneconomic to attempt on-site reliquefaction of such gases.

On-site mechanical plant

The components include a compressor and prime mover, condenser, chiller/evaporator, intercooler, pumps, float valves, oil separator, cooling tower, instrumentation and controls. An evaporative condenser may replace the condenser and cooling tower arrangement. The prime mover is usually an electric motor, but in remote situations, or those with unreliable electricity supply systems, a diesel engine drive may be substituted.

The ammonia or freon charged compressor system, illustrated diagrammatically in Fig. 5.1, provides chilled brine, usually a calcium chloride solution of Specific Gravity (SG) 1·24 to 1·28, which is then pumped around the closed circuit freeze-tube arrangement via distribution mains at temperatures within the range $-20\,°C$ to $-40\,°C$. In most circumstances a lowest brine temperature of $-25\,°C$ will achieve the desired ice-wall thickness reasonably quickly at the optimum mechanical efficiency of two-stage compressors. Figure 5.2 shows the freezing point

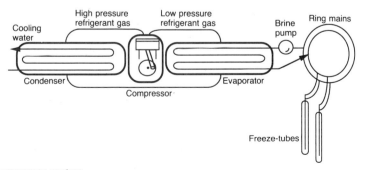

Fig. 5.1. Circulatory compressor system

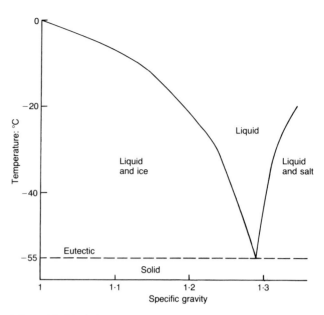

Fig. 5.2. Characteristics of calcium chloride brine

characteristics of calcium chloride brine; it will be seen that too dense an SG is potentially as serious as a weak solution: either could freeze in the evaporator.

Two-stage compression of ammonia (refrigerant R.717) is common in the UK and US; single-stage ammonia compressors with a rotary booster are also used in North America to achieve the desired low brine temperature, while some European contractors use freon (refrigerant R.22) as the primary refrigerant. Ammonia, though toxic, has attractive thermodynamic properties which are understood and well documented, is low in cost, and does not contribute to the greenhouse effect. The unit price of freon is five times that of ammonia, and the cost of charging a system with freon taking into account density differences is up to seven times more. Taking into account the difficulty in recognizing leakages and the ozone depletion potential of CFC gases (e.g. freon), ammonia is staging a recovery in popularity. Fuller discussion on ammonia and its safe use appears in Chapter 8.

Reciprocating and screw-type compressors are suitable. The latter have controlled adjustment (reduction) of output as the load falls, once the ice-wall has formed and reached its design thickness.

Users of freon sometimes employ it as a single refrigerant, i.e. it is circulated directly around the freeze-tubes instead of brine. Elimination of the double temperature step results in a gain of 15% in thermal efficiency and a refrigeration temperature of $-45\,°C$ is possible, but a larger pump is needed (unless direct expansion at 12 bar self motivating pressure difference is employed) together with hermetical glands. As it has no smell, leak detection is difficult and expensive.

REFRIGERATION SYSTEMS

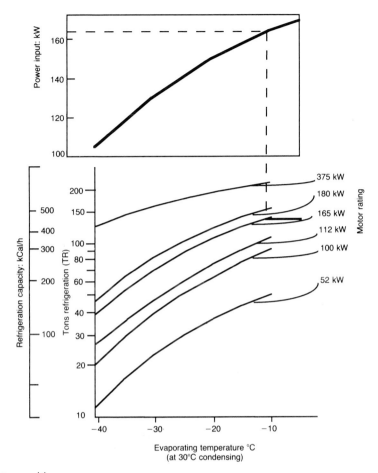

Fig. 5.3. Typical plant capacities

Typical plant capacities and characteristic output curves are shown in Fig. 5.3. It should be noted that utilization of the full potential output of a compressor is dependent on the size of motor fitted; at the warmer evaporating temperatures the power available effectively acts as a cut-off and the actual capacity curve becomes a horizontal line, as illustrated for the 165 kW (220 hp) unit. This can be particularly significant during the early part of the primary freeze-period — the 'pull-down' stage — since the activity will be on the critical path of the contract. Similarly, the condensing capacity must be adequate for the full range of operating conditions. A heat rejection capacity of up to 400 kcal/h per meter of freeze-tube at evaporating temperatures of −30°C is needed.

For mobility and ease of erection most ground freezing contractors use purpose-built package units (the main components are permanently mounted on skid-frames), in containers or on trailers. Typical package units are shown in Figs 5.4 and 5.5.

Mechanical plant requires power and water services and, if in a sensitive area, acoustic shrouds to minimize noise disturbance.

Fig. 5.4. Self-contained, medium size, two-stage reciprocating compressor plant of 84 TR (250 000 kcal/h) capacity (at $-30\,°C$ evaporating). Photo courtesy of BDF Co Ltd.

Fig. 5.5. Self-contained, large size, screw compressor plant of 170 TR (500 000 kcal/h) capacity (at $-30\,°C$ evaporating). Photo courtesy of BDF Co Ltd.

Building the ice-wall may typically take from three to twelve weeks according to the conditions being tackled and the capacity of the freeze-plant installed. Actual plant selection will take into account programme constraints and safety in the event of breakdown.

As a general rule the maximum capacity needed during the 'pull-down' stage should be supplied from two or more units. Then, at the construction/ice-maintenance stage when the security is most important and the refrigeration duty is lowest, the unit(s) no longer needed will provide standby backup against breakdowns, as required by British Standards. When the scale of work is small and the duty can be satisfied by a single unit of plant, a second unit must be kept available on standby for service at short notice.

Off-site produced liquefied gas

Bulk delivery of LN by road tanker into Vacuum Insulated Storage Tanks (VIT), a commercial reality throughout the UK, and in most industrialized countries, from the 1950s, offers a powerful, simple and practical refrigerative agent to the ground freezing process. Its self pressure, while trying to vaporize in the VIT, is sufficient to force the liquid through the circuit; its latent heat is given up to the freeze-tubes at $-196°C$; the resultant gas is exhausted to the atmosphere — expendable after one cycle, still at a low temperature; nitrogen is inert and constitutes approximately 80% of the air. A typical arrangement is shown diagrammatically in Fig. 5.6; a VIT is illustrated on site at Chelsea, in Fig. 7.44.

A freezing system based on LN has no moving parts, is silent,

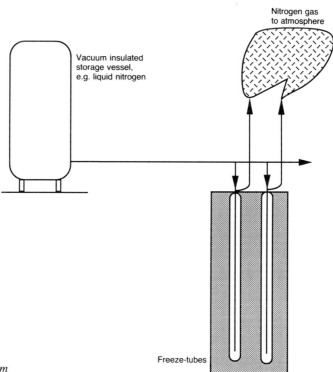

Fig. 5.6. Expendable system

and requires no major services although an electrical supply of domestic scale for instrumentation is useful; refrigeration is fast, typically taking two to six days to form a contiguous ice-wall. The scale of works that can be accommodated is limited by the capability of the supplier to deliver the product to the site at an adequate rate, a function of the location of the site relative to that of the air separation production plant. However, as the product is lost to the atmosphere after only one cycle, the cost of ice-maintenance for lengthy construction works may be excessive, even though the scale may be small enough to permit or justify its use. This effectively restricts LN freezing to small-scale exercises or emergency applications. See also Chapter 9.

Distribution systems

The common denominator in all ground freezing systems is the pipework arrangement which permits circulation of a cold medium within the volume of ground to be frozen.

Brine systems

The components needed are a brine balance tank, an insulated supply or delivery main, the freeze-tube assemblies, a return or exhaust main, and appropriate pumps, valves and instrumentation. Brine systems rely on the pumped circulation of large volumes of brine and a small temperature differential between flow and return. Typically the differential falls progressively from around 15°C to about 5°C over the first ten to fifteen days of the PFP, and eventually to only a degree or two at the ice-maintenance stage (see Fig. 5.7). Depending on the freeze-tube spacing, the plant capacity, and the refrigeration load (i.e. the scale of the project), the latent heat stage, from commencement of refrigeration to ice-wall closure, may occupy between three weeks and three months.

The brine mains will be sized on normal hydraulics principles to achieve laminar flow, typically of 150 to 250 mm diameter, with 50 to 80 mm insulation to minimize extraneous warming. Distribution of brine from the mains to the freeze-tubes is most efficiently achieved via ring mains (typically of 75 to 120 mm diameter pipe), since this ensures an equal pressure drop across each freeze-tube, and therefore similar flow-rates and heat

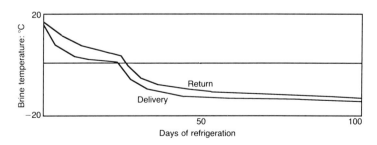

Fig. 5.7. Brine temperature characteristic 'pull-down' curves

REFRIGERATION SYSTEMS

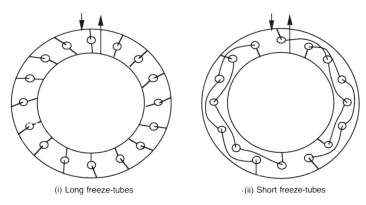

(i) Long freeze-tubes   (ii) Short freeze-tubes

Fig. 5.8. Alternative circuit connections (plan views)

extraction (other conditions being equal). Fine tuning of each freeze-tube is possible with the control/isolating valves. Bleed valves are needed at each freeze-tube header to release air pockets.

The freeze-tubes may be connected to the ring mains either all in parallel with each other or as parallel groups of two or more in series. Figure 5.8 shows a parallel and a 4-series × 4-parallel arrangement.

Hydraulic considerations, i.e. head loss against pump characteristics, will determine whether series connections are practical; if so such arrangements will be the most economical. Deep shafts, where freeze-tubes are longer than 50 m or so, will usually adopt parallel systems; small, shallow pits of say 3 m diameter and 6 m depth could, from hydraulic considerations, be connected as a fully series system, but this is undesirable in that should there be failure of one freeze-tube the whole system will be ineffective until the repair is located and rectified, or the errant freeze-tube is bypassed. For this reason there should be not less than two parallel circuits, and the connections to the supply and return mains should be so arranged that no pair of neighbouring freeze-tubes is in the same series circuit. Then, if one freeze-tube fails, the worst spacing between functional freeze-tubes during the repair work is double, which should just be sufficient to maintain the status quo.

The 'normal' direction of brine flow is down the inner pipe and up the annulus. This delivers the coldest brine to the bottom where ice-growth will therefore be greatest, resulting in elongated pear shaped ice-columns. This is of advantage when an increase in ice-wall thickness with depth is wanted — the norm. Occasionally, if the thermal conditions near the surface are more severe than at depth, flow in the reverse direction to invert the pear shape, either initially or throughout the exercise, may be desirable.

Since the heat transfer rate for turbulent flow is significantly greater than for laminar flow, the difference in diameters between inner and outer pipes of the freeze-tube assembly should be small to achieve a low Reynolds number, i.e. laminar flow in the inner pipe and a high R_e in the annulus, particularly for deep shaft freezings.

During their working lives the outer elements of the freeze-tubes have to withstand severe stresses generated mainly by differential contraction between themselves and the frozen ground to which they become bonded, and to earth/rock pressures. The brine for which they act as conductors must not be lost to the strata since it would then constitute an anti-freeze at the temperatures in hand. The material of the freeze-tube outers must therefore be of good quality, as must be the means of connection of successive lengths. Seamless steel tube of high quality, e.g. J55 or N80, is frequently specified for deep shaft freezings, together with either self sealing taper-screw threaded connections of equal strength to the pipe itself (UK practice), or welded joints. Records show that, where freeze-tubes have fractured in service, failure was usually at a joint.

The inner tube may be standard steel tube with threaded or welded couplings, or continuous lengths of polythene tubing. Apart from the ease of installation of single lengths of polythene, the possibility of breakage at an intermediate level (e.g. at a threaded joint) is removed; such a breakage at an intermediate level, which may not be detected until some weeks after the event, would result in a short circuit of the flow path and much reduced refrigerative effect at the lower end of the freeze-tube. A breakage at the point of suspension at the top of the freeze-tube results in total loss of refrigerative effect in that freeze-tube. This condition is easily observed since the temperature differential between flow and return for that tube is also lost. Once recognized a repair can be undertaken. Polythene, however, exhibits a high coefficient of contraction with reduction of temperature and is buoyant. These characteristics can be accommodated by fitting an appropriate length of sliding steel weight pipe at the lower end.

Cryogenic systems

Off-site produced liquefied gas (cryogenic) systems operate at very high temperature differentials and relatively low flow rates. Thus a delivery main is usually only $c.25$ mm diameter; it will be of stainless steel and constitute the inner component of up to 8 m long, 80 mm diameter sections of a prefabricated twin tube, vacuum-insulated assembly. With 'conventional' insulation around copper pipe, the overall diameter would need to be at least 300 mm to even approach the thermal efficiency of vacuum insulation, an essential feature if early vaporization (due to heat input) is to be avoided, and control of pipeline pressure to reasonable levels, is to be achieved.

As with the delivery main, the inner element of the freeze-tubes will also be small, rarely larger than 20 mm. Consequently the outer need only be 40 mm/50 mm diameter.

The direction of flow will be chosen on the same basis as for brine freezing; connection of freeze-tubes in series, however, is less appropriate and often undesirable since the rapidly falling temperature gradient along the circuit is reflected in lesser growth of the frozen ground column around it.

It is relatively simple to control the delivery rate of LN such that the primary (or only) freeze-tube in a circuit is always effectively fully charged with *liquid* which, when evaporating, releases the latent heat and induces the desired uniform *ice* growth along the length of the freeze-tube. When the now gaseous nitrogen leaves that freeze-tube its temperature is still colder than $-100\,°C$, but the potential energy represented by that temperature is the sensible heat, and the heat absorbed per unit length of flow path is lower and falling. In fact when lost to the atmosphere the nitrogen gas will still be too cold for comfort. Any attempt to utilize the cold gas in second or later series freeze-tubes will, in the same time-scale, lead to small ice-columns of ever decreasing size towards the exhaust, which are rarely of practical use.

With horizontal arrays of freeze-tubes in a 'cover' pattern as in Fig. 1.9(ii), it is inevitable that some tubes will be directed upwards. This can lead to pressure build-up and loss of flow if the rate of evaporation allows too much *liquid* nitrogen to reach the (low) header, thus trapping *gaseous* nitrogen behind in the main body of the freeze-tube. This problem can be dealt with by installing a cascade system; semicircular nylon clip-on weirs are attached to the inner pipes at 1 m intervals as they are installed [Maishman (1988)]. This system also encourages more even refrigeration over the length of the freeze-tube, and should therefore be used in each horizontal or sub-horizontal freeze-tube whether it is level, or inclined upwards or downwards.

References

BRITISH CRYOGENICS COUNCIL 1991. *Cryogenics safety manual.* 3rd edition. Butterworth-Heinemann, Oxford.

MAISHMAN D. 1988. A short tunnel in Seattle using liquid nitrogen cascades. *Proc. 5th ISGF*, Nottingham, **2**, 561–562.

STANFORD W. and HILL G. B. 1972. *Cooling towers, principles and practice.* Carter Industrial Products Ltd.

WESTAWAY C. R. and LOOMIS A. W. (eds) 1977. *Cameron hydraulic data.* Ingersoll-Rand, New Jersey.

6. Construction using AGF

With contribution by Dr F. Alan Auld

The practical elements of executing works based on the ground freezing method involve the installation and commissioning of the refrigeration system, generation of the necessary ice-wall, excavation of the sub-surface space while maintaining the integrity of the ice-wall, and construction of the permanent lining following which the ice can be allowed to thaw. Each stage demands careful monitoring and control for safe and efficient completion.

The first three stages require consecutive time slots in a bar-chart programme, while the fourth stage — installation of the lining — may proceed concurrently with advance of the excavation, or as a further consecutive and separate activity as soon as the excavation is complete. A delay between stages one and two has little effect on contract efficiency, but once refrigeration has commenced any interruption of associated activities may lead to significantly increased costs. Phased programmes are cost-effective when two or more excavations are to be made at the same location.

Freeze-tube placement

As we have seen, the prime requirement is that a series of freeze-tubes be placed to an accurately controlled pattern. It is of little concern to the final result whether the outer element of the freeze-tube assembly is forced into the ground or lowered into a pre-formed hole, so long as it performs reliably and durably. Various methods are available, and selection will depend on speed of installation and costs.

Vertical and angled freeze-tubes

Rotary mud-flush drillhole
Undoubtedly the majority of freeze-tubes are installed into drillholes formed by rotary drilling, any loose surface strata being supported by casing and the remainder (most) of the depth supported by the conventional bentonite mud-flushing system. Nearly all drilling contractors include this method of open-hole or cored drilling in their repertoire. Air-flush, which tends to create or enlarge voids in loose strata, is not recommended for ground freezing projects.

Advance of the borehole is accomplished by means of a power driven rotary table through which the drill-pipe passes. The bottom-mounted bit, weighted with drill collars (heavy lengths of drill-pipe), cuts the strata as it is rotated, the cuttings being carried to surface by the circulation of drilling mud down the drill-pipe and up the annulus. The pressure of the thixotropic mud ensures that the walls of the hole remain stable.

Accuracy of drilling to depths of 250 m or so is achieved by adopting the 'pendulum' drilling technique, i.e. a drill column of greater weight than that needed for cutting purposes is provided, the excess weight being 'suspended' from the rig mast. In this way verticality to within 0·5% is usually possible. When excessive inaccuracy occurs even when practising pendulum drilling, or when the design includes angled holes, the technique of directional drilling will be needed. This method relies on the flow of the drilling fluid to achieve rotation of the bit, the rotary table being redundant. By mounting a turbo-drill or Down-Hole Motor (DHM) immediately above the bit, drill-pipe rotation above that level is avoided, and the orientation (azimuth) of drilling can be controlled. This affords the opportunity to install a bent sub and non-magnetic

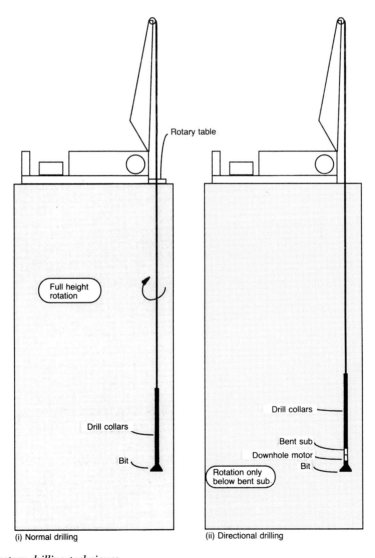

Fig. 6.1. *Alternative rotary drilling techniques*

collar (non-rotating) just above the DHM. The differences between the normal and directional drilling set-ups are illustrated in Fig. 6.1.

By directing the angle of the bent sub diametrically opposite to the direction of deviation, the borehole will tend to recover its alignment. This is checked and controlled by carrying out single-shot down-hole gyro surveys at regular intervals, typically 9 m when drill-rods are added. The single-shot film disc can be developed and read, and computations made and plotted, as soon as the instrument is retrieved from the borehole.

On completion of drilling a multi-shot survey on a single traverse of the instrument down the borehole provides an as-completed record of the course of the hole in sub-surface space. A suite of surveys for the complete freeze-tube installation provides the control for monitoring of the subsequent growth of the ice-wall. See also *freeze-tube survey* later in this chapter.

Rotary drill-and-case drillhole (overburden drilling)
By a combination of percussion, rotation and high pressure water flushing, a drill-rod and enclosing pipe are sunk concurrently through the overburden. The pipe is tipped with a ring bit and the drill-rod with a cross bit, both of tungsten carbide cutting faces. The cross bit projects beyond the ring bit to act as a pilot.

Independent and instant control of percussion, rotation, feed rate and flushing allows adjustment to suit penetration of any type of overburden as it is encountered. The rod or the pipe may be advanced independently as required.

If drilling is to be continued after reaching bedrock the ring bit is first allowed to collar a short distance into bedrock; the remainder of the hole is drilled by the drill-rod alone. When the full depth is reached the rods are withdrawn while the pipe is left in the overburden section of the hole as casing. Maximum depths of about 15 m at 125 mm diameter and 30 m at 90 mm diameter limit the usefulness of this method for ground freezing purposes.

Shell and auger borehole
The standard percussive 'site investigation' rig offers a very convenient means of installing 200 mm or 150 mm diameter temporary casing through surface deposits for shallow freezings, or to permit open hole rotary drilling to greater depths, prior to installing the freeze-tubes. The rig is light, manoeuvrable and free of subordinate tanks, pumps and so on which form part of the complement of rotary rigs; as such it offers advantages of speed and low cost in the right conditions.

Drilled-in tube with 'lost' sealable cutting shoe
Although practical, this method is more commonly used for (and is described under the heading of) horizontal freeze-tubes — see below.

CONSTRUCTION USING AGF

Driven/vibrated tube
In soft ground of limited depth it may be practical to drive a sealed tube of appropriate diameter speedily and without damage. Damage limitation will be a function of the quality of the tubing and the drive anvil, elements which tend to increase costs. Although the appraisal may be favourable, this method is rarely used, if only because few ground freezing practitioners carry suitable equipment; they would have to improvise or hire in.

Horizontal and sub-horizontal freeze-tubes

Rotary drill-and-case drillholes (overburden drilling) have already been described.

Drilled-in tube with 'lost' sealable cutting shoe
Drilling through the wall of a pit which is performing as a cofferdam, or through ports in the tailskin of a Tunnel Boring Machine (TBM), can lead to the ingress of water and soil, an inconvenient and potentially hazardous situation. Control of such inflows is effected quite simply by drilling through a gland or stuffing box which offers a tight fit to the passage of the drill-pipe.

The arrangement illustrated in Fig. 6.2 was undertaken at *Runcorn* (Chapter 7) where the 9 m diameter working pit had been constructed of 1·2 m diameter secant bored piles. Cored holes 300 mm deep were drilled at each freeze-tube position ready to receive casing-stubs fitted with a flange and gate-valve; the stubs were set securely in place with epoxy resin. The low pressure stuffing box assembly, incorporating a bleed-valve and (second) gate-valve, could then be mounted on each valved stub successively while the outer element of the freeze-tube was drilled into place as if it were the drill-pipe. The bleed-valve allowed controlled 'circulation' of drilling fluid (water) balanced with limitation of the loss of fines from the strata to the volume of the hole being drilled.

Drilling was effected with a sacrificial cutting shoe which permitted normal flow of drilling fluid through a non-return valve;

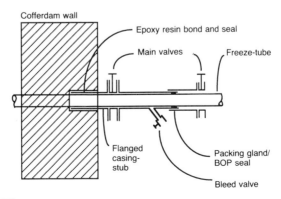

Fig. 6.2. Stuffing box arrangement

on completion of drilling, and before installing the inner tube, a sealing plug was screwed into place in the shoe so that brine would not be lost to the strata during the refrigeration stage. See Garnett (1981).

For deep excavations subject to high hydrostatic heads, the stuffing box needs to be designed to the principles of a blow out preventer.

Accuracy of freeze-tube placement

The gap between diverging tubes may exceed the capacity of the system to generate a competent ice-wall, leaving a 'window' or a weakness which at the excavation stage could result in flooding and/or collapse. Accuracy targets are set according to the tolerances of the design with respect to uniformity of ice-wall growth and programme time constraints. If the sole function of the ice-wall is a waterstop (impermeable barrier) the relative uniformity of the ice-wall cross-section is of low importance, whereas fusion of *all* neighbouring columns is paramount; a low level of individual accuracy is acceptable in this case provided that the course of each hole with respect to its neighbour stays within acceptable limits. When the function of the ice-wall is structural (retaining wall) the uniformity of cross-section assumes primary importance, and becomes very significant with increase in depth.

Typically a target from which drilling should not stray may be a rectangle of 1·5 m (tangentially) × 2·5 m (radially), centred on a point on a radius through a freeze-tube 0·25 m outside the freeze circle. A rigorous target is to control the course of the borehole within 0·5 m radius of the plumb line.

When freeze-tubes diverge beyond acceptable limits, either redrilling at one or more identified locations, or the drilling of a supplementary freeze-tube becomes necessary to close wide gaps and restore a regular pattern at all depths (see Fig. 6.3). At the *Three Valleys* site the previous grouting works proved sufficient to tolerate unmeasured drilling inaccuracies. At *Asfordby* it was possible to measure the accuracy of freeze-tube placement at depth, and thereby verify the surveying results. See Chapter 7.

Freeze-tube survey

We have seen that unequal spacing of freeze-tubes is undesirable and its avoidance may be crucial. Various devices are available to survey the course of the borehole; they include the following.

Pajari (mechanical plumb)
This instrument contains a sphere which pivots independently with respect to both inclination and azimuth, and incorporates a time-delay which can be pre-set to lock the pivoting mechanism. Thus, by allowing the instrument to stand at the desired level during the locking action, the inclination angle and plan direction can be read when it is recovered. Readings are accurate only to about 0·5°.

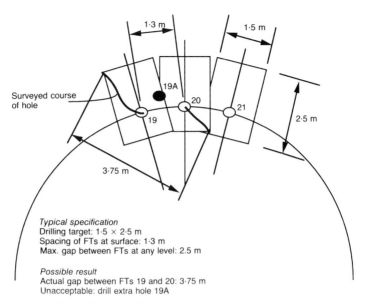

Fig. 6.3. *Typical drilling target for shaft freeze-tubes*

Inclinometer (electronic dipmeter)
Inclinometers were designed to detect deformation, usually of slopes. They are based on a probe, containing two gravity sensing transducers sensing planes at right angles, which is lowered on a graduated six-conductor, steel-cored, waterproof cable (standard length 30 m) connected to the surface readout unit; to control the orientation (azimuth) the probe is fitted with spring loaded wheels which follow tracking grooves in a purpose made guide tube. System accuracy is claimed to be within 3 mm at 25 m penetration.

These devices will operate at any angle and so are also suited to angled and horizontal freeze-tubes. A disadvantage is that the largest standard guide tube size at 85 mm is much smaller than the 125 mm bore of many freeze-tubes; annular stabilizers are therefore required to hold the guide tube steady and permit the making of accurate and repeatable surveys. See Dunnicliffe and Green (1988).

Strain gauge deflectometer
The functional length of this instrument is 2 m with a strain-gauged hinge at the mid-point. Fixed wheels at 1 m centres on the lower side, with matching spring-loaded wheels opposite, retain the unit in a uniform position relative to the casing as it is advanced along a horizontal borehole. Readings are taken at metre intervals based on calibration of the strain gauge signals. By repeating the traverse with the instrument rotated by 90° a full spatial survey is obtained.

Fotobor
This reflex-optical dip and direction indicator is a precision multi-shot instrument particularly suitable for surveying the course of

non-vertical boreholes. The instrument comprises a 45 mm diameter probe, 6·1 m to 13·3 m long, is watertight when assembled, and contains a camera and three reflective rings set at intervals of 1·5 m or 3 m. To ensure a snug fit within internal-flush casing, annular stabilizing collars are fitted externally. As the probe is advanced by drill-pipe or wireline, by the same increments as the spacing of the reflective rings, it flexes slightly as it closely follows the course of the drillhole. This flexure deflects the rings from the axis of the instrument which is then recorded by the camera (older versions) or electronically (current versions), from which coordinates and offsets are easily calculated or computed to yield a complete survey. The results can be presented graphically as orthogonal plots of deviation versus penetration or as a 'birds eye view' along the axis of the hole.

As the advance matches the ring spacing, successive readings are made at each station; this ensures great accuracy and enables back corrections to be made. The short instrument can negotiate curves of 40 m radius, while maximum accuracy is achievable for radii of up to 500 m. An accuracy of 0·2% is claimed. A penetration range of 1800 m in one run is possible.

Gyro — e.g. Eastman, Sperry Sun
Single-shot directional surveys are accomplished by photographing the position of an angle indicator, referenced to a compass card with a calibrated angle unit. The film disc records the inclination (drift angle) and the magnetic direction of the borehole. Other components of the instrument are a timing device, camera, and battery, all enclosed within a waterproof sleeve. The camera is pre-focused, and the disc trap design permits daylight loading and unloading.

A multi-shot instrument, carried within the lowest drill collar, and fitted with an advancing film carrier permits successive readings to be made within a completed hole as each length of drill-pipe is added or removed. The result of a continuous multi-shot survey is more accurate than can be obtained by compiling a borehole survey plot from a series of single-shot readings.

Monitoring

Refrigeration system

Plant performance
Monitoring of refrigeration plant power, pressures, temperatures, levels and flows follows normal routines established by the refrigeration industry to ensure efficient and safe performance. Modern plants are normally equipped with fail-safe auto shut-down systems and alarms in the event of break-downs and abnormal pressures — high head pressure and low suction pressure — or liquid levels/hydraulic pressures, according to the requirements and specification of each operator.

Brine flow and temperatures
It is essential that all sections of the circuit continuously receive their design flow of brine. The ring main distribution system ensures an equal pressure drop over each freeze-tube, thereby encouraging equal distribution of 'cold'. Total flow of brine is monitored by an in-line flowmeter, and the same can be applied elsewhere in the circuit when desirable.

The temperature difference between flow and return of brine in each freeze-tube is recorded at regular intervals, and the similarity of such readings with respect to each other is normally sufficient evidence of the proper functioning of each freeze-tube; an anomalous reading would merit investigation and rectification of the cause. Flow in individual circuits (e.g. each freeze-tube) can be checked at any time with ultrasonic flowmeters held for a short time against a convenient portion of the tube.

Loss of circulation (of coolant)
Apart from (outer) freeze-tube fracture (as described below), an infrequent occurrence is breakage of the inner pipe or its suspension device. When this happens the flow path is short-circuited, and that freeze-tube becomes ineffective. The result of prolonged malfunction of this kind is similar to inaccurate placement of the freeze-tube — a window or weakness in the ice-wall. Detection is normally easy and quick if flow and return brine temperatures are being regularly monitored.

Loss of coolant
Brine in the wrong environment, i.e. outside the circuit, constitutes an anti-freeze. It is therefore most important that fractured freeze-tubes or broken brine mains are detected quickly, and brine circulation through the affected parts of the circuit stopped while a repair is effected, to minimize loss of brine to the strata being frozen.

Ice-wall growth and integrity

Pressure relief hole
Once ice-wall closure has been completed its thickness increases both internally and externally so long as the heat extraction exceeds the rate of supply from the strata. Unless the refrigerative capacity is reduced at this stage it is inevitable that said growth will continue, with the consequence that shafts and tunnels of limited diameter may freeze solid to their centre in a relatively short time-span. The effect on the unfrozen core of the volume increase of water as it is frozen, is an increase of pressure in that zone; this affects the water trapped therein. Unless relieved, this pressure could reach a value greater than that which the developing ice-wall can sustain. The solution is simple: the provision centrally within the shaft space of a pressure relief hole, i.e. a hole furnished with perforated tubing through which the water can be expelled to surface. In addition to this essential function, such a hole also acts as a tell-tale monitor

of the fact of closure of *all* the freeze-tubes enclosing the space — it proves that the ice-wall has integrity, and confirms (or enables adjustment of) the interpretation of the effective ice-wall thickness based on the temperature observation holes (see below).

If closure is not achieved, or if it is lost, e.g. due to seepage flow of groundwater, further refrigerative effort will be needed to close the 'window', as it is termed. Locating a window may be possible by study of water level/pressure readings from observation wells or piezometers, but it is more usual to undertake mapping of borehole temperatures or cross-hole ultrasonic measurements. In either case the system has to be closed down in whole or part to permit access to the freeze-tubes to run the appropriate device. Readings can be made at short intervals of depth over the full length of each freeze-tube quite quickly, and they will soon indicate where the problem lies.

Temperature pull-down
It is important to measure the delivery and return temperatures of the brine at the plant and, if used, the brine balance tank; they are fundamental parameters for assessing the effective output of the plant and the overall heat extraction from the ground. The aim is for a quick reduction of the delivery temperature through zero to the design operating temperature, during which period the temperature difference between flow and return also falls, as discussed in Chapter 5.

Temperature observation holes
Checking the formation, growth and integrity of the ice-wall is facilitated by measuring the temperature of the strata at strategic points. It is usual to provide two or more observation holes furnished with a series of thermocouples at depths corresponding with known strata and groundwater conditions. Practice varies as to their positioning: commonly they will be sited on radii which bisect the gaps between adjacent freeze-tubes, and some $0 \cdot 5 - 2$ m outside the freeze-tube circle. By selecting the two most divergent freeze-tubes to locate one of the observation holes, or the upstream side of the installation where underground water movement is known or suspected, the 'worst' scenario will be monitored.

A variant in China, where inward ice-wall growth is curtailed (see Chapter 4), it is also usual to include an observation hole between the excavation line and the freeze-tube circle.

Piezometers
Standard piezometers may be required to monitor the water level/pressure in an aquifer outside the freezing zone which may be affected by or during the works.

Ground levels and pressures
The frost heave phenomenon may be of concern to the owners or occupiers of neighbouring property, and level measurements are recorded at and near structures, and on the sites of anticipated future works. See *Frost heave* in Chapter 3.

Uniformity of ice-wall thickness
Geophysical methods have been used to assess the thickness of frozen ground actually generated around the circumference of an excavation. It is a useful way of confirming the uniformity (competence) of the ice-wall, particularly when groundwater flow across the site is likely.

Thawing
Natural thawing commences soon after cessation of active refrigeration; it will be relatively slow depending on the thermal characteristics of the lining, and the temperature and ventilation regime in the excavation. The structural component will be lost first as the ice-wall becomes thinner, followed by the cofferdam component as 'windows' appear, but discrete masses of frozen ground may take six months or more to disappear completely. See also Back-wall grouting.

Side-effects

There do not appear to be any documented accounts of structural (strength) failure of an ice-wall. There are, however, several identifiable occasions when performance, for various reasons, has not been as intended. The usual result is flooding or collapse.

Correcting the effects of under-performance is usually time consuming and costly, and generates adverse publicity which reflects on the reputation of the method. The majority of the causes fall under such headings as inadequate site design information, insufficient monitoring or human errors.

Examples that have arisen in practice are discussed, sometimes with reference to case descriptions and other chapters, but otherwise without identification of the site, or of the parties involved. See also *Risk*, in Chapter 9.

Groundwater

Groundwater quality (salt or oil contamination)
The groundwater may be saline: if so the freezing-point will be depressed several degrees below zero (see Chapter 3), or it may be contaminated with hydrocarbon waste — oil or tar — which does not freeze solid. The salinity transition is usually gradual with depth, whereas oily residues float and will usually be concentrated at or above the groundwater table.

The presence of salt(s), usually dissolved in the pore water, inhibits the freezing process by depressing the freezing-point temperature, lengthens the time taken to reach the latent heat stage, and yields a lower frozen strength compared to that exhibited by

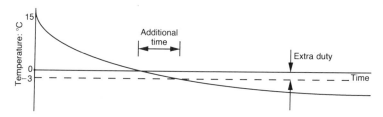

Fig. 6.4. *Effect of depressed freezing-point on pull-down*

similar non-saline soils. A colder average ice-wall temperature and/or a thicker ice-wall may therefore be needed if distress is to be avoided. The effect is most apparent in clays, which also exhibit low thermal conductivity.

Salinity is defined as the ratio of weight of salt to weight of pore water; sea water typically has a salinity of 25–35 parts per thousand, and a depressed freezing-point of about $-2 \cdot 5°C$. For a typical ground temperature of 15°C in the UK this adds some 16% to the 'pull-down' temperature range; thus allowance is made for more energy when allocating refrigeration capacity and a longer time period when assessing the length of the primary freeze-period. This can be appreciated by reference to Fig. 6.4.

Characteristic (eutectic) curves are published for individual salts — e.g. sodium chloride and calcium chloride (see Fig. 5.3) — from which the freezing-point depression can be estimated, but as most salts are impure in nature it is better (and cheaper) to verify the freezing-point by test, rather than to rely on interpretation of this value from chemical analyses.

Oily conditions are best dealt with either by displacement by pumping clean water through the zone (possibly followed by grouting) prior to freezing, and by the introduction of oil-absorbent materials as the excavation advances, according to the severity. Sloughing of the face of the excavation will occur where oil remains, but the severity will be limited if adequate precautions are taken. If infiltration of heavy oils into permeable strata above the groundwater table has taken place, ground freezing may be ineffective even if the suggested measures are tried. See *Gibraltar*, in Chapter 7.

Flow/movement

The most common source of heat input, which counters the refrigerative effort, is due to natural or induced groundwater flow, and due account must be taken in the design. Slowness in creating the desired ice-wall, or in achieving an adequate thickness, may be due to pre-existing unidentified or under-estimated flow; deterioration of an existing ice-wall may be due to (re)commencement of intermittent pumping of water to or from the strata. Counter measures, as discussed in Chapter 4, include increasing the refrigerative capacity and/or reducing/eliminating the flow.

Dewatering of adjacent ground may be required to progress other works. Careful consideration of the hydraulics will be required if it is desired to carry on the two activities at the same time: the distance between the ice-wall and the nearest well, the permeability of the strata, the quantity of water to be pumped, the extent of the drawdown, and the relative durations of the two processes will all need to be evaluated. It is unlikely that the two activities could proceed unless either the ice-wall is created before dewatering commences or the dewatering has reached a stable condition before refrigeration commences, and then the dewatering is continued in a uniform manner throughout the life of the ice-wall. See *Rogers Pass*, Chapter 7.

Multiple aquifers
The drilling mud and/or the groundwater will normally act as a heat conducting 'bond' to the strata, allowing the freeze-tubes to perform efficiently. However, in some hydrological conditions the annuli between the freeze-tubes and the strata may act as unwanted drains if not filled with solid material.

Artesian, or even sub-artesian, pressures from confined aquifers, could give rise to water flow to the surface, or between aquifers, after completion of the ground freezing scheme. Such flows could cause contamination or depletion of a water resource, or flooding, or apply hydrostatic pressures to sections of shaft lining designed for non-hydrostatic conditions. To prevent this situation the freeze-tubes can either be recovered at the end of the exercise and the holes cemented, or they can be permanently abandoned by filling the annuli with cement at the time they are installed (pre-abandonment).

Basal cut-off
Unless all sides of the excavation, including the base, are to be frozen it follows that if the frozen ground is to function as a cofferdam, the stratum in which the freeze-tubes terminate must be strong and impermeable to act as a cut-off.

Even when the site investigation data is sufficient to enable a reliable assessment to be made, it is good practice to treat the drilling of the central pressure relief hole as a further opportunity to verify the condition of the basal stratum. The recovery of continuous cores permits examination of the material for fissures or other discontinuities, when the safe minimum penetration of the freeze-tube can be decided. On occasions the evidence may suggest that the interpretation be checked in some of the freeze-holes as well. Lack of borehole evidence of weathering or discontinuities can never be *totally* reliable — the old maxim that the *actual* construction is the ultimate investigation remains true.

See Chapter 4 and the descriptions of *Belle Isle* and *Milwaukee*, in Chapter 7.

Low moisture content
There must be sufficient moisture present, generally greater than 10%, to bond the soil grains when frozen if the ice-wall is to act as a reliable retaining wall/cofferdam. Various techniques have been used to ensure this when the natural moisture content was too low; see the descriptions of *Brussels* and *Mannheim* in Chapter 7.

Thermal factors

Thermal conductivity
The rate of ice growth varies according to the thermal conductivity of the stratum (Chapter 3). Thus, in mixed strata, the ice-wall may be formed at all depths yet vary in thickness. It is essential that the monitoring, and interpretation thereof, establishes that the thickness/strength is adequate at all levels prior to the excavation reaching critical horizons. When the applied stress exceeds the supporting strength, the resultant over-stress could lead to collapse and flooding of the excavation, with attendant lengthy delay while the situation is recovered. This characteristic is noted in the descriptions at *Boulby*. See *groundwater quality* above, and *Salinity* in Chapter 3.

Frost heave
As we have seen in Chapter 3, the development of ice lenses may result in frost heave of relatively large dimensions. The effect is generally restricted to the plan area of the frozen body, and therefore only influences ground/structures in the immediate vicinity of vertical excavations, or freezings directly above tunnels.

Many works have been described where ground freezing for vertical shafts and pits has been conducted in close proximity to sensitive buildings and services, and for tunnels below buildings, without detrimental effect to the property. Jones reviewed the available UK shaft records (1982) (see Table 6.1) and international tunnel records (1993) (see Table 6.2), and emphasized the need for quality field records to validate the predictive methods being developed. Some projects particularly sensitive to heave deformation are described in Chapter 7 (e.g. *Blackpool, Bromham Bridge, Burgos* and *Lowestoft*).

At *Selby Wistow* mine (see Chapter 7) freezing was carried out for the air and fan drifts as well as for the shaft sinking. The drifts were box section reinforced concrete ducts at 30° inclination supported by the foreshaft construction and piles. The shallow strata comprised two silty clay zones separated by a water bearing sand layer near the surface, 8·8−9·2 m depth, and below 18 m depth. A frozen cofferdam terminating in clay was an attractive alternative to sheet piling, since it would offer a self-supporting boundary without the need for strutting, and utilize refrigeration plant already on site. Maximum heave of 124 mm at the fan drift and 51 mm at the air drift was recorded; this was followed by some thaw settlement. Of itself the movement was not severe, but it was noted that interaction between two frozen bodies presents problems of control and stress due to lateral expansion of the ground during freezing. See Auld (1988).

Thaw settlement
Settlement can occur during the thawing stage, usually but not necessarily following heave, and has only occasionally been significant. The extent can be controlled to fine limits by careful grouting during the early stages of the thaw period. See Chapter 3.

Rapid freezing during the PFP, followed by minimal refrigeration during the SFP, has been shown to reduce frost heave by an order of magnitude. This was achieved at *Milchbuck* (see Chapter 7), the first recorded site at which cyclic freezing during the SFP was intentionally and successfully practiced for the express purpose of minimizing frost heave when frozen-ground tunnelling was undertaken beneath a built up area of a city.

Settlement is not unusual with non-frozen tunnelling works, even with the utmost care. When such tunnelling has to pass below sensitive buildings at shallow depth, the introduction of a frozen zone between the foundations and the tunnel crown can be particularly beneficial in minimizing settlement.

Creep
Highly stressed ice-walls, particularly for shafts in deep clayey strata, may suffer inward deformation of the exposed ice-wall while it is being prepared for placing temporary or permanent lining. Being time-related, the phenomenon must be recognized and handled expeditiously, i.e. support must be positioned within a short, controlled time span. In these conditions it is inevitable that some deformation will occur; advantage can be taken of this in the lining design. See Chapter 4.

With the hanging permanent-lining method favoured in UK and North America, the excavation should include sufficient over-break to allow for the anticipated creep deformation, thus preserving the design thickness of the concrete lining. If deformation occurs more quickly than anticipated it may be necessary to reduce the height of each increment of sinking advance. In Germany it is common to cast a compressible thin lining, offering a structural component while sinking, then erect a permanent seal and full structural lining from sump to surface.

An alternative, as practised in China, is to install a temporary lining incorporating an outer skin of compressible material and an inner skin offering some measure of structural strength as the shaft is sunk, then construct the permanent structural lining from the bottom upwards. See Chapter 4.

Table 6.1 Details of major British shafts sunk with artificial ground freezing 1947–1976 [after Jones (1982)]

Ref/location and date		Depth to water-table: m	Ground conditions		Finished internal dia.: m	Depth frozen: m	Freezing circle pitch dia.: m	Freeze tube spacing: m	Freeze tube dia.: mm	Initial brine temp: °C	To closure: days	Total freezing: days	Heave: mm	Notes
			Depth below surface: m	Strata										
1 Calverton (Notts) 1947–49	No 2	19	0–105 105–196 >196	Bunter sst Permian Marls and Limestones Coal Measures	6·3	125	10·0	1·22	76–150	−20	60	265	0	No detailed observations but movement insignificant
2 Bevercotes (Notts) 1952–56	No 1		0–3 3–6 6–15 15–204 204–360	Drift K. Waterstones Green beds Bunter Sst Permian	7·3	248	13·4	1·23	125	−20	183	478	152	Heave began after 70 days
	No 2		>360	Coal Measures	7·3	252	13·4	1·23	125	−20	203	423	200	?
3 Lea Hall	No 1	2	0–20 20–69	Alluvium Keuper Sst	7·3	218	12·2	1·13	125	−19	116	534	NR	Below 143 m Bunter is known as Littleworth Sst
	No 2		69–185	Bunter Sst		222	12·2	1·13	125	−19	137			
4 Wearmouth (Tyne & Wear) 1953–55			21 21–95	Boulder Clay Permian Marls and Lst Coal Measures	7·3	108	12·2	1·13	125	−20	116		NR	
5 Cotgrave (Notts) 1955–57	No 1		0–128 128–162 162–247 247–259	Keuper-Marl K. Waterstones Bunter Sst Permian Marl	7·3	268	13·4	1·24	125	−20	153	410	429	Many fractured tubes No. 2 sunk first

CONSTRUCTION USING AGF

Project	Shaft	Depth (m)	Geology									Remarks
6 Kellingley (W. Yorks) 1958–60	No 2	>259	Lst and Breccia Coal Measures	7.3	268	13.4	1.24	125	−20	154	554	380	
	No 1	5.5 / 0–13 / 13–149	Bunter Sst / Permian Marls and Limestones	7.3	195	14.0	1.22	125		189	662	100	Excavation dia = 9.75 m. Some heave associated with grout injection
	No 2	>149	Coal Measures	7.3	195	14.0	1.22	125	−18	201	671	106	
7 Ely-Ouse (Cambs) 1968–69 (Near surface)													
Blackdyke		0–22 / >22	Chalk / Gault	4.5	24.4	7.0	0.92	114	−22	37	105	85	
Little Ouse		0–12 / 12–23 / >23	Superficial / Chalk / Gault	4.5	25.6	7.0	0.96	125 and 76	−27	31	105	49	Superficial = peats silts, sands and gravels
Lakenheath		0–15 / 15–28 / >28	Superficial / Chalk / Gault	4.5	31.7	7.0	0.92	114		37	75	24	
Lark		0–12 / 12–42 / >42	Superficial / Chalk / Gault	4.5	45.1	7.3	0.96	114	−33.7	36	83	53	
Worlington		0–58 / >58	Chalk / Gault	4.5	60.4	7.3	0.96	114	−29.7	29	140	144	55 mm settlement subsequently
Kennett		0–13 / 13–71 / >71	Superficial / Chalk / Gault	7.31	74.3	10.9	0.96	125	−22	53	200	43	

Note: The table excludes shafts at Hawthorn, Co. Durham and Boulby, Cleveland which were not frozen to the ground surface. NR = not recorded.

1 Hill (1957), Wadsworth (1957). 2 Wadsworth (1957). Smith *et al.* (1973) 3 Wood (1956). 4 Wood (1958). 5 Redfern and Pinder (1964), Wood (1957). 6 Maishman (1959), Firth and Gill (1963). 7 Collins and Deacon (1972).
These sources were checked and supplemented where possible with Foraky Limited records.

Table 6.2 Surface movements associated with tunnels constructed with AGF [after Jones (1993)]

Ref/location	Strata	z_w: m	Cover: c/m	Dia: D/m	Coolant	Type	Duration: days G	Σ	Movements H_m	S_m	z_o/R^2: m^{-1}	Notes
1 Charratt	Silt, sand and gravel	2	3·5	1·5	LN	HC	3		0	0	7·56	Fluctuating groundwater level
2 Tokyo	Loam, clay, sand Silt, sand	12	25	3·14	B	hC	45	160+	0	0	10·78	Freeze tubes 15° to horizontal
3 Washington	Clayey sand Sand and silt	1·5	2·7	3·8	B	HC		31	125	130	1·27	Closure probably earlier
4 Essen	Fill, silt	2·7	5·7	8	B	IA		21	0	60	0·61	Tunnel 7 × 9 m
5 Milchbuck	Gravel-sand-silt		6	13	B	hA			105 40	45 60	0·30	Horseshoe tunnel 12·1 × 14·2 m. GW lowered first. Heave less than 10 mm on later stages. 2–4 mm under buildings
6 Japan			13	5·2	B	HC	18	110	6		2·31	Precautions taken otherwise 90 mm heave expected
7 Syracuse	3 m loose fill (cinders and sand) 76 m loose sand and silt	3·6	2·1	3·2	B	HC		≈47	101 40	45 60	1·45	Water injected above water-table. Location subject to regional settlement
8 Ueno, Japan	Sands and silts		18	12·8	B	vA		≈150	13		0·59	Displacement absorption holes provided. Dewatering, although original GWL unclear
9 Vienna, TCB	Sandy gravel Clay/silt Sand		1·5*	6·5	B	HS			13	14	0·45	Seasonally varying perched water table
10 Vienna, U3	Silt and clay		3*	6·5	LN	hA		—	1	10	0·59	Water-table about 1 m above tunnel crown. Intermittent freezing

Notes: z_w = depth water-table, * to underside foundations. Coolant: LN = liquid nitrogen, B = brine. Type: H = horizontal, h = inclined (near horizontal), v = inclined (near vertical), C = complete cylinder, A = arch, S = slab. Duration: G = time to closure (cylinders only). Movements: H_m = maximum heave, S_m = maximum settlement, Σ = total. z_o = depth to centre (=C+D), R = radius (=C+D), R = not available or not applicable.

1 Stoss and Valk (1979). 2 Takashi et al. (1978). 3 Jones and Brown (1978). 4 Valk (1980). 5 Mettier (1985), Bebi and Mettier (1979), Aerni and Mettier

The permanent works

Mining contractors have developed various methods of mineshaft construction, which can all accommodate the frozen state and are described in the mining literature. These include

- sink and line in stages, often referred to as the hanging-lining method, commonly used for deep shafts in the UK and North America
- sink to the sump after (forming a shaft collar), followed by slipforming from the base to the collar, often used for shallow shafts in the UK, e.g. *Ely Ouse* (Chapter 7)
- sink and place a hanging primary lining, then install the final structural lining upwards to the surface, a common technique in Germany
- raise drill from an existing mine complex where there is already an infrastructure for muck disposal, e.g. *Kentwood, USA* (Chapter 7).

Tunnel construction may be by hand, with an auger or thrust bore, or by machine.

Recovery by ground freezing following flooding or collapse is usually dealt with by hand techniques. The strong, watertight conditions generated by a frozen enclosure offer an extremely safe working environment which can proceed with the minimum of supplementary support — struts and props are rarely necessary, as long as the cold atmosphere can be controlled to safeguard the integrity of the ice-wall.

Major tunnel works will usually be undertaken by some form of mechanical advance, e.g. TBM. If the tunnel core is to be frozen, the available torque and the cutter types will need to be in the range able to deal with rock, e.g. *Gascoigne Wood* (Chapter 7). For TBMs designed for soft ground conditions the frozen ground will need to be generated outside the profile of the TBM.

Shaft linings may be cast in situ in mass concrete, sometimes reinforced, or pre-cast concrete or cast iron bolted segments, or combined concrete and steel. Tunnels are generally lined with segments, not necessarily bolted. Concrete placement in proximity to frozen ground often occurs, whichever method of lining is adopted; this topic is discussed below.

Shaft sinking and tunnelling methods are all dealt with in the literature, and can each be utilized in association with ground freezing.

Excavation and lining

Excavation of the hard frozen soil can be likened to dealing with concrete but in a cold atmosphere. Pneumatic tools or explosives can be used.

Pneumatic tools are the obvious first choice for small schemes and awkward places. Care must be taken to minimize the moisture content of the air lines and choice of lubricants: sharpening facilities should be available for the tools.

It is common in shaft sinking to use explosives to break out the sump. The quantity of explosive will be determined according to the volume to be broken, the nature of the strata with respect to shock wave propagation, and the radial proximity of the freeze-tubes and their specification. Typical limits for a 'pull' of 2 m effective will be of the order of drilled depth 2·4 m, trimmer hole charge 350 g, cut hole charge 1·5 kg, total charge 250 g/m^2 of shaft cross-section. A shot firing sequence using delays, which first removes a central relief wedge and successively increases the broken diameter, with the boundary trimming holes fired last, is essential. Muck removal is easily effected with cactus-type grabs into skips, or (mainly in Canada) by wall-mounted Cryderman grabs.

Concreting against frozen ground

The design of concrete to be cast in situ in a sub-zero environment, and particularly when the frozen ground is to act as one element of the 'shutter', must be undertaken carefully. The serviceability of concrete which is to sustain a long-term sub-zero temperature has also to be considered.

Marsh (1959) and Wood (1956) described the installation of mass concrete shaft linings in sub-zero conditions at British coal mines. The main considerations were

- the possible effect of the frost on the ultimate quality, strength and watertightness of the linings
- the necessary precautions to be taken in placing concrete against frozen strata
- the safety factor to be allowed.

It was realized that, as the specified concrete strength of 34·5 N/mm^2 was more than the average achieved at that time, good quality control of the mix and rigorous supervision during placing would be essential. The Industrial Research Laboratories of the City of Birmingham were engaged to carry out laboratory tests on concrete against frozen soil, which proved that there would not be any deleterious effects.

Geddes (1955), at Newcastle University, produced a theoretical analysis of the development of temperature for concrete shaft linings cast directly against frozen ground. The predictions were subsequently checked at Murton-Eppleton shafts. When the Dutch Statemine Maurits #III [see Weehuizen (1959)] was sunk in 1955–7, in situ tests also confirmed no problems in this sphere. It became clear that, provided certain simple principles are followed, a finished concrete product which is homogeneous, undamaged by the freezing action, and durable throughout its design life, will be achieved.

The heat balance at the point of contact is fundamental. That is, the volume of concrete placed must generate sufficient heat of hydration to dominate the adjacent freezing action long enough to allow the initial set to proceed at a positive temperature.

Weehuizen (1959) concluded that the minimum thickness should be 600 mm, and this has been a general guideline adopted for deep shaft concrete linings in the UK mining industry [see Auld (1989)]. Thinner thicknesses have been used for shallow shafts, e.g. at Ely Ouse a 380 mm design thickness plus 80 mm safety margin was adopted successfully.

Another fundamental principle to observe is the use of the correct mix temperature of 19–20°C at the time of placing. Too high an initial temperature plus peak heat of hydration leads to an excessive temperature gradient when refreezing takes place, and instigates thermal contraction cracking; too low a temperature will allow the freezing action to retard and dominate the heat of hydration process, resulting in poor quality concrete.

Even in ideal conditions early release of the formwork minimizes the prospect of tangential (tensile) cracking of circular shaft walls, since all 'movement' takes place radially inwards. The regular construction joints, normally at intervals of 6 m, prevent axial thermal cracking. [This was found by Bagheri (1993) to be true for linings placed against unfrozen ground.]

Bagheri et al. (1993) showed that higher characteristic strengths could be achieved with micro-silica concrete, typically 70–85 N/mm^2, compared with about 45 N/mm^2 for normal concrete; this allows thinner sections to be used which offer greater tensile strength and lower risk of thermal contraction when placed against frozen ground. Chinese authors report similar findings [see Zhang and Hu (1991)].

If the heat of hydration dominates, it follows that there must be some thawing of the inner face of the ice-wall, albeit of small dimension and for a short time. Only with very thin ice-walls will this temporary deterioration be significant in terms of overall stability. In this situation it is helpful to accelerate the setting rate of the newly placed concrete; this can be achieved by adding a mixture of antifreeze, e.g. $NaNO_2$, and an accelerator, e.g. Na_2SO_4. So-called rapid hardening cements, i.e. those with admixtures which either accelerate the release of the heat of hydration — and incidentally counter sulphate degradation — or which reduce the freezing-point of the mixture, are useful beyond cement enrichment for fine-tuning the operation. Other typical additives include $NaCl$, $CaCl_2$, K_2CO_3, and $C_6H_{15}O_3N$.

Confined space (as in a mining situation) incurs access, handling and placement limitations, while the effects of blasting on the recently placed and maturing concrete (often part of the underground excavation technique) must be allowed for.

Placement
The concrete may be poured between shutters, one of which can be the exposed face of the ice-wall, in stages or sections. The junctions of successive pours — construction joints — are potential ingress or egress leakage paths for water under pressure according

to the groundwater conditions and the purpose of the eventual construction, i.e. to exclude or to contain liquid(s). To counter such eventuality it is necessary to incorporate water seals (shallow shafts) or high pressure grout seals (deep shafts) into the joints during the pouring stages, with provision for subsequent injection should such measures not be totally effective.

An alternative technique for shaft linings, which eliminates cold joints, is to use a slipform. Continuous advance rates of 0·3 m/h are practical, and produce very serviceable structures. It is often necessary with this procedure to safeguard the exposed ice-wall above the stage by placing insulation blankets against it while sinking proceeds. Below the stage, which restricts the ventilation flow, very high temperatures can develop during what becomes a continuous period of hydration, and it has proved necessary to mix the concrete with chilled water to compensate.

In frozen tunnel works, it is often desirable in modern practice to apply shotcrete as a quickly applied primary 'structural' lining-cum-insulation rather than leave the face of the ice-wall exposed for any length of time. Again the temporary advance of the zero isotherm into the ice-wall affords sufficient delay for the initial set to occur. Harvey (1993) reported that tests on cores from the Du Toits Kloof tunnel lining shotcrete, compared with those on non-frozen cores, showed no adverse effects.

Frozen concrete performance
As with frozen ground, the compressive strength and the elastic modulus of concrete increase with depression of sub-zero temperature, to a maximum of 250% and 150% respectively at $-100°C$ compared with ambient values; the tensile strength increases by approximately 200% maximum at $-50°C$, according to Marshall (1982). There is little change in the thermal conductivity below 0°C. Thermal contraction is more complex: it is generally recognized that gradual contraction occurs to $-4°C$, then an expansion to $-40°C$, then a resumption of contraction. The scale of this pattern varies from negligible for dry concrete to very significant for wet concrete, i.e. it is a function of the pore-ice. The lower the water/cement ratio, the less the influence of the pore-ice.

Back-wall grouting

The long term performance of the completed structure can be compromised if post-thaw deformation is possible. Such deformation — usually settlement — can arise when water from thawing of pockets of ice or frost is expelled by consolidation processes as the overburden pressure again becomes dominant.

The problem can be avoided, or at least minimized, by conducting 'back-wall' grouting. Grout is injected under controlled pressure through the lining on a chase-through principle, i.e. with a free-draining outlet some 3 m or so ahead of each grout injection point, during the thaw stage. This should be timed when the inward growth has receded from the outer face of the lining but before the integrity of the ice-wall is lost, usually between two and five weeks from cessation of refrigeration. The water will be seen to be expelled and all significant voids will be filled. With sensitive structures it is prudent to repeat the exercise three to six weeks later, thus ensuring a 'tight' condition. Additionally, such injections serve to penetrate and seal any potential leakage paths in the lining. See *Selby* in Chapter 7.

Freeze-tube recovery or abandonment

There may be technical reasons requiring that the freeze-tubes are recovered from the ground once they have served their purpose. If there is no such need the decision will be determined by economic considerations, i.e. the cost of recovery versus the second-hand value of the tubes.

Recovery is best effected within the first few weeks following termination of refrigeration while the frozen ground is relatively intact. Circulation of warm brine/water through the system for a short time will thaw the bond between frozen ground and the tube, create a water-lubricated annulus, and enable the tubes to be withdrawn by jacking, and crane or drill-rig.

Whilst it is unlikely that freeze-tube material will be lost in the ground, there is always a possibility, however small, that fracture will occur during pulling, leaving the remotest section in the ground at or near to its working location. 'Fishing' for this lost section will be difficult as the frozen bond will soon re-establish, and the ability to rewarm the tube has now been lost. If such a loss in the path of the works will pose difficulties in advancing the machinery, consideration must be given to using alternative materials for the outer element of the freeze-tubes; e.g. aluminium or copper could be milled away or cut through with greater ease, causing less damage to drilling cutters.

Following thorough cleaning to rid the recovered metal of salt deposits from the brine, which is corrosive, the tubes may be stored for future re-use. It should be noted that traces of salt will remain, however good the cleaning process, and deterioration will occur more quickly than for virgin material.

Abandonment in straightforward geological conditions may simply require filling of the tubes with sand or cement, to avoid cavitation when the metal eventually rots. When confined or multiple aquifers have been penetrated the abandonment must totally seal the connection from the aquifer to ground surface, or between aquifers, to prevent unwanted and potentially hazardous loss or contamination. Permanent abandonment may be carried out when the freeze-tubes are installed, and is termed 'pre-abandonment', or following freezing when it is termed 'post-abandonment'.

In either case the aim is to fill the annulus between the drillhole and the outer element of the freeze-tube with cement grout, and various techniques are used to achieve a good seal. Pre-abandonment methods usually rely on the oil-field 'wiper-plug' technique — that is an injection tube mounted in an inflatable packer is advanced down the drillhole casing (in this case the outer element of the freeze-tube); following flushing the drilling mud with water, cement is injected in a continuous operation until all remaining mud and water is expelled as the cement reaches the surface. The success of the operation can be checked by running a down-hole bond-log (another oil-field practice). Post-abandonment is effected by forcing grouting injection tubes down the annulus, and gradually withdrawing them as the grout is injected. It is more difficult to achieve a continuous seal by this method.

References

AERNI K. and METTIER K. 1980. Ground freezing for the construction of the three-lane Milchbuck road tunnel, Zurich. *Proc. 2nd ISGF*, Trondheim, 1, 889–895, 2, 115–10.

AULD F. A. 1982a. Concrete in underground works. *Proc. NS on Concrete in the Energy Industry*, Concrete Soc., Runcorn.

AULD F. A. 1982b. Ultimate strength of concrete shaft linings and its influence on design. *Proc. NS on Strata Mechanics*, Newcastle, 134–140.

AULD F. A. 1983. Design of temporary support for frozen shaft construction. *Proc. 1st NSGF Progress in AGF*, BGFS, Nottingham, 1–8.

AULD F. A. 1988. The application of AGF to the construction of the ventilation drifts at Selby Wistow mine. *Proc. 5th ISGF*, Nottingham, 2, 513–524.

AULD F. A. 1989. High-strength, superior durability, concrete shaft linings. *Proc. IC on Shaft Engineering*, IMM, Harrogate, 25–37.

AULD F. A. 1995. Casting concrete against frozen ground. *Proc. 7th ISGF*, Nancy, 2, (in press).

BAGHERI A. R. *et al.* 1993. The use of microsilica concrete in the construction of mine shaft linings. *The Mining Engineer*, May, 307–317.

BEBI P. C. and METTIER K. R. 1979. Ground freezing for the construction of the 3-lane Milchbuck road tunnel, Zurich. *Tunnelling 79*, IMM, London, Ch 16, 245–255.

COLLINS S. P. and DEACON W. G. 1972. Shaftsinking by ground freezing in Ely-Ouse Essex scheme. *Proc. ICE*, 7506S, May, 129–156, 319–336.

DEIX F. and BRAUN B. 1988. Vienna subway construction — use of brine freezing in combination with NATM under compressed air. *Proc. 5th ISGF*, Nottingham, 1, 321–330.

DUNNICLIFFE and GREEN 1988. *Geotechnical instrumentation for monitoring field performance*. Wiley.

FIRTH G.W. and GILL J.J. 1963. The sinking of Kellingley shafts. *Mining Engineer (Proc. IME)*, 123, Dec., 147–166

GARNETT G. A. 1981. Tunnelling on development of first crossing of Manchester Ship Canal. *Cooling Prize presentation*, BGS.

GEDDES J. D. 1958. The prediction and measurement of temperatures in concrete structures with particular reference to some recent construction in frozen ground. *Bulletin 11 University of Durham*, Dept of Civil Engng, 43pp.

HARVEY 1993. Effective control of groundwater by use of ground freezing. *Proc. IC on Groundwater Problems in Urban Areas*, London.

HARVEY S. J. and MARTIN C. J. 1988. Construction of the Asfordby mine shafts through Bunter Sandstone by use of ground freezing. *Proc. 5th ISGF*, Nottingham, **1**, 339–348, and in *Mining Engineer,* **148**–323, 51, 53–58.

JONES J. S. 1980. Engineering practice in AGF — the State of the Art. *Proc. 2nd ISGF*, Trondheim, **1**, 837–856.

JONES J. S. and BROWN R. E. 1978. Soft ground tunneling by ground freezing. *Tunnelling and Underground Structures*, US Trnsptn Research Board, **684**, 28–36.

JONES R. H. 1982. Ground movements associated with artificial freezing. *Proc. 3rd ISGF*, Hanover NH, **1**, 295–304.

JONES R. H. 1993. Control of ground movements in AGF works, BGFS, *Technical Memorandum TM4*, 12pp.

KIRIYAMA S. *et al.* 1980. Artificial ground freezing in shield work. *Proc. 2nd ISGF*, Trondheim, **1**, 940–951.

LACY H. S. *et al.* 1982. A case history of a tunnel constructed by ground freezing. *Proc. 3rd ISGF*, Hanover NH, **1**, 389–396.

MAISHMAN D. 1959. Shaft sinking using the freezing process — Kellingley Colliery. *Iron and Steel Trades Review*. 30 Oct., 707–715.

MAISHMAN D. and POWERS J. P. 1982. Ground freezing in tunnels — 3 unusual applications. *Proc. 3rd ISGF*, Hanover NH, **1**, 397–410.

MARSH F. 1959. Shaft sinking in Great Britain since 1947. *Proc. IS on Shaft Sinking*, IMM, London, 215–237.

MARSHALL A. L. 1982. Cryogenic concrete. *Cryogenics*, Nov., 555–565.

MARSHALL A. L. 1983. Cryogenic concrete and the marine storage of LNG. *Proc. 2nd IC on cryogenic concrete*.

MARTAK L. V. 1988. Ground freezing in non-saturated soil conditions using liquid nitrogen. *Proc. 5th ISGF*, Nottingham, **1**, 367–376.

METTIER K. 1985. Ground freezing for the construction of the Milchbuck road tunnel in Zurich — an engineering task resolving between theory and practice. *Proc. 4th ISGF*, Sapporo, **2**, 263–269.

REDFERN A. and PINDER B.F. 1964. Shaft lining and grouting problems associated with Cotgrave and Bevercotes shafts. *Colliery Guardian*, **209**, Dec., 817–826, Jan., 14–21.

SCHMID L. 1981. Milchbuck tunnel: application of the freezing method to drive a three-lane highway tunnel close to the surface. *Proc. 5th RETC*, San Francisco, May, **1**, 427–445.

SHUSTER J. A. 1982a. Ground freezing failures — causes and prevention. *Proc. 3rd ISGF*, Hanover NH, **1**, 315(10pp).

STOSS K. and VALK J. 1978. Chances and limitations of ground freezing and liquid nitrogen. *Proc. 1st ISGF*, Bochum, **1**, 303–312, and in *Engng Geol.*, 1979, **13**, 485–494.

TAKASHI T. *et al.* 1978. Effect of penetration rate of freezing and confining stress on the frost heave ratio of soil. *Proc. 3rd IC on Permafrost*, Edmonton.

TAKASHI T. *et al.* 1982. Artificial ground freezing in shield work. *Proc. 3rd ISGF*, Hanover NH, **1**, 415–422.

TURNER F. H. 1979. *Concrete and cryogenics*. Cement and Concrete Association, 100pp.

VALK J. 1980. The successful application of an unusual ground freezing method to secure tunnel excavation. *Proc. 2nd ISGF*, Trondheim, **2**, 79–93.

WADSWORTH A. 1957. Recent shaft sinking developments in the East Midlands. *Proc IME*, **117**, Apr., 397–416.

WEEHUIZEN J. M. 1959. New shafts of the Dutch State mines. *Proc. IS on Shaft Sinking*, IMM, London, 28–65.

WOOD A. F. 1956. Placing mass concrete shaft lining in frozen ground. *Trans IME*, **116**, 713–727.

WOOD A. F. 1958. Shaft sinking at Wearmouth. *Colliery Engng*, **35**, May, 188–195; June, 234–238; July, 280–284.

YANG Y. 1982. Strength development of concrete placed in frozen soil and the thermal effects. *Proc. 3rd ISGF*, Hanover NH, **1**, 375–382.

ZHANG C. and HU D. 1991. Technical innovations in freeze shaft construction of Jining #2 coal mine, PRC. *Proc. 6th ISGF*, Beijing, **1**, 385–390.

7. Case examples

There are many examples of the use of ground freezing for open excavations from the surface, and for tunnelling through adverse ground conditions beneath or adjacent to sensitive structures, which have been undertaken around the world. The main features of a number of these, many of which the author has personal experience or knowledge of, are briefly described in this chapter. Selected references are given where fuller information of particular projects has been published.

Given that AGF has often been thought to be the 'method of last resort', the list of projects where it has been employed, reported to some degree in the literature and given as Appendix C, is surprisingly lengthy. The data associated with individual projects forms the subject of a database in the course of preparation.

The examples which follow have been grouped under shafts, tunnels and miscellaneous; shafts and tunnels have been sub-divided according to whether the ground freezing works were incorporated by design or were resorted to following discovery of a construction difficulty; these groups have been further sub-divided into instances when the freeze-tubes were installed from the surface or from below ground. Scheme diagrams are included to illustrate the principles of the freezing design, but are not to scale and do not necessarily indicate the actual number of freeze-tubes.

The principal participants in each project [the Engineer, the Main Contractor (MC) and the Specialist Ground Freezing Contractor (SGFC)] are referred to under the trading title appropriate at the date of contract award; some which have since changed are National Coal Board to British Coal, North Thames Gas Board to British Gas, JD & DM Watson to Watson Hawksley, Mott Hay & Anderson to Mott MacDonald, and Foraky Limited to British Drilling and Freezing Co Ltd. (*NR* is entered where the names of the participants are not reported in the primary literature.)

Shafts, pits and open excavations

As the chosen design method of temporary works

Frozen from the surface
   Belmont (Australia) 1981
Near Newcastle (NSW), ground freezing was employed to sink a new 5·8 m diameter ventilation shaft through 30 m of superficial deposits. Unusually, the site first required the construction of an island and access causeway, as it lay in the middle of a lagoon. The freeze-tube drilling therefore had to pass through the recently placed, saturated fill before penetrating the natural ground. See Fig. 7.1.

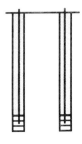

Fig. 7.1. Mine ventilation shaft on man-made island at Belmont, Australia. Photo courtesy of BDF Co Ltd.

Point of interest A straightforward freezing in an unusual situation.

Engr: *Owner*; MC: *Cementation (Australia)*; SGFC: *Foraky Limited*.

See Scholes (1982).

Blackpool 1969

A disused aircraft hangar was to be put to industrial use. The conversion required a number of pits, to house metal treatment tanks, to be sunk close to stanchion bases within the building. The upper strata were sand over peat to some 6 m depth, over a further 6 m of soft clay; the tanks were to be 6 m deep coinciding with the top of the soft clay which was too weak to support the new structure. An open pit is shown in Fig. 7.2.

Bored cast-in-place concrete piles, with their cap level corresponding with the top of the soft clay, were installed prior to freezing to receive and support the final structure. To avoid interference with the stability of the existing stanchions the pit was then sunk within a circular ice-wall taken to the base of the soft clay. The work was completed satisfactorily with brine plant.

Point of interest Carried out without disturbance to the stanchion bases alongside the ice-wall.

Engr: *GKN*; MC: *GKN*; SGFC: *Foraky Limited*.

See Maishman (1971).

Canvey Island 1967–1968

In 1964 a Liquefied Natural Gas (LNG) import terminal was constructed alongside the Thames Estuary, to receive LNG at −160°C by sea from an LNG export terminal in Algeria. The terminals had conventional above ground metal storage tanks. In

Fig. 7.2. Open pit inside old aircraft hangar at Blackpool. Photo courtesy of BDF Co Ltd.

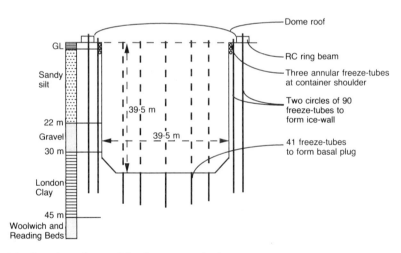

Fig. 7.3. Inground LNG tank at Canvey Island (cross-section)

1967–8, four additional inground tanks were constructed at Canvey (see Fig. 7.3), and one in Algeria. It was considered that advantage could be taken of the refrigerating effect of the LNG to dispense with structural components below ground level, since the frozen soil would form a wall and base and act as insulation, and the only additional structure would be an insulated roof of adequate mass to minimize boil-off and prevent loss of gas to the atmosphere. Thus pre-freezing the ground as a construction aid was logical and effective.

The strata comprised 22 m of sandy silt, over 8 m of gravel, over 15 m of London Clay; below the clay, the Woolwich and Reading Beds supported a sub-artesian head almost to ground level. Each container was 39·5 m diameter by 39·5 m deep, thus terminating within the London Clay. These conditions called for twin circles of 90 freeze-tubes around each container, and 41 further freeze-tubes within their cores to freeze the basal clay which, although strong, would not have withstood the sub-artesian uplift pressures associated with the underlying Woolwich and Reading Beds. The core freeze-tubes were fitted with twin inner tubes to minimize ice creation at shallow levels and ease excavation. See Chapter 5. The construction is shown in Figs 7.4 and 7.5.

The original concept was that refrigeration would be provided by means of a methane/propane heat exchanger; however, at the time there were concerns (which turned out to be unfounded) that the political crisis in Algeria might lead to delays in supply of liquid methane; conventional mechanical refrigeration machines were therefore installed for this stage of the operation.

Points of interest The freezing was successfully carried out within a water-table affected by tidal variations.

The very close proximity of the four tanks constrained drainage/egress of groundwater from the centre of the site; as a result heave amounting to a maximum of 1·5 m was noted. Ground heave at the boundary of the frozen zone was less than 20 mm.

Prior to construction it had been anticipated that the ice-wall thickness, due to the very cold product being stored, would increase to an estimated 30 m. In the event this dimension was exceeded, and by 1977 had reached 45 m at 30 m depth with a declining growth rate of 1·5 m/year. The underestimate was accounted for by (a) a higher thermal conductivity of frozen soil at very low temperatures than had been measured in the laboratory, and (b) development of some fissures in the inner (coldest) section of the

Fig. 7.4. Overview of preparations for ground freezing at Canvey Island. Photo courtesy of British Gas.

Fig. 7.5. The insulated excavation at Canvey Island just prior to erection of the roof. Photo courtesy of British Gas.

ice-wall thereby increasing the available heat transfer surface area and the rate of freezing.

It was established that the fissuring occurred when the tensile stresses produced on cooling from AGF temperatures (around $-25°C$) to LNG temperatures (around $-160°C$) exceeded the tensile strength of the frozen soil, and that fissuring was confined to the walls since there was no penetration of LNG into the base. Fissuring ceased once the major stresses were relieved and the tank temperatures stabilized; the cracks did not penetrate to the (warmer) outer section of the ice-wall which is subject to a continuous freezing regime and a state of high compressive stress.

To avoid possible adverse effects (from continued growth of the ice-wall) on neighbouring surface tanks supported on shallow piles, a series of heating-tubes were installed in 1977 at the 47 m bound. These were of similar design to freeze-tubes, but through which hot water ($+45°C$) could be circulated to act as a thermal barrier for the remaining life of the containers. The line of heating-tubes crossed the protective bunds associated with two of the surface tanks, and this posed installation problems. Operation of the thermal barrier caused the ice-wall boundary to recede slightly and stabilize at about 45 m.

Problems associated with excessive boil-off led to operation of the tanks at less than theoretical capacity, but they nevertheless made a valuable contribution during the currency of the base-load supply contract. In 1982 the inground installation had served its purpose and a decision was made to reduce the function of the terminal to 'peak-shaving' during periods of high gas demand; the inground caverns then became redundant and were filled in, an exercise which required careful planning and execution.

Ground freezing techniques were successfully used to help construct these large cylindrical excavations, although the concept of using them to store cryogenic liquid without an internal liquid-tight lining was shown to be unsound.

Engr: *North Thames Gas Board/Conch Methane*; MC: *Sir Robert McAlpine*; SGFC: *Foraky Limited*.

See Miller and Gordon-Brown (1967); Manning (1973); Findlay et al. (1982), Graham et al. (1983).

Carcroft 1979

A replacement culvert drain, being laid within a sheet-piled open cut kept dry by sump pumping, had to be connected into an existing operational pumphouse. It was expected that flow of water and running sand/silt into the excavation could not be controlled without risk of major damage to the pumphouse. By freezing twin walls from surface to a 7 m deep underlying impermeable clay layer, to link the sheet-piles with the pumphouse structure, and across both ends of the section, it was possible to isolate the area from the groundwater table. Being a relatively small, short duration exercise at a site remote from a suitable electric power supply, refrigeration was by LN.

Engr: *National Coal Board*; MC: *Camm*; SGFC: *Foraky Limited*.

Ely Ouse–Essex 1968

This major scheme enabled surplus water from the fenland drainage system to be transferred to Essex via a tunnel and river courses. It arose from recognition at the (tunnel) site investigation stage that the sinking of a series of five permanent and up to five temporary shafts through the chalk aquifer to the underlying Gault Clay, without affecting the level or quality of the groundwater, could easily be achieved by the ground freezing method. Coupled with acceptance by the Contractor of penalty clause time constraints and quality of shaft lining determinable by miniscule (inward) leakage limits, various arguments against using freezing were countered, and the works for the five permanent shafts were awarded as a main contract to the ground freezing specialist contractor. Subsequently the Contractor for the tunnel in the Gault, which would link the shafts, required one further (temporary) shaft

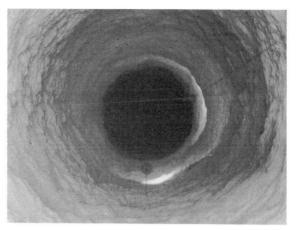

Fig. 7.6. Shaft excavation in frozen chalk at the Ely Ouse–Essex scheme. Photo courtesy of Binnie & Partners.

to optimize his programme, and this was undertaken as a sub-contract.

Apart from the southernmost up-shaft of 24 ft (7·3 m) finished diameter to house the pumps, all shafts were to be 15 ft (4·6 m) diameter. The up-shaft was constructed on the sink-and-line principle, whereas the common-size shafts were sunk consecutively to full depth, for programmed successive use of a slipform. The latter method, which avoids 'cold joints' between successive pours of in situ concrete, easily met the in-leakage criteria, and all such shafts were completed within their respective time schedules. Only the larger shaft over-ran by three weeks, mainly accounted for by surface flooding at the commencement of the works. The shaft excavation is shown in Fig. 7.6.

Points of interest The whole of the works was completed at a cost below budget.

Stringent time and quality specifications were met.

Engr: *Binnie & Partners*; MC: *Foraky(5)/Nuttall(1)*; SGFC: *Foraky Limited*.

See Collins and Deacon (1972).

Gibraltar 1976, 1977, 1979, 1981
Diesel driven generating sets supplying electric power to the naval garrison were sited on reclaimed land which, it was said, had been laid to aircraft runway compaction specification; however, excessive differential settlement had occurred over a number of years. When each set was due for renewal/replacement the opportunity was taken to place stable load supports founded in natural rock. Each generating set required two 3 m diameter 'piles' or columns passing through 6 m of rock fill, of which 4·5 m was saturated with sea water. Each pile was constructed within a circular ice-wall, much as a shaft is.

Points of interest The 'groundwater' was sea water (fill

deposited in the sea dock), and its surface was contaminated with oil as in any repair dockyard. The placing of oil-absorbent bolsters against the excavation proved to be adequate in this instance.

Engr: *PSA*; MC: *J Mackley (3), Shand (1)*; SGFC: *Foraky Limited*.

See Harris and Woodhead (1978).

Kentwood, Michigan (USA) 1988
An interesting constructional variation was employed at the Kentwood minesite when freezing from the surface was combined with raise-boring from an existing underground roadway (Fig. 7.7). Prior to drilling a 0·3 m diameter pilot-hole 64 m deep from surface into the mine, a 6 m internal diameter ice-wall was generated through the 43 m of heavily water-bearing mixed overburden strata. The pilot-hole was then reamed to 2·7 m diameter by raise boring through the relatively soft but competent (unfrozen) shales and gypsums, then into and through the protective ice-wall to ground level. A 2·1 m diameter hydrostatic steel lining complete with escape manway was then grouted into place to secure the works, following which the freezing operation could be terminated.

Engr: *Georgia-Pacific Corp (Owner)*; MC: *Dynatec Mining Corp*; SGFC: *FreezeWALL Inc*.

See Walsh *et al.* (1991).

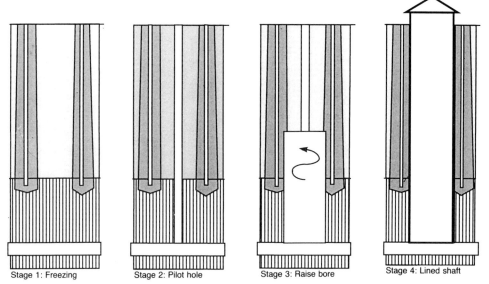

Fig. 7.7. *Raise-bore concept used at Kentwood (cross-section)*

Belle Isle, Louisiana (USA) 1970
The target here was to reach and exploit a sub-surface salt dome, one of a series of domes that occur across the Gulf of Mexico off the Louisiana coast and into the US. The particular site is located on an island within the coastal limits of the Mississippi delta. The

strata comprise 46 m of estuarine silt, clay and sand with some gravel towards its base, then massive high purity salt. The salinity of the groundwater falls from sea water concentrations at the surface until near the gravels, when high values equivalent to a freezing point of $-21\,°C$, associated with brine solution from the bedrock, are found; the 5 m diameter by 80 m deep shaft freezing therefore had to be designed to cope with these conditions.

This was handled by a combination of closely spaced freeze-tubes, generous freeze-plant capacity to achieve brine temperatures of near $-40\,°C$, plentiful brine pumping capacity, and a penetration into the salt of 34 m.

Points of interest The surface of the dome dipped across the shaft diameter by 6 m. As this was the most critical horizon, the 0·6 m thick in situ concrete shaft lining over the whole zone had to be concreted in a single pour of 7·5 m height — any lesser dimension would have imposed severe temperature differentials between newly poured lining and the subsequent excavation.

Some time after completion of the shaft lining into the salt dome, a flooding and collapse, identified as being due to a geological fissure passing very close to the shaft, resulted in a spectacular surface crater with the shaft lining and headframe lost into the mine. Later a repeat rehabilitation exercise was undertaken, again involving a frozen shaft sinking but with a much thicker ice-wall.

Engr: *Cargill (Owner)*; MC: *Cementation (America)*; SGFC: *Foraky Limited*.

Lowestoft 1967

A shallow shaft was required alongside the dock entrance at Lowestoft, located beside the swingbridge traverse area regarded by the Docks Engineer as 'very sensitive'. He had therefore precluded any method of construction which the Contractor could not 'guarantee' to be free from adverse side-effects on the neighbouring ground, particularly heave or settlement. Shaft sinking within frozen granular strata was successfully completed using the brine system.

Point of interest There were no adverse effects.

Engr: *JD & DM Watson*; MC: *King*; SGFC: *Foraky Limited*.

Milwaukee, Wisconsin (USA) 1985

The 50 m depth of soft water-bearing silts, sands and clays immediately under Milwaukee presents difficulties for shaft sinkers, and over the years many methods — e.g. dewatering, caisson sinking and compressed air — have been used with varying degrees of success to a maximum depth of 26 m. Several contracts let in the mid-eighties as part of a major city-wide deep-level sewerage scheme included freezing to over 50 m depth as the ground treatment technique.

One of the sites included three separate 6 m diameter shafts, 46 m apart, to enable and serve an in-line pump station 104 m

GROUND FREEZING IN PRACTICE

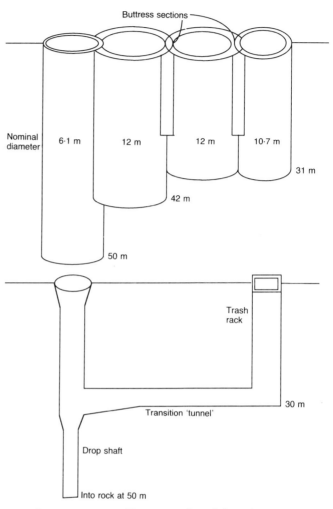

Fig. 7.8. (upper) frozen ground structures to enable construction of (lower) storm water surface-to-tunnel transfer systems in Milwaukee

below surface in the dolomite bedrock; another required three closely spaced excavations of varying shape and depth. The surface deposits were 52 m thick at the in-line pump station site, and could be characterized as two layers for the purpose of the ground freezing design: an upper layer 25 m thick of post-glacial saturated silty sand, organic silt and fat clay; and the remainder dense glacial or lacustrine clays with bands of silt, sand and gravel. This led to a two-level freeze-tube pattern, 34 at 1·5 m spacing to 30 m depth and 17 at 3 m spacing i.e. alternate tubes) to 55 m. The refrigeration was carried out sequentially to minimize the plant and power demand.

At the second site (Fig. 7.8) a 30 m deep, 10·7 m internal diameter shaft (trash rack) was to be connected by a sub-level transition tunnel and cascade to a 6 m diameter drop shaft extending into the rockhead, the whole complex being some 42 m long. The cost of encompassing all these elements within a single elliptical ice-wall was prohibitive; to minimize the length of the ice-wall, and the required refrigeration capacity, and the volume of excavation, a series of four cells (linked circular/elliptical ice-walls) was created. First the drop shaft was frozen and constructed to a depth of 55 m, thus providing an 'anchor' for the remaining cells. The intermediate cells over the tunnel length were elliptical, though approximating to circles of 12 m diameter, and the final cell 10·7 m diameter. Analyses indicated the need to counter tensile stresses at the link points between the cells. This was dealt with by concreting steel beams into boreholes prior to freezing. When the linked cells were frozen, openings were cut through the common parts of the ice-walls to form the tunnel and link the two shafts.

Point of interest Flooding occurred at one of the inline shafts when the excavation entered the bedrock; the reason could have been either a 'window' in the ice-wall or a water-charged fissure extending below the base of the ice-wall in the bedrock. An independent assessment concluded that the ice-wall itself was intact.

Engr: *CH2M*; (1) MC: *Cementation (America)*; SGFC: *Foraky Limited*. (2) MC: *NR*; SGFC: *Geofreeze Inc*.

See (1) Doig (1985), Sopko *et al*. (1991); (2) Shuster and Sopko (1989).

Saskatchewan (Canada) 1963 – 1967
Exploration for oil in the 1950s disclosed vast deposits of Devonian potash beneath Saskatchewan and Manitoba in central Canada, at depths of 900 m or more at the Manitoba end of the field. The deposit is overlain by younger beds, mostly shales, but invariably including the Blairmore, a formation of up to 120 m thickness containing zones of uncemented sand, saturated with saline water. To reach and exploit the rich potash deposit required mineshafts penetrating the Blairmore and able to withstand an active pressure of 60 bar. Freezing depths of 600 m and more were handled by twin banks of ammonia plants, with shaft linings of thick plates of welded or cast iron, machined to accurate fitting tolerances, which were purpose made for the various projects.

Points of interest Factory inspection procedures and inspections equivalent at least to today's BS5750 quality assurance requirements were instituted at Canadian and British foundries; oil field drilling techniques were adapted to deal with the task of quickly and accurately inserting 700 m deep freeze-tubes; and an effective ice-wall thickness of 5·5 m with an average temperature of $-15\,°C$ was created through the Blairmore.

Engr: *Owners (various)*; MC: *Cementation (Canada)*; SGFC: *Foraky Limited*.

See Kelland and Black (1969).

Selby Project 1977 – 1982
Five pairs of shafts (and twin drifts reported separately). These

major works to further the development of a major new coal mining complex required ground freezing on a large scale through the Bunter Sandstone for all shafts, and the Basal Sand (Permian) at one of them. The Selby shafts at Wistow, Riccall, Stillingfleet, North Selby and Whitemoor were all frozen from ground surface to depths of 148/273* m, 253 m, 165 m, 280 m and 307 m respectively (* included the Basal Sand).

Note: The drifts at *Gascoigne Wood* are described on p. 154 as part of the Section on tunnels.

Points of interest At one site with particularly shallow groundwater, difficulty was experienced with closing the ice-wall at a very shallow level even though closure had already occurred at depth, thus unduly delaying the start of excavation. Slowness of this kind can occur due to the warming influence of the atmosphere combined with the returning warmer brine temperature during the pull-down at the tops of the freeze-tubes, as described in Chapter 1. Here, however, the principal cause was identified as being due to extensive stone fill across the site — an old access road which acted as a 'french drain'. A solution in such circumstances is to install up to three, or even more, horizontal annular freeze-tubes to hasten closure of the 'Vs' between the tops of the vertical freeze-tubes.

At another site, flooding of the shaft took place after the shaft sinking had passed below the frozen zone. This occurred five months after the termination of refrigeration, i.e. following the thaw stage: groundwater entered the shaft via a construction joint of the cast in situ lining, within the depth which had been frozen, prior to the carrying out of back-wall grouting. This topic is discussed in Chapter 6.

Engr: *National Coal Board*; MC: *Cementation (3), Thyssen (GB) (2)*; SGFC: *Foraky Limited*.

See Wild and Forrest (1981).

Timmins, Kidd Creek Mine (Canada) 1974
A 'bonanza' orebody was discovered in northern Ontario by aerial geophysics, at the location of a long-abandoned gold prospector's cabin. Muskeg and soft clay totalling some 12 m in thickness covered (and hid) the sub-outcrop of the 1·5 km diameter orebody which haded at around 80°C. Initially the ore was won from an open pit, but while this was in hand a 900 m deep shaft was sunk alongside to continue recovery by deep mining methods; this shaft was sunk at the only point on the site where rock outcropped. Thus, as the open pit was reaching its depth limit, a second shaft to 1000 m depth was commenced.

The shallowest founding level for the headframe, and equally the most convenient level to commence sinking, was at the base of the soft clay. The rectangular excavation was made within an elliptical ice-wall encompassing the 16 m × 11 m excavation for the eventual 12 m × 7 m rectangular headframe structure. The

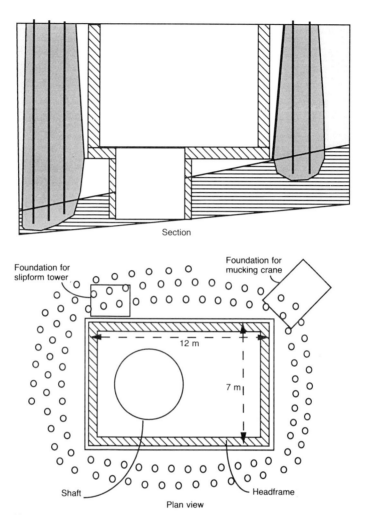

Fig. 7.9. *Scheme for headframe foundation at Timmins*

bedrock dipped from 7 m to 19 m depth across a diagonal of the pit. Clays and peats have low thermal conductivity and offer the least gain in strength on freezing, so two rings of freeze-tubes were provided around the shallower sides of the pit to 12 m depth, and three rings around the deeper section. See Figs 7.9 and 7.10.

The thick ice-wall thereby generated could be used additionally as a foundation for the Alimak slipform tower, and for trucks reversed to receive spoil recovered by crane-bucket from a dozed spoil heap in the excavation. To minimize melting of the ice-wall as it became exposed to atmospheric degradation — by heat and heavy rainfall — the whole establishment was well insulated.

Point of interest A large relatively shallow, rectangular excavation was undertaken during the season of hot days and frequent half-hour periods of late afternoon heavy rain. Such conditions dictate the provision of generous refrigerative capacity.

Fig. 7.10. Slipformed headframe rising through the freeze-tube headers and brine distribution system at Timmins. Photo courtesy BDF Co Ltd.

But such capacity alone can be dwarfed by the melting and erosive effects of direct warming and wetting which causes sloughing of exposed surfaces as the excavation is advanced, thus quickly reducing the ice-wall thickness to the point when its integrity is in doubt. This is quite easily controlled by applying waterproof insulation.

At Timmins this was achieved by nailing blanket insulation, attached to a waterproof foil backing, to the ice-wall immediately following its exposure to the elements, and by stringing hawsers across the excavation to form a square grid supporting tarpaulin sheets, thus shading and isolating the excavation which thereby became an intermediate, essentially constant temperature, thermal barrier.

Engr: *Texas Gulf Inc*; MC: *Cementation (Canada)*; SGFC: *Foraky Limited*.

Wrexham 1971

The route selected for a new road bypassing Wrexham passed over abandoned mineworkings; at one point an open and flooded old mineshaft lay within the boundary fences and had to be dealt with to the satisfaction, *inter alia*, of the National Coal Board. The combination of flooded old workings, a second similarly abandoned old shaft 30 m away outside the fenceline, and flooded glacial deposits some 30 m thick at the surface dictated the construction of a sub-surface reinforced concrete cap in solid rock; this would ensure that backfill placed below and above the cap would remain in place and not be syphoned into the workings with later fluctuations of the groundwater system.

The procedure adopted was (a) place hardcore backfilling until the shaft was filled to 60 m below surface, (b) install twin welded steel tubes of about 1 m diameter (large diameter waterwell casing, sealed at their base) within the top section, (c) dispose twelve freeze-tubes in a regular pattern between said casings and the shaft wall, and then (d) fill the remaining annular space with sand, illustrated in Figs 7.11–13.

The saturated sand could then be frozen until the frozen mass bonded with the brick shaft lining and the surrounding strata. Once frozen the casings could be used as access shafts to permit excavation at the competent rock level by first linking them to yield a working chamber, then enlarging the excavation until it extended sufficiently into the rock to provide the space and circumferential seating needed to cast the cap itself.

Point of interest When excavating for the cap it was found that the shaft had been relined off-centre at some unrecorded time in its life, necessitating a redesign of the cap, and the reordering of reinforcement to cater with a larger diameter than expected.

Engr: *Borough Engr, Wrexham/National Coal Board*; MC: *Foraky Limited*; SGFC: *Foraky Limited*.

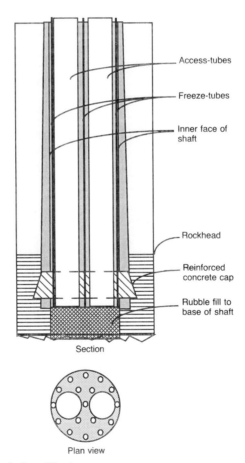

Fig. 7.11. Abandonment of old mineshaft at Wrexham

GROUND FREEZING IN PRACTICE

Fig. 7.12. Sand being tipped into the shaft between the access casings and the freeze-tubes at Wrexham. Photo courtesy of BDF Co Ltd.

Fig. 7.13. Man-riding kibble arriving in the cap excavation, with freeze-tubes exposed and reinforcement in place. Photo courtesy of BDF Co Ltd.

See Gregory and Maishman (1973).
Similar work, effected horizontally from nearby underground roadways, was undertaken in Belgium.
See Buttiens (1978).

CASE EXAMPLES — SHAFTS

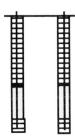

Sub-surface freezing
  Asfordby 1985–1987

Two pairs of shafts. The Bunter Sandstone lies from 300–390 m below surface. Here the freeze-tube drilling was undertaken from the surface which, on account of the significant depth, also required the use of directional drilling techniques to maintain an exceptionally high degree of vertical accuracy in their placement. On completion of each drill-hole survey the freeze-tube was lowered into place with the upper end, temporarily blanked off, at a level 25 m above the Bunter Sandstone.

When shaft sinking reached that level an annular chamber was excavated to expose these 'lost' heads and house the ring mains delivering and recovering brine coolant to and from the freeze-tubes (Fig. 7.14). Although this procedure meant an interruption to the sinking cycle, the cost saving from not freezing the shallower horizons was considerable; the permanent headframes were erected during the primary freeze period.

Point of interest There are very few occasions on which it is possible to verify the accuracy, not only of the position of the freeze-tubes themselves, but also of the borehole surveys on which the accuracy of freeze-tube placement is usually assessed. On this occasion corroboration of both criteria was possible when the underground freeze cellar was excavated: the heads of the freeze-tubes were located in the cellar floor and their positions accurately determined. Comparison with the target pattern showed that all the freeze-tubes were within their 0·5 m radius target circles, and the maximum error of the multi-shot gyroscopic borehole surveys at 275 m depth was 233 mm. The claimed accuracy of such surveys

Fig. 7.14. Underground freeze-chamber at Asfordby shaft showing distribution mains and instrumentation. Photo courtesy of BDF Co Ltd.

is 0·05°/100 m, which equates at 275 m penetration to 240 mm deviation.

Engr: *British Coal*; MC: *Cementation*; SGFC: *British Drilling & Freezing Co Ltd*.

See Harvey and Martin (1988).

Boulby 1974

Unusual features of this project included (a) two 5·5 m diameter shafts sunk by different techniques, the first by freezing with a double steel permanent lining, and the second by grouting and tubbing; (b) directionally drilled freeze-holes to intercept the intended underground freeze-cellar. It was expected that the freezing method would be quicker though dearer than the grouting method, and a steel lining would be advantageous when rigid winding guides were installed, two features significant to shaft 1.

Of the total shaft depth of 1150 m, the zone from 610 m to 945 m was frozen, the brine coolant being pumped via insulated mains in the upper shaft to the ring mains installed in a freeze-cellar 590 m below surface. The freeze-hole drilling commenced at surface on a circle of 33·5 m diameter, reducing to 12·5 m diameter at and below the freeze-cellar, thus permitting concurrent shaft sinking operations. There was, of course, a delay to sinking during the primary freeze period.

Additional freeze-plants were installed during the works when it was noted that the ice-wall growth rate was slower than expected, and supplementary works were needed when fissure-water entered the shaft at 859 m below surface (see below and Chapter 9).

Point of interest The progress achieved in the frozen shaft, notwithstanding the interruption due to flooding, vindicated the choice and (with hindsight) suggested that if the economy of scale had also been considered, a two-shaft freezing would have compared favourably, in cost and time, with the scheme adopted.

Engr: *Whitby Potash*; MC & SGFC: *Thyssen (GB)*.

See Cleasby *et al.* (1975).

London, St Pauls 1966

Apart from small scale trials, this was the first site in the UK at which freezing was undertaken with LN. A shallow shaft had to traverse Thames gravels to reach an existing tunnel in the London Clay. Being a small operation it was considered an ideal opportunity to gain experience with the 'new' refrigerant, while providing a time advantage to the client.

The 4·6 m diameter shaft was first sunk to 8 m, just above the water table in the gravel. Timber decking over the upstanding shaft collar acted as a drilling platform from which angled freeze-tubes were installed into the clay at 0·4 m centres around the perimeter of the shaft sump. Ice-wall closure was achieved in 2·5 days, but continued for a further 3·5 days before excavation was resumed, to generate sufficient volume of frozen ground to permit completion

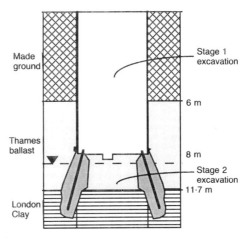

Fig. 7.15. Shaft deepening at St Pauls (cross-section)

of the shaft without refrigeration. Sinking and lining through the gravels was completed in three days. The construction is shown in Fig. 7.15.

Point of interest It will be noted that if the shaft excavation is to be continued at the same diameter, the freeze-tubes and associated distribution mains will have to be dismantled thus precluding any further freezing to preserve the ice-wall. It was expected that with controlled minimal ventilation within the shaft the integrity of the ice-wall would survive for up to four days in granular soil, and this proved to be reasonable in granular strata; at another site dealt with some years later in a similar manner but in soft silty clay, 2·5 days proved to be the safe limit in cohesive soil.

Engr: *Sir William Halcrow*; MC: *Mitchell Bros*; SGFC: *Foraky Limited*.

See Miller and Gordon-Brown (1967).

Unity, Saskatchewan (Canada) 1960s
Ground freezing was selected to deal with a saturated sand layer extending from 235 ft to 280 ft (72–85 m) below surface for a 12·5 ft (3·8 m) diameter shaft being sunk to exploit potash deposits. The concept adopted involved sinking the shaft to 100 ft (30 m) below surface where an enlarged chamber was excavated to facilitate sub-surface freeze-tube drilling. Inclined drilling to pass 6 ft (1·8 m) outside the shaft excavation was undertaken. All of the 36 holes were surveyed for angle only and, being deemed to be sufficiently straight, 1·5 in (38 mm) internal diameter freeze-tubes (EX casing) were duly installed and furnished with 0·5 in (13 mm) copper inner pipes.

Freezing was undertaken with brine plant, and with the shaft flooded to ensure static groundwater conditions. After ten days the shaft was dewatered and no inflows were observed indicating that ice-wall closure had been attained. On the following day all the freeze-tubes broke, releasing brine to the formations, shortly followed by flooding as the ice-wall was eroded.

Each freeze-tube was then 'salvaged' by running 0·5 in plastic freeze-tube outers and copper inners. A central pressure relief/observation hole was then installed which indicated ice-wall closure after a further 60 days of freezing. Shaft sinking to the bottom of the frozen zone was completed 90 days later.

Point of interest The initial freeze-tube sizes used here are, to the author's knowledge, the smallest ever used for a brine freezing exercise, though typical for LN freezing. The event-dictated change to even smaller freeze-tubes reflects the much increased primary freeze-period, and the eventual success was undoubtedly achieved with a relatively thin ice-wall.

Engr: *Western Potash*; MC: *NR*; SGFC: *Winston Bros*.
See Dawson (1954).

As an aid to recovery

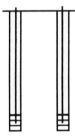

Frozen from the surface
Aberdeen 1980
The access shaft, for the construction of a long sea outfall, was to be 7·3 m diameter and 66 m deep. The upper half of the sinking had to pass through soft ground comprising boulder clays to 20 m, clay with sand lenses to 23 m, sands and gravels to 30 m, and the lower half through rock.

From the outset it was appreciated that ground treatment would be needed between 20 m and 30 m, and allowance for grouting was made in the contract documents. Doubts were expressed by the Contractor at the time of award but the cost and duration of grouting were expected to be lower than those for ground freezing.

On reaching 20 m, the top of the grouted zone, a major inflow of water occurred, and sinking was brought to a halt. After several unsuccessful attempts (including regrouting) to seal the water make which by now was 1400 l/m, ground freezing was assessed to be the most feasible of the several alternatives considered, even though the prime cost was still considered high.

As the design diameter of the shaft was to be maintained, the freeze-tubes were drilled around the shaft and freezing took place from the surface into the bedrock. A brine temperature of $-35°C$ was applied to create an average ice-wall temperature of $-18°C$. Excavation was resumed seven weeks after commencing refrigeration, and continued into competent rock without interruption.

Point of interest Comparing prices based on the original tender price of ground freezing = 100, the range of grouting bids was 70−125, and of dewatering 105−115. Comparing completion costs to the same base, grouting accounted for 180 and ground freezing 150, i.e. more was actually spent on abortive grouting than was eventually spent on ground freezing, even allowing for associated main contractor delay and attendance costs. See Chapter 9.

Engr: *Donald & Wishart & Partners*; MC: *Balfour Beatty Ltd*; SGFC: *Foraky Limited*.

CASE EXAMPLES — SHAFTS

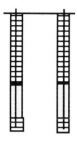

Sub-surface freezing
Crewe 1960

A 40 ft (12 m) diameter casting pit being sunk within a circular ring of steel sheet piling 44 ft (19 m) diameter and 60 ft (18 m) deep, was excavated to 40 ft (12 m) and concreted with a 6·5 ft (2 m) thick base plug/floor. Following a short period during which other works were undertaken, it was found on inspection that the uplift pressure on the base due to the shallower groundwater table had raised the floor slab by about 1 ft (0·3 m) from the as-cast level. A re-examination of the strata details showed that the very variable glacial drift of boulder clays, lake clays and sand, silt and sandy silt, achieved a thickness locally of 82 ft (24·6 m) before the Keuper Marl bedrock was reached. It was now urgent that the plug be safely replaced and the casting pit completed. To eliminate any possibility of recurrence of the uplift failure, it was decided that the excavation be taken at a gradually reducing diameter to the Marl as a foundation horizon, and backfilled with concrete to support the floor slab. The excavation is shown in Fig. 7.16.

By installing freeze-tubes on a circle 39 ft (12 m) diameter, i.e. just within the existing pit walls as Fig. 1.6(iii), to reach into the Marl, a cofferdam was created; the inward growth of the ice-wall, which increased with depth, acted as a stable excavation boundary. The works were completed without further incident.

Point of interest For a duplicate facility (proposed but not

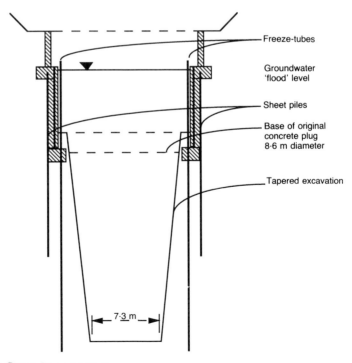

Fig. 7.16. Casting pit at Crewe (cross-section)

proceeded with), the client intended to specify the ground freezing method based on his experience.

Engr: *Sir Alexander Gibb & Partners*; MC: *Sir A McAlpine*; SGFC: *Foraky Limited*.

Rogers Pass (Canada) 1986

Deep-well dewatering was the method adopted to sink a shaft through 95 m of water-bearing silty boulder clay with extensive beds of interlocked quartzitic boulders which infilled a high-level valley over the mid-point of a 14·7 km long rail tunnel being driven below the Canadian Rockies. Of the three wells installed only two developed their full potential, achieving an effective pulldown of 77·7 m.

Inflows of less than 0·04 l/s, which were experienced in reaching a depth of 62·5 m, caused instability in parts of the exposed wall; concern was felt that conditions would worsen and problems become too severe below the drawdown level. It was decided to implement ground freezing for the remaining 32 m of sinking. Two overlapping freeze-tube covers, each approximately 22 m long, were used as illustrated in Fig. 7.17. The first cover comprised 42 freeze-tubes at a sub-vertical angle of 6°, and the second cover 30 freeze-tubes at 8·5°; in the latter case five central freeze-tubes were installed into the base to resist the upward groundwater

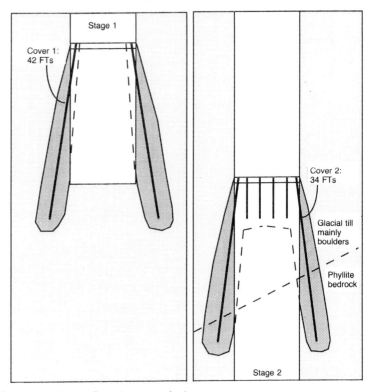

Fig. 7.17. Two-stage freezing at Rogers Pass (cross-section)

pressure. The deep-well depressed water level was maintained throughout the freezing operation.

Point of interest It is unusual for groundwater lowering to remain operational concurrent with ground freezing operations.

Engr: *Canadian Pacific Railway*; MC: *Cementation (Canada)*; SGFC: *British Drilling & Freezing Co Ltd..*

See Macfarlane *et al.* (1988).

Tooting Bec 1991

As the 2·5 m diameter, 40 m deep, water ring main tunnel approached Tooting Bec the London Clay/Woolwich and Reading Beds/Thanet Sand interfaces rose above floor level, and an inrush of sand and water resulted.

To recover the situation the first task was to sink an access shaft just ahead of the trapped tunnelling machine. Ground freezing was employed over the lower section of the shaft as illustrated in Figs 7.18 and 7.19. The 'recovery' is described later in this chapter.

Engr: *Thames Water*; MC: *Fairclough CE*; SGFC: *British Drilling & Freezing Co Ltd.*

See Clarke and Mackenzie (1994).

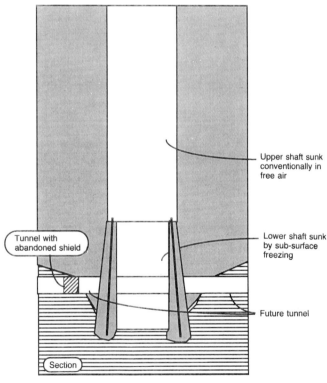

Fig. 7.18. Scheme for rescue shaft at Tooting Bec (see Fig. 7.37 for scheme to recover shield and resume tunnelling)

GROUND FREEZING IN PRACTICE

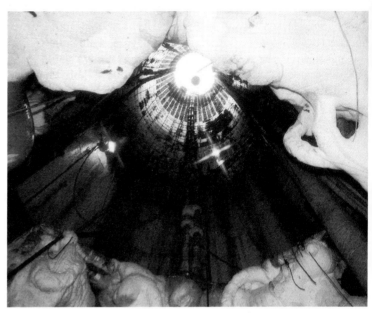

Fig. 7.19. Looking up the rescue shaft at Tooting Bec. Photo courtesy of BDF Co Ltd.

Tunnels

As the chosen design method of temporary works

With horizontal freeze-tubes

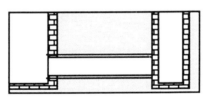

Blackwall Tunnel 1967
Completion of a second road tunnel under the river Thames at Blackwall afforded the opportunity to renovate and upgrade the 19th century tunnel which included two sections with sharp reverse curves at the ventilation shaft locations. To improve traffic flows, when the two tunnels would eventually operate as separated dual carriageways, the tunnel eyes in these ventilation shafts were to be enlarged thus permitting an increase in the horizontal radii of the reverse curves.

Following a spectacular collapse and inrush of soil into the tunnel as the first eye was being dismantled, it was decided that protective measures were needed at two of the remaining locations. The original construction comprised cast iron bolted segments backed by 300 mm or so of in situ concrete; the latter was found to have decomposed with time and was no longer structurally useful. A scheme using horizontal freeze-tubes drilled through the 0·9 m thick wall of the ventilation shaft around the tunnel eye to form

CASE EXAMPLES — TUNNELS

Fig. 7.20. Cavity at Blackwall Tunnel from initial loss of strata on dismantling the tunnel eye. Photo courtesy of Mott MacDonald.

Fig. 7.21. Freezing system at the second tunnel eye at the Blackwall Tunnel. Photo courtesy of BDF Co Ltd.

a frozen collar, embodying the defunct concrete and some of the surrounding gravels, was devised and undertaken. This enabled the original concept to proceed at the remaining eyes.

Engr: *Mott Hay & Anderson*; MC: *Balfour Beatty Ltd*; SGFC: *Foraky Limited*.

Cairo (Egypt) 1990

Ground freezing was used to form the InterConnecting Tunnel (ICT) between the 39 m diameter main pumping station and the 18 m diameter distribution chamber at Ameria, Cairo, and to effect the connection from the pre-arrived spine tunnel into the distribution chamber once the latter had been sunk to depth.

Horizontal freezing was adopted in both cases. Some holes were slightly inclined under the invert of the ICT because the sump level of the shaft matched that of the tunnel invert; similarly for the spine tunnel, to accommodate the smaller external diameter of the tunnel compared with the starting diameter of the freeze-tube circle around the tunnel eye in the shaft wall. The installation of the freeze-tubes through stuffing boxes, and the tunnelling operations, were undertaken from within the distribution chamber in both cases. Conventional ammonia mechanical refrigeration plant was supplied from the UK. (See Figs 7.22 and 7.23.)

A partial collapse of the frozen roof of the ICT immediately adjacent to the pumping station, when preparations were being made in the main pumping station shaft for the imminent arrival of the ICT, led to the rupture of a shallow-level force-main with consequent influx of water and soil. Both shafts were then flooded to stabilize the water pressure outside and inside the caissons while a number of vertical freeze-tubes were installed to reinstate the frozen ground over the tunnel crown. An explanation for the ice-wall collapse has not been discussed in the literature to date.

Fig. 7.22. Face excavation with protective insulation at Cairo. Photo courtesy of BDF Co Ltd.

Fig. 7.23. Excavation of the invert below floor level at Cairo. Photo courtesy of BDF Co Ltd.

Engr: *Ambric (Binnie-Taylor)*; MC: *Christiani*; SGFC: *British Drilling & Freezing Co. Ltd.*

Dundee 1981
A sewer tunnel was to be driven beneath the railway east of Dundee Station from vacant ground between the main four-track railway line and the docks, to a target sheet-piled reception pit in the parallel main road. Horizontal freeze-tubes were installed to reach the sheet pile face, and a hand-dug tunnel lined with bolted segments executed within the brine refrigerated cylindrical ice-wall.

Point of interest Monitoring of the railway track levels was undertaken throughout the operation and for three months beyond conclusion of freezing; no changes of level were detected.

Engr: *Tayside Regional Council*; MC: *J Mowlem & Co Ltd*; SGFC: *Foraky Limited*.

Du Toits Kloof (South Africa) 1981
Following the need to salvage a parallel pilot tunnel by local freezing with LN, the designers specified that construction of the main road tunnel through the initial incompetent ground (talus and decomposed granite) covering the granite bedrock be undertaken within the protection of a 2 m thick canopy/cylinder of frozen ground (Figs 7.24 and 7.25).

About 155 m of the drive (plus cut-off) required protection; it was divided into 32 m stages, each surrounded by a series of sub-horizontal freeze-tubes and terminated by vertical freeze-tubes from the surface to form a vertical frozen bulkhead. Enlargement of the tunnel prior to the bulkhead provided chambers for drilling the next-stage freeze-tubes completely outside the tunnel space, thus ensuring adequate ice-wall thickness and integrity.

Points of interest Even with careful drilling techniques

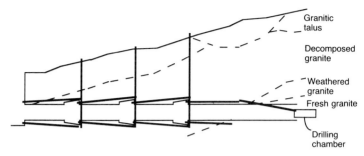

Fig. 7.24. Tunnel at Du Toits Kloof (long-section, frozen zones omitted)

Fig. 7.25. Breaking through the contiguous piles to form the tunnel eye at Du Toits Kloof. Photo courtesy of BDF Co Ltd.

incorporating stabilized drill-rods, accurate drilling over the 32 m penetration proved to be very difficult, and several extra holes were drilled to compensate for divergences deemed to be unacceptable.

To avoid the need for a very deep vertical bulkhead between stages four and five, a drilling chamber was excavated in the fresh granite, via a heading constructed from the parallel pilot bore. This chamber also served as a collecting point for draining water from the weathered/fresh granite interface, thus reducing the flow from this source through the zone to be frozen.

Engr: *Van Niekerk Kleyn & Edwards*; MC: *LTA*; SGFC: *Foraky Limited*.

See Cockcroft *et al.* (1982), Harvey (1983).

Helsinki (Finland) 1981
This project concerns the passage of twin 6·1 m diameter tunnels 35 m below the surface through a buried channel comprising glacial

water-bearing soils, a cleft lying between two hard rock bodies which hosted the rest of the tunnel drivages. The area, in the centre of Helsinki, is a major focal point for surface transport systems, sub-surface concourses and many buildings on shallow timber piles, all of which had to continue in uninterrupted service without adverse effects.

It was decided that an overall length of 50 m, encompassing the 30 m section in soil, should be treated to permit initiation of the excavation and lining works within stable rock. After consideration of space and possible side-effects horizontal sub-surface freezing was commissioned. Two shafts for access, supplies/services and ventilation were constructed to link with the tunnels on either side of the cleft; drilling was undertaken from chambers constructed from the existing tunnel drives on each side of the cleft. Freeze-tubes were installed on twin circles of 8·4 m and 9·8 m diameter (16 to each), and three within the central core, to generate a 2·5 m thick ice-wall around the excavation. The brine method with freon compressors achieved a primary freeze in twelve weeks for one tunnel and seven weeks for the other.

Ground movements were monitored, mainly from shallow settlement plates, and by taking levels on tram lines and by buried magnetic extensometers. More than 80% of the settlement noted occurred during the pre-freezing operations including the installation of freeze-tubes, but satisfactory explanations were not obtained. No significant ground movements occurred attributable to either the freeze or the thaw, and no movements were observed in any of the adjacent buildings.

Point of interest   A compressed air facility, provided as a fail-safe back-up, remained unused.

Engr: *Mott Hay & Anderson with K Hanson & Co Oy*; MC: *Lemminkainen Oy*; SGFC: *Kankkija*.

See Lake and Norie (1982).

Mannheim (Germany) 1988–1989
Twin tunnels to carry a dual carriageway bypass beneath eleven busy rail tracks required ground protection both below the water table and in the low moisture content strata above the water table. Sub-horizontal freeze-tubes 92 m long from both ends of the 184 m tunnel length had to be very accurately drilled, using directional drilling techniques, to establish the dumb-bell cross-section shape, shown in Fig. 7.26. Advantage was taken of the intended directional control, i.e. elimination of deviation, to space the freeze-tubes at the unusually large dimension of 1·8 m. A vertical frozen bulkhead was created at the mid-point.

Pre-grouting was carried out to reduce the permeability, and thereby minimize water flow problems. Additionally it was necessary to increase the moisture content of the strata above the water table, to ensure sufficient strength when frozen to support the active railways above. An auxiliary gallery was driven above

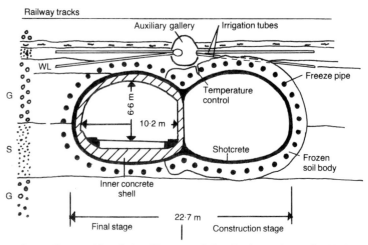

Fig. 7.26. Cross-section of tunnelling at Mannheim (Germany) [after Borkenstein et al. (1991)]

the centreline of the twin tunnels, from which transverse horizontal irrigation wells were installed.

Point of interest The design was supported by laboratory experimental work devoted to the mechanical properties of the frozen soils at various moisture contents, and the viscosity of the irrigation water which would ensure a controlled seepage rate during the primary freeze period.

MC: *Bilfinger & Berger, Hochtief, Diringer & Scheidel, Klee, Sax & Klee JV and others*; SGFC: *Deilmann-Haniel*.

Milchbuck (Switzerland) 1978

A 1·3 km highway tunnel in Zurich was driven from the south portal through a 350 m length of morainal deposit containing pockets of gravel, sand and silt with groundwater under sub-artesian pressure, before reaching solid rock comprising alternating layers of marl, sandstone and limestone. Ground freezing was used in

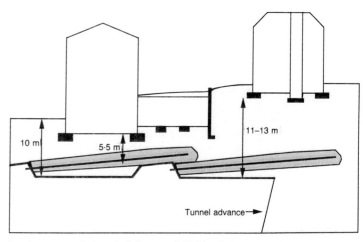

Fig. 7.27. Overlapping freezing stages below buildings at Milchbuck (cross-section)

the moraine section in horizontal bays with freeze-tubes typically 34 m in length (but without vertical bulkheads at the end of each bay). Brine was circulated at −40°C. Following multi-level excavation the exposed frozen ground was continuously shotcreted, and the invert concreted. The working face, also, was protected by shotcrete whenever it was not being advanced.

Enlarged drilling chambers 8 m long were excavated at the end of each bay to permit the setting up of the drilling template and machines for introducing the freeze-tubes for the succeeding bay. The construction is shown in Fig. 7.27.

Point of interest Apart from the first bay the tunnel route passed below multi-floor business and domestic property sensitive to ground movements. Careful observations made during the freezing of the first bay indicated a maximum heave of 105 mm, which could not be tolerated under the built-up area. It was found, by operating the freeze-plants on a cyclic basis during the ice-maintenance stage, thus restricting further growth of the ice-wall, that the heave was virtually eliminated — 5 mm or less.

Engr: *Bundesamt fur Strassenbau*; MC: *Brunner/Locher/Prader*; SGFC: *Electrowatt*.

See Bebi and Mettier (1979), Huder (1981).

Runcorn 1980

The syphon-like connection of tunnels carrying three water mains under major waterways — the Manchester Ship Canal and the estuary of the River Mersey — was to be eliminated by linking the existing tunnels in-line to form a single continuous aqueduct; this would enable the resulting tunnel to be relined and utilized full-bore, thus quadrupling its capacity. To safeguard against instability and flooding from either waterway the Consultant had considered many methods of undertaking this task, the safest and most effective being to construct a large diameter temporary shaft at the mid-point of the 38 m gap, and then form two short tunnels from it to the original terminal shafts of the existing tunnels (Fig. 7.28).

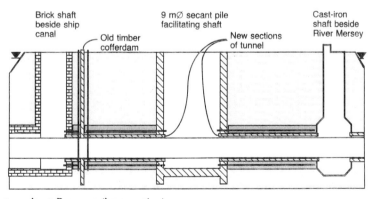

Fig. 7.28. Connecting tunnels at Runcorn (long-section)

GROUND FREEZING IN PRACTICE

Fig. 7.29. Old timber caisson exposed in tunnel excavation at Runcorn. Photo courtesy of North-West Water.

Fig. 7.30. Tunnel breakout through secant pile wall at Runcorn. Photo courtesy of Mott MacDonald.

The 9 m internal diameter shaft was excavated within bored secant piles of 1·2 m diameter, and provided an adequate working space from which to install horizontal freeze-tubes to carry out 'brine' ground freezing for the 13/14 m long connecting tunnels. To prevent loss of water and more importantly loss of soil fines (ground support), the freeze-tube drilling was undertaken from an adjustable platform through a stuffing box securely mounted onto flanged pipes pre-set into the secant piles with epoxy resin. While freeze-tube drilling was in hand for the north tunnel, exploratory work being undertaken to establish the quality of backfill behind the brick shaft associated with the south tunnel had shown that a timber obstruction lay in the path of the south tunnel freeze-tubes 1 m short of the shaft.

The timber could not be dealt with by the drilling method already provided. The design was therefore varied to include short additional horizontal freeze-tubes drilled from within the original shaft to the 'rear' side of the timber obstruction, found later to be part of the cofferdam (Fig. 7.29) within which the brick-lined tunnel and shafts had been built in open cut before the canal was flooded. Programme delays were minimized by using LN as the refrigerant for the short length. The tunnel eye is illustrated in Fig. 7.30.

Engr: *Mott Hay & Anderson*; MC: *FJC Lilley Ltd*; SGFC: *Foraky Limited*.

See Harris and Norie (1982), Lake and Norrie (1982).

Tokyo (Japan) 1985
This example is of tunnel junctioning following drivage by TBMs from both portals to a meeting below water at or near the midpoint. Two merged cones of frozen ground were formed by installing freeze-tubes from each tailskin at angles between 20° and 25°. To achieve a maximum spacing of 0·8 m, 108 freeze-tubes were drilled into place. In addition the inside face of the tailskin was also refrigerated, via pipework installed at manufacturing stage, to ensure good frozen adhesion. Each cone was served by a refrigeration plant of 96 000 kcal/h to supply brine at about $-25°C$. A total of 34 temperature sensors were installed to monitor the ice-wall.

With the upper part of the tunnel in sand and the invert in silts, frost heave and thaw settlement of the invert were expected. To minimize the effect the upper level refrigeration was carried out ahead of that of the invert to limit the period of refrigeration in the lower part to a minimum; similarly hot water was circulated through the lower freeze-tubes in place of brine to accelerate the thaw.

Engr: *Tokyo Sewage Works Agency*; MC: *NR*; SGFC: *NR*.
See Numazawa *et al.* (1988).

Frozen from the surface
  Born (Switzerland) 1976

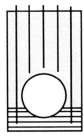

Commencement of twin-track tunnelling from the north portal had to contend with historic landslide debris comprising clayey silts up to 30 m thick with a moisture content of up to 25%. Detailed technical and cost assessments indicated to the Railway Authority that ground freezing was the correct choice.

The freeze-tube drilling pattern from the surface followed the profile of the tunnel roof, and extended into the rockhead on either flank. To minimize the volume of frozen ground the freeze-tubes were fanned as illustrated in Fig. 7.31, and only the lower part

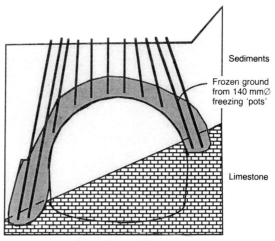

Fig. 7.31. Born railway tunnel (cross-section) [after Wind (1978)]

of each was designed to operate as a fully refrigerating element; the upper section acted principally as flow and return pipes. The frozen arch was maintained for six weeks during the excavation and lining works.

Engr: *NR*; MC: *NR*; SGFC: *Deilmann Haniel*.
See Wind (1978).

Selby, Gascoigne Wood 1980
At the Gascoigne Wood drifts the Basal Sand stratum, which was to be frozen, occurred at a depth of 170–180 m. The drifts, sloping at 1 in 4, required freezing for just over 100 m (horizontal) of their advance. The novel method selected involved drilling freeze-holes from the surface along the centreline of each drift, initially vertically then, from a depth between 84–100 m, intentionally steered alternately to pass either side of the intended drift to form an 'inverted-Y' cross-section pattern, to create eventually a tent-like frozen canopy through the Basal Sand horizon.

Engr: *National Coal Board*; MC: *Cementation*; SGFC: *Foraky Limited*.
See Wild and Forrest (1981).

Tottenham Hale 1966, Vauxhall Cross 1968, Pimlico 1970
A series of arch freezings were required to protect the inclined escalator drives at new stations on the London Underground Victoria Line. In each case the station concourse area was constructed as an open excavation within a perimeter wall through the gravels into the London Clay foundation stratum.

The roof of the escalator drive, starting from leaving the perimeter wall until the drive was totally within the London Clay, was protected by a frozen arch, formed by brine freezing through vertical/sub-vertical freeze-tubes (Fig. 7.32). Careful monitoring confirmed that there was no ground heave, which could have affected the nearby buildings. The site is shown in Fig. 7.33.

CASE EXAMPLES — TUNNELS

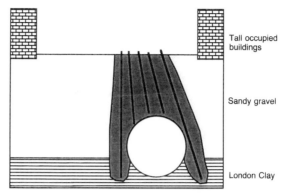

Fig. 7.32. *Escalator drivage at Pimlico (cross-section)*

Fig. 7.33 *Freeze-plant and system alongside medium-rise buildings just prior to excavation of an inclined escalator drive for Pimlico Station.* Photo courtesy of Mott MacDonald.

Engr: *Mott Hay & Anderson*; MC: *A Waddington at Tottenham Hale, Kinnear Moodie at Vauxhall Cross, Balfour Beatty at Pimlico*; SGFC: *Foraky Limited*.

See Lee (1969), Megaw and Bartlett (1981).

As an aid to recovery

With horizontal freeze-tubes

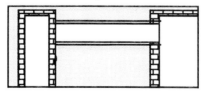

155

Edinburgh, Craigentinney 1974

Here a trunk sewer tunnelling contract involved passing below a golf course and a hospital at relatively shallow depth. The works were commenced in compressed air at 34.5 kN/m^2, the maximum that the thin overburden cover would tolerate. A thin lens of saturated fine sand was encountered just below axis level when the face was beneath the hospital; the stratum was unstable and yielded water even against the air pressure. The face was boarded to safeguard the works while supplementary measures were instituted. A series of freeze-tubes was driven forward in a fan-disposition, so that refrigeration would solidify the weak stratum on either side of the next excavation stage and across the face.

The LN refrigerant was fed from a storage vessel installed alongside an intermediate manhole already constructed and sealed off (to contain the compressed air). Twin holes were drilled through the seal to accommodate the insulated delivery pipeline which fed a distribution manifold at the face, and the exhaust pipeline used to discharge the resultant nitrogen gas to the atmosphere. Each stage averaged 3 m of freezing, permitting excavation for and erection of four 0.6 m rings at a time. Six advances totalling 15 m were completed in 21 working days, and the facility was retained for the remaining two weeks drivage under compressed air against the eventuality of further instability.

Engr: *City Engr, Edinburgh*; MC: *J Mowlem & Co Ltd*; SGFC: *Foraky Limited*. See Harris (1974).

Renfrew 1975

Following the collapse of a sheet-piled trench excavation through the west-bound carriageway of the A8 road, it was decided that the remainder of the new sewer construction below the east-bound carriageway be undertaken in tunnel. Access to and working space at the foot of the embankment was available, so horizontal drilling for 14 freeze-tubes was undertaken. Only 9 m penetration was practical, so the length was treated in two stages, and frozen with LN. Bolted segmental lining was erected as the excavation advanced. The excavation was slowed by the presence of timber, steel pipe, RSJs and sheets, some being old fill but mainly the aftermath of the collapse of the sheet-piling.

Engr: *W A Fairhurst & Partners*; MC: *L Fairclough Ltd*; SGFC: *Foraky Limited*. See Harris (1975).

River Medway 1976

Remoteness of the tunnel face collapse from the Isle of Grain access shaft (1 km), and 50 m below the estuary of the River Medway towards Chetney Marshes, posed problems of access and handling of the refrigerant. The original 2.4 m diameter face was abandoned and a diversion, enlarged to 3.3 m diameter, was constructed to within 10 m of the weak zone. Concurrently a 2.4 m diameter tunnel was advanced from the opposite shore, on a matching

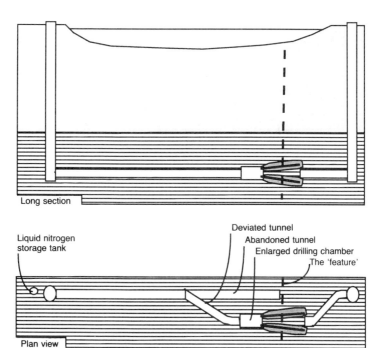

Fig. 7.34. Tunnel recovery under the River Medway

alignment, also to within 10 m of the weak zone, thus leaving a 20 m connection to be made. See Fig. 7.34.

Interpretation of exploratory drilling from the diversion face suggested the presence of a sand and sea-shell filled fissure (Fig. 7.35), in the London Clay, of 150–200 mm thickness subject to fluctuating water pressure of the order of up to 350 kN/m^2. A fan-shaped pattern of 40 freeze-tubes was installed to receive LN refrigerant (Fig. 7.36). The usual LN storage vessel was installed near the shaft-head at a point accessible by road tankers. The LN was then transferred by vacuum insulated pipeline to rail-mounted 500 litre insulated tanks at the shaft bottom, ready for hauling to the face on a shuttle principle. After circulation within the freeze-tubes, the resultant nitrogen gas was collected in a manifold and vented to atmosphere via a 200 mm diameter pipeline running along the tunnel and up the Grain shaft.

Difficulties were experienced with the upward inclined freeze-tubes in achieving regular ice-growth along their length. This arises because the liquid flows to the lowest point which tends to 'starve' the upper (remote) section of the freeze-tube. Rudimentary 'weirs' were provided within the freeze-tubes to alleviate this problem. [Simple and very effective cascade designs were used on a similar exercise in Seattle — See Maishman (1988).] The whole operation to complete the 20 m connection occupied 13 weeks.

Engr: *C Haswell & Partners*; MC: *J Mowlem & Co. Ltd*; SGFC: *Foraky Limited*.

See Harris and Woodhead (1977).

GROUND FREEZING IN PRACTICE

Fig. 7.35. The frozen sea-shell filled feature being approached by the tunnelling shield under the River Medway. Photo courtesy of Charles Haswell & Partners.

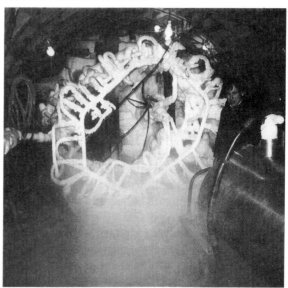

Fig. 7.36. Freezing system at the tunnel face below the River Medway, showing the rail-mounted LN supply tank in foreground. Photo courtesy Charles Haswell & Partners

CASE EXAMPLES — TUNNELS

Tooting Bec 1991
Having gained access to the tunnel level (via the shaft described earlier in this chapter) it was possible (a) to connect with the abandoned heading, and (b) to construct a launch chamber for a replacement tunnelling machine able to cope with the severe conditions, by conducting horizontal freezings from the shaft sump (Figs 7.37 and 7.38).

Engr: *Thames Water Authority*; MC: *Fairclough CE Ltd*; SGFC: *British Drilling & Freezing Co Ltd*.

See Clarke and Mackenzie (1994).

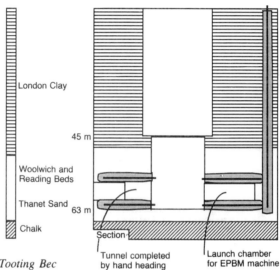

Fig. 7.37. Tunnel recovery at Tooting Bec

Fig. 7.38. Launch chamber to recover the TBM at Tooting Bec, Thames Water ring main. Photo courtesy of BDF Co Ltd.

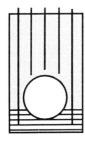

Frozen from the surface
  Alexandria (Egypt) 1987
A small shield tunnel, being driven under a canal by controlled intrusion of plastic clay, started rising uncontrollably. It was bulkheaded to limit the collapse and effect temporary security, following which recovery of the face was achieved by creating a frozen canopy with LN-fed vertical freeze-tubes. This enabled the central door, omitted at the outset, to be fitted to the shield.

  Engr: *City Sewerage Authority*; MC. *Egyptian Arab Contracting/Mini Tunnels Intl*; SGFC: *British Drilling & Freezing Co Ltd*.

Isle of Grain (1974–1975)
A collapse occurred at the face of a 4·65 m diameter tunnel being machine driven 52 m below surface in the London Clay. The collapse led to funnelling of overburden material into the working space and a large depression at the surface. The tunnelling machine was engulfed.

  Recovery measures included the immediate application of compressed air to minimize further soil and water ingress, followed by regrading and concreting of the ground surface to permit freeze-tube drilling in a pattern as Fig. 1.7(i) over and ahead of the collapse zone. Once the ice arch was formed, using a bank of ammonia compressors, the air pressure could be partially reduced to enable an underground inspection. Normal air pressure was then resumed and the debris cleared. The tunnelling machine could then be repaired prior to restarting the tunnelling advance under the protection of the ice arch and into undisturbed ground.

  Engr: *LG Mouchel*; MC: *John Laing*; SC: *Cementation*; SGFC: *Foraky Limited*.

  See Harris (1976).

Sheffield 1981
The Don Valley trunk sewer being driven with a hard-rock shield in coal measure mudstones beneath 5 m of drift, encountered a declining clay aquiclude in the crown with associated weathered unstable rock. Even after grouting an inleak of groundwater at 3 l/s was experienced. These conditions could have been safely handled with a soft ground hood on the face of the shield.

  An arch, frozen with LN (illustrated in Fig. 7.39), under which the shield could be advanced and long enough to permit a chamber to be excavated ahead of the advanced shield, was therefore provided. A total of 144 freeze-tubes at 0·6 m centres along the boundaries and on a 1·5 m grid within, was installed by drilling with light top-drive drilling rigs. A total of 10^6 m^3 of LN (measured as gas) was consumed in creating and maintaining the arch. A conventional soft-ground hood was then attached to the face of the shield within the chamber, and the works resumed as a soft-ground tunnel.

CASE EXAMPLES — TUNNELS

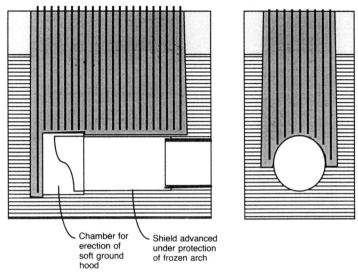

Fig. 7.39. *Shield chamber at Sheffield (sections)*

Engr: *City Engr, Sheffield*; MC: *Don Valley Contractors*; SGFC: *Foraky Limited*.

Stirchley (Birmingham) 1974
Following collapse of a 2 m diameter heading 35 m short of its target shaft, it was established that the drive had encountered a buried channel containing weak saturated soil. Of this 25 m could be adequately protected by grouting, leaving a 10 m gap immediately up to the shaft which would have to be frozen.

Being within a street of terrace housing with nominal front gardens where it would be difficult to provide power supplies, and round-the-clock noise would be unacceptable, ground freezing with LN was chosen. Freeze-tubes were installed in vertical boreholes sunk with shell and auger rigs.

After five days refrigeration, tunnelling was resumed by hand digging, and erection of bolted segmental lining was completed over a period of 15 days by two eight-hour shifts. From this heading the remaining 25 m was installed by pipejacking under the grouted cover to link with the pre-installed main drive.

Engr: *City Engr, Birmingham*; MC: *FJC Lilley Ltd*; SGFC: *Foraky Limited*.

See Harris and Reed (1975).

Stonehouse 1986
Soft ground tunnelling commenced in competent clays but encountered a buried channel of weak soil; a bulkhead was erected to safeguard the works already completed. A frozen canopy was then established across the buried channel from an array of vertical freeze-tubes fed with LN.

Engr: *Strathclyde Regional Council*; MC: *Farrans Civil Engineering Ltd*; SGFC: *British Drilling & Freezing Co Ltd*.

Thamesmead 1987

Attempts to make a sub-surface connection between two adjacent, newly constructed manholes were frustrated by water and soil runs when the breakout was started. The strata were not amenable to injection methods, and groundwater lowering was rejected by the client. A small frozen block, using LN, was successful.

Engr: *Wimpey/PSA*; MC: *Balfour Beatty Ltd*; SGFC: *British Drilling & Freezing Co Ltd*.

Three Valleys 1984

A 2·8 m diameter, mechanized shield, open face, single drive tunnel was being driven in and near the base of the London Clay (35 m below the surface) when it encountered a geological anomaly containing sands and silts subject to a water head of over 30 m. On further investigation a diapiric structure was detected where the superficial deposits were significantly deeper than elsewhere over a tunnel length of 55 m, and extended in depth to a local peak of the underlying Woolwich and Reading Beds which had risen above the basement level of the London Clay; over this length the London Clay was missing: see Fig. 7.40.

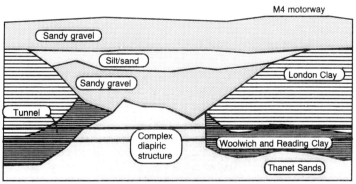

Fig. 7.40. Generalized structure of anomaly (long-section)

Fig. 7.41. Nitrogen venting from the tunnel freezing at the Three Valleys site. Photo courtesy of Binnie & Partners.

Fig. 7.42. Excavating the frozen tunnel at the Three Valleys site. Photo courtesy of Binnie & Partners.

Table 7.1. Financial assessments, Three Valleys (in thousands of pounds)

Option	Lowest quoted cost	Estimated delay cost	Total cost
Jet grouting	350	210	560
Curtain grouting	535	165	700
Claquage grouting	257	138	395
Dewatering	350	600	950
Brine freezing	300	330	630
LN freezing	500	150	650

To control the ground and water pressures by compressed air was considered to be untenable. Due to the high groundwater pressures proposals for grouting and ground freezing solutions were then sought, and a three-stage bentonite−cement/claquage + sodium silicate/bentonite−cement grouting scheme was chosen and implemented. This proved to be inadequate due to problems with the sodium silicate grout.

Time delays were by now becoming intolerable, and further requests for ground freezing proposals stipulated that, for speed and maximum security against groundwater inflow, LN freezing was required. This was conducted via five rows of freeze-tubes placed in vertical holes drilled from the surface, and refrigeration was applied on a progressive basis in phase with tunnelling advance underground (Figs 7.41 and 7.42).

Points of interest The initial choice of grouting over freezing is stated to have been based on cost, time, reliability of the method (safety), and the experience of the contractor. The financial assessments, based on lowest quoted cost plus estimated delay cost at £30 000/week, were stated to be as shown in Table 7.1. In the event £598 000 was spent on grouting and £650 000 on ground freezing (including main contractor's charges). No comparisons under the other headings have been given, but the presence of flowing spring water into a nearby borrow pit was a factor against brine freezing in the reliability assessment.

The logistics of this exercise centred on

(a) the need to drill approximately 250 boreholes quickly and accurately to a depth of 40 m: it is recorded that drilling did not meet the 1 in 100 specification, and many of the boreholes drifted up to 1·7 m off plumb (3·8 in 100), leading to the possibility of unfrozen windows unless corrected by placement of additional freeze-tubes (two unfrozen windows were encountered nevertheless, where the grouting had stabilized the ground; this was fortunate as regards the safety of the tunnel) and

(b) the provision of the largest uninterrupted supply of LN demanded by the construction industry from the producers: a total of $5·5 \times 10^6$ m^3 (gas vol.) ($7·96 \times 10^6$ litres of liquid) were consumed, requiring peak deliveries by road of up to 1×10^5 m^3/day equivalent to six truckloads.

Engr: *Binnie & Partners*; MC: *Thyssen (GB)*; SGFC: *Stent Foundations/Rodio*.

See Baker and James (1990).

Miscellaneous

Gap freezing

Duisberg (Germany) 1978

The balance of groundwater flow and the level of the water table in the glacial and quaternary sands and gravels below much of the city of Duisberg, can easily be adversely affected by major 'linear' inground construction work. If said works are founded in the underlying impermeable tertiary silts and clays they can act as barriers with a consequent rise in groundwater level on the upstream side, and depression on the downstream side. The effects would be flooding of basements or drying of wells/settlement of structures respectively. If the angle of incidence of groundwater flow with the axis of the works exceeds 20° it is necessary, as part of the design, to ensure that flow continues throughout and after the construction works.

Diaphragm walls with 'gap freezing' were introduced for cut and cover metro tunnel construction through such strata conditions, within which groundwater flows of 0·5 to 15 m/day, at 45° to the tunnel line, were common. See Fig. 7.43.

Following studies which included FE analyses and a full scale trial, the following design was adopted.

(a) Install 5·4 m lengths of diaphragm wall extending 2 m into the basal cut-off, leaving gaps of 1·35 m (20%) between panels.

(b) Construct cross walls at each end and at the midpoint, thus creating two 100 m blocks.

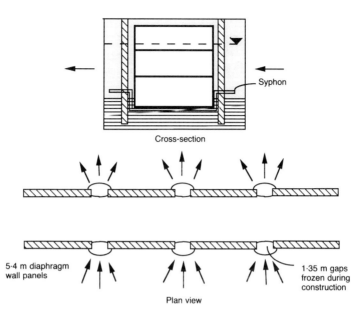

Fig. 7.43. Gap freezing at Duisberg

(c) Install the freeze-tubes.
(d) Freeze the gaps, first on the upstream side of block 1, then on the downstream side.
(e) Excavate within the cofferdam so created, install horizontal filter wells just above basal cut-off level on up- and downstream sides linked to a syphon below tunnel floor level (wherever the floor level occurred below the basal cut-off level), cast the floor slab, and complete the tunnel walls to achieve full security.
(f) After allowing thawing, repeat for block 2.

On the upstream side freeze-tubes were included in the end of each diaphragm wall panel plus five across the gap; on the downstream side four freeze-tubes were placed across the gap. A plant of 600 000 kcal/h capacity supplied brine at temperatures down to $-40\,°C$ to achieve closure times of up to seven and eleven days.
Engr: *City Engineer, Duisberg*; MC: *Hochtief*; SGFC: *Hochtief*.
See Weiler and Vagt (1979, 1980, 1981).

Fosdyke 1971 and Cheltenham 1972 — sealing parted sheet piles
At both these sites excavation of access pits within sheet-piled cofferdams was frustrated when water and soil 'flowed' between sheet-pile clutches which had parted during driving. The gaps were sealed by carrying out local LN freezing, thus re-establishing the cofferdam and allowing work to resume safely.
Engr: *British Gas*; MC: *Shand Ltd*; SGFC: *Foraky Limited (Fosdyke)*. Engr: *Septimus Willis*; MC: *Esply-Tyas Ltd*; SGFC: *Foraky Limited (Cheltenham)*.
See Harris (1972b).

Inset *London, Chelsea Harbour 1987 — junction of shaft with crown of sewer*

Sewage disposal from a new prestige development at Chelsea Harbour was to be through an active 19th century brick sewer which lay beneath one corner of the site. Difficulty was experienced sinking the shaft, and injection treatment of the sands forming the cover to the connection required between shaft and tunnel was

Fig. 7.44. Compact LN site layout at Chelsea Harbour. Photo courtesy of BDF Co Ltd.

Fig. 7.45. The excavated chamber over the old sewer tunnel crown at Chelsea Harbour. Photo courtesy of BDF Co Ltd.

CASE EXAMPLES — MISCELLANEOUS

unsuccessful. The Grouting Contractor then recommended that freezing be used. Freezing with LN from the surface was carried out, the surface layout being shown in Fig. 7.44, and the resultant excavation over the old sewer in Fig. 7.45.

Engr: *Thames Water Authority*; MC: *J Murphy Ltd*; SGFC: *British Drilling & Freezing Co Ltd*.

Protection

Bromham Bridge 1991 — safeguarding an ancient structure
It was feared that excavation close to the foundations of a 14th century multi-arch bridge would lead to loss of support and unacceptable structural damage. Normal river flow is through the main arch, the remaining arches being dry except in flood conditions. A new sewer was (to be) installed through one of the flood arches.

The headroom of just over 2 m dictated the use of a mini-drill to install the freeze-tubes which were placed to create a 35 m cofferdam through the gravels, centred on the bridge. Liquid nitrogen refrigerant was used, and the works were completed without detriment to the bridge. See Fig. 7.46.

Engr: *North Beds Borough Engineer*; MC: *C Gregory Civil Engineering Ltd*; SGFC: *British Drilling & Freezing Co Ltd*.

Fig. 7.46. Protecting the 14th century foundations of Bromham Bridge by LN freezing. Photo courtesy of BDF Co Ltd.

Temporary roadway

Green River, Kentucky (USA) 1985
A frozen silty clay roadway of three 6 m wide tracks, each 700 m long within a 25 m wide strip, enabled a 3000 ton dragline to 'walk' across an alluvial flood plain in Kentucky in one day. To achieve the frozen state some 12 600 m of 90 mm diameter horizontal freeze-tubes were laid in a series/parallel array, and connected to a central 600 TR (2000 kcal/h) multi-unit freeze-plant. The ground surface was insulated with 100 mm straw covered with

sheet polythene over the full width to minimize heat losses to the atmosphere and shed rain water to either side.

After 46 days of refrigeration the thickness of frozen ground in each strip had reached the design thickness of 1·83 m, and the decision to move the dragline on day 49 was taken. This unusual exercise was the subject of careful monitoring and led to a post-completion study and design methodology proposal.

Engr: *Owner, Peabody Coal Co*; MC & SGFC: *FreezeWALL Inc.*
See Maishman *et al*. (1988).

Underpinning

Burgos (Spain) 1986
To enable construction of an 11 m deep underground car park below a 15th century palace, while preserving the fabric, it was necessary to underpin the shallow foundations. Part of the perimeter could be dealt with by diaphragm wall techniques, but to complete the boundary beneath the main facade a chain of frozen small diameter shafts was formed following the principle illustrated in Fig. 1.10; a detail at a corner is reproduced as Fig. 7.47.

Sands, gravels and sandy clays of 6–10 m thickness rested upon marl, with the water table at 2·5 m. Once a water-tight peripheral wall into the marl had been established, other excavation works within the boundary could be begun. By freezing alternate mini-shafts in two phases, the 3 m long sections of underpinning could be completed and linked to form a continuous wall.

Engr: *NR*; MC: *NR*; SGFC: *Rodio*.
See Rojo and Novillo (1988).

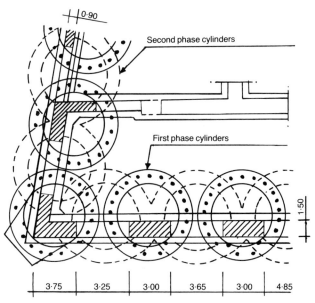

Fig. 7.47. Plan of overlapping frozen soil cylinders and pier foundation panels at Burgos (Spain) [after Rojo and Novillo (1988)]

CASE EXAMPLES — MISCELLANEOUS

São Paulo (Brazil) 1942
Rapid settlement occurred as a 26 storey building was nearing completion in 1942, and urgent measures became necessary to arrest the movement, eliminate tilt and restore the building to its original elevation. Successive layers of clays, sands and silty sands and clayey sands of varying thicknesses, continuity, saturation, strength and compressibility, and decomposed gneiss and granite, overall some 90 m thick, rested on solid rock. A stiff clay layer identified over most of the site served as the foundation stratum for bulbous Franki-piles, commonly used in the area. Their installation proceeded as expected and the building rose to its full height without incident.

Some five months later, when setting out the lift guides it was realised that a 1 in 500 tilt towards a corner, and a differential settlement of 60 mm, had occurred. The whole building was moving monolithically. Investigations established the presence of a previously unrecognized wedge of compressible soft silty clay, and this was deemed to be the cause.

A ground freezing scheme, comprising 162 freeze-tubes drilled through the heavily reinforced concrete floor slab to a depth of 14 m at the affected corner, was instituted to arrest the tilting movement. The rate of settlement increased alarmingly to a peak of 6 mm/day during the early drilling operations, which led to commencement of freezing as soon as possible. Drilling and freezing continued concurrently for 15 weeks, and the settlement gradually lessened as the refrigerative effort increased. After another four weeks settlement had ceased and a 10 mm rise was recorded at the point of previous maximum settlement.

Jacking was then used to eliminate the tilt and restore the building to its original elevation. A number of freeze-tubes were then decommissioned to leave a balanced system of ice-maintenance, and to release capacity for additional freezing while two additional columns were installed.

Several of the piles within the frozen zone were, by now, under tension and had to be cut, while others were becoming overloaded. Additional 'mega-Hume' piles were introduced to deal with this situation. Belled pits were excavated through the frozen ground into dense sand for cast in situ supplementary foundations to absorb any thaw settlement loadings and provide reaction for jacking wedges. All movements ceased over a further 18 months.

Engr: *NR*; MC: *Severo & Villares SA/Estacas Franki SA*; SGFC: *Foraky SA*.

See Dumont-Villiares (1956).

Vienna (Austria) 1986
A section of the Vienna subway considered to pose particularly difficult construction constraints required twin single-track tunnels to pass below a major telecommunications facility with only 1·5 m cover between tunnel crown and building foundations. The general

conditions already dictated the use of the New Austrian Tunnelling Method (NATM), supplemented by compressed air and partial dewatering, for the main tunnel drives.

Extensive laboratory testing of frozen alluvial deposits and verification of design data by conducting freezing at two test sites was carried out as part of the design for the 64 m long section, to ensure minimal heave and settlement during and following the freezing activity. The freezing was limited to forming a horizontal slab, 1 m thick, to act as a groundwater control, structural support and air-loss/blow-out barrier over the tunnels. The design criteria were to choose freeze-tube spacings which optimized creation of the ice-wall, and the proportion of on/off refrigeration cycles to minimize frost heave activity during the maintenance stage.

The length of individual LN-fed freeze-tubes was limited to 32–33 m by installing from both ends of the section with an overlap at the mid-point. Continuous freezing was carried out for five days, following which maintenance freezing occupied periods of 2–7·5 h/day. Maximum soil movements recorded were 13 mm heave during active freezing, 6 mm settlement due to tunnelling, 18 mm settlement due to thaw consolidation/secondary settlement when tunnelling/depressurizing the tunnel, with a final maximum total settlement of 14 mm.

Engr: *Magistratsabteilung*; MC: *JV Arge U6/3*; SGFC: *Gruen, Bilfinger & Deilmann-Haniel*.

See Deix and Braun (1988).

A similar scheme was undertaken in *Brussels*, Belgium, where in addition the dry sand needed irrigation to increase the moisture content.

Engr: *Lipski SA*; MC: *Egta SA*; SGFC: *Foraky SA*.

See Gonze *et al.* (1985).

References

BAKER A. C. J. and JAMES A. N. 1990. Three Valleys Water Committee: tunnel connection to Thames Water reservoirs. *Proc. ICE*, Part 1, **88**, 929–954.

BEBI P. C. and METTIER K. R. 1979. Ground freezing for the construction of the 3-lane Milchbuck road tunnel, Zurich. *Proc. Tunnelling '79*, IMM, 245–255.

BORKENSTEIN D. *et al.* 1991. Construction of a shallow tunnel under protection of a frozen soil structure. *Proc. 6th ISGF*, Beijing, **2**, 481–487.

BUTTIENS E. 1978. Conversion of abandoned collieries in south Belgium to LPG storage with description of special plugging of various shafts. *Proc. 1st ISGF*, **1**, 349–356. (Also *Proc. Rockstore*, Sweden, **1** (1979), 101–105.)

CLARKE A. P. J. and MACKENZIE C. N. P. 1994. Overcoming ground difficulties at Tooting Bec. *Proc. ICE*, **102**–2, 60–75.

CLEASBY J. V. *et al.* 1975. Shaft sinking at Boulby Mine, Cleveland Potash Ltd. *Trans. IMM*, **84**, 7–28 and 147–148.

COCKCROFT T. N. *et al.* 1982. Construction of a section of the Du Toits Kloof tunnel by use of ground freezing. *Proc. Tunnelling '82*, IMM, 105–116.

COLLINS S. P. and DEACON W. G. 1972. Shaft sinking by ground freezing in Ely-Ouse Essex scheme. *Proc. ICE*, Supp. vii & xx, 7506S, May, 129–156, 319–336.

DAWSON A. S. 1954. Quicksand frozen for shaft sinking. *Cndn Mining J.*, Oct., 68–70.

DEIX F. and BRAUN B. 1988. Vienna subway construction — use of brine freezing in combination with NATM under compressed air. *Proc. 5th ISGF*, Nottingham, 1, 321–330.

DOIG P. J. 1985. Grouting and freezing for shaft water control in Milwaukee. *Proc. RETC*, 2, 1211–1224.

DUMONT-VILLAIRES A. 1956. The underpinning of the 26 story 'Companhia Panlista de Segmos' building, Brazil, *Géotechnique*, 6, 1–14.

FINDLAY A. J. et al. 1982. Recent developments in the design and operation of LNG facilities in the UK. *Proc. IC on World Gas*, B34, 1–23.

GONZE P. et al. 1985. Sand ground freezing for the construction of a subway station in Brussels. *Proc. 4th ISGF*, Sapporo, 1, 277–284.

GRAHAM E. B. et al. 1983. The British Gas Canvey Island LNG Terminal: a review of developments. *Proc. IC LNG7*, Jakarta, May, 17pp.

GREGORY O. and MAISHMAN D. 1973. Motorway construction meets an unusual problem in old shaft treatment. *Trans. IMinE (Manchester)*, Feb.

HARRIS J. S. 1972b. Ground freezing seals sheet-pile leaks. *Contract J.*, 10 Feb.

HARRIS J. S. 1974. Cryogenic treatment of shafts and tunnels. *Tunnels and Tunnelling*, Sept. 67–70.

HARRIS J. S. 1975. Freezing relieves bad ground conditions. *Underground Services*, 3-2, 22–23.

HARRIS J. S. 1976. Ground freezing — some uses of this geotechnical process in civil engineering applications during the last decade. *Proc. C. on the Next Decade*, Jan.

HARRIS J. S. and NORIE E. H. 1982. Construction of two short tunnels using artificial ground freezing. *Proc. 3rd ISGF*, Hanover NH, 1, 383–388.

HARRIS J. S. and REED R. J. 1975. Ground freezing at Stirchley. *Ground Engng*, Sept., 46–48.

HARRIS J. S. and WOODHEAD F. A. 1977. Ground freezing deals with tunnel instability. *Ground Engineering*, Sept., 47–48.

HARRIS J. S. and WOODHEAD F. A. 1978. Ground freezing for large diameter foundation piers. *Consulting Engineer*, Jan.

HARVEY S. J. 1983. Ground freezing successfully applied to the construction of the Du Toits Kloof tunnel. *Proc. 1st NSGF*, Nottingham, 51–58.

HARVEY S. J. and MARTIN C. J. 1988. Construction of the Asfordby mine shafts through Bunter Sandstone by use of ground freezing. *Proc. 5th ISGF*, Nottingham, 1, 339–348; and in *The Mining Engineer*, 148–323, 51, 53–58.

HUDER J. 1981. Milchbuck tunnel. *Proc. 10th IC SM & FE.*.

JACKSON D. 1989. Contractual aspects of AGF. *Proc. 5th NSGF*, Nottingham, 63–69.

KELLAND J. D. and BLACK J. C. 1969. Cominco's Saskatchewan potash shafts. *Proc. 9th Commonwealth Mining Congress*.

LAKE L. M. and NORIE E. H. 1982. Application of horizontal freezing in tunnel construction — two case histories. *Proc. Tunnelling '82*, IMM, 283–289.

LEE J. 1969. Escalator tunnel at Tottenham Hale Station, Victoria Line. *Proc. ICE*, 7270S, 423–429.

MACFARLANE I. M. et al. 1988. Application of ground freezing at Rogers Pass, Canada. *Proc. 5th ISGF*, Nottingham, 2, 563–564.

MAISHMAN D. 1971. Ground freezing. *Civil Engng and Public Works Rev.*, Oct.

MAISHMAN D. 1988. A short tunnel in Seattle frozen using liquid nitrogen cascade lances. *Proc. 5th ISGF*, Nottingham, 2, 561–562.

MAISHMAN D. and POWERS J. P. 1982. Ground freezing in tunnels — 3 unusual applications. *Proc. 3rd ISGF*, Hanover NH, **1**, 397–410.

MAISHMAN D., POWERS J. P. and LUNARDINI V. J. 1988. Freezing a temporary roadway for transport of a 3000 ton dragline. *Proc. 5th ISGF, Nottingham*, **1**, 357–365.

MANNING G. P. 1973. In-ground storage of LNG and its effect on the surrounding ground. *Underground Services*, June, 15–23.

MEGAW T. M. and BARTLETT J. V. 1981. Ground treatment (by freezing). *In Tunnels — planning, design and construction*, **2**, 112–114.

MILLER H. W. and GORDON-BROWN T. P. 1967. Recent developments in ground freezing. *Proc. IoRef*, Nov.

NUMAZAWA K., TANAKA M. and HANAWA N. 1988. Application of the freezing method to the undersea connection of a large diameter shield tunnel. *Proc. 5th ISGF*, Nottingham, **1**, 383–388.

ROJO J. L. and NOVILLO A. 1988. Recent applications of soil freezing techniques in Spanish construction works. *Proc. 5th ISGF*, Nottingham, **2**, 525–532.

SCHOLES W. A. 1982. Freezing an island to sink a shaft. *Tunnels and Tunnelling*, Jan.

SHUSTER J. A. and SOPKO J. A. 1989. Ground freezing to control ground water and support deep storm sewer structural excavations. *Proc. 9th RETC*, Los Angeles, 149–155.

SOPKO J. A. *et al.* 1991. Frozen earth cofferdam design. *Proc. 6th ISGF*, Beijing, **1**, 263–272.

WALSH A. R., HART D. E. and MAISHMAN D. 1991. Shaft construction by raise boring through artificially frozen ground. *Proc. 6th ISGF*, Beijing, **1**, 369–378.

WEILER A. and VAGT J. 1979. The Duisberg method of metro construction — an example for establishing large underground civil engineering projects in flowing groundwater. *Proc. 7th European Conf. on Soil Mech. & Fndn Engng*, Brighton, **3**, 299–305.

WEILER A. and VAGT J. 1980. The Duisberg method of metro construction — a successful application of the gap-freezing method. *Proc. 2nd ISGF*, Trondheim, **1**, 916–927.

WEILER A. and VAGT J. 1981. Gap freezing solves groundwater problems. *Tunnels and Tunnelling*, Dec., 31–34.

WILD W. M. and FORREST W. 1981. The application of the freezing process to 10 shafts and 2 drifts at the Selby project. *Proc. IME: The Mining Engineer*, June, 895–904.

WIND H. 1978. The soil freezing method for large tunnel constructions. *Proc. 1st ISGF*, Bochum, **2**, 119–126; and in *Engng Geol.*, 1979, **13**, 417–423.

8. Standards and safety

with contribution by Alan Clough

The health and safety of freeze plant operatives, and others who may be working close by, is covered in the UK by various pieces of legislation, the most important of which are given below.

Health and Safety at Work etc. Act 1974

Although this piece of revised legislation became law almost 20 years ago, the requirements are still rigorously enforced today. Section 2 places responsibility on the employer to ensure the health, safety and welfare at work of all employees. This means that the employer must

- provide and maintain plant and systems that are safe and without risks to health
- arrange to remove any risks to health and safety in connection with the use, handling, storage and transport of articles and substances
- provide information, training and supervision as is necessary to ensure the health and safety of employees
- ensure that all places of work, within the employer's control, are safe and without risks to health, and also that access to, and egress from them are safe and without any risk
- provide and maintain a working environment that is without risk to health, and adequate for the employees' welfare at work.

The above items are equally important but emphasis is usually placed on the provision of information and training. This must be given immediate consideration when new operatives are taken on. Training must be given to an agreed programme, usually to the Company's Safe System of Work procedures, which must contain all legislative requirements; appropriate records must be kept.

In January 1993 a comprehensive set of regulations was introduced in the UK in an attempt to improve the health, safety and welfare conditions of all employed persons. The most important, and indeed the one upon which the other five pieces of legislation hinge, is the *Management of Health, Safety and Welfare Regulations 1993*.

Regulation 3 sets out the requirements for Risk Assessment. All items within a person's job must be assessed and any significant findings recorded. When making the assessment it is necessary to be aware of the many different meanings given to the terms

Hazard, Risk and Safe in the occupational health sense

- *hazard* — something with the *potential* to cause harm
- *risk* — the likelihood and effect of that *potential* being realized
- *safe* — something is safe if there is a general acceptance that the risk involved in its use is a reasonable one.

Regulation 4 states that an employer shall have a system which in terms of Health and Safety has built into it

- planning
- organisation
- control
- monitoring and review.

A health surveillance procedure must be established to be followed in the event of impending serious danger to persons working in the factory or on the works. The documentation must identify a named person to implement this procedure. Again the risk assessments should identify foreseeable events that need to be covered in these procedures.

Employers should make available to their employees all relevant information about all relevant risks. While this information is limited to what the employees need to know to ensure their health and safety, it must be easily understood. It would be advisable to check the knowledge of employees shortly after any such information has been given.

Other new sets of regulations which need to be addressed, and which the HSE will check on routine visits and in the event of a serious injury, include

- Personal Protective Equipment at Work Regulations 1993
- Work Equipment Regulations 1993
- Workplace Health, Safety and Welfare Regulations 1993
- Manual Handling Regulations 1993
- Display Screen Equipment at Work Regulations 1993.

These regulations are very comprehensive and in many instances repeal certain sections of both the Factories Act 1961 and the Offices Shops and Railway Premises Act 1963.

Note: Particular attention should be paid to requirements relating to working in confined spaces and gaseous atmospheres.

Control of Substances Hazardous to Health (COSHH) 1988

Under this legislation, as described in HSE publication EH40, suppliers of any chemical or hazardous substance must provide a Hazard Data Sheet to the purchaser, which must contain all relevant information to enable the user to make a fair and accurate assessment of any hazard associated with this material, assess all chemicals including oils, greases, paints and gasses, and any chemical used in an office environment. In fact any sort, type and preparation which is to be used by employees in their work situation must be covered by a hazard data sheet and risk assessment.

Regulations, standards and codes of practice

The following have particular relevance to ground freezing in construction.

The Noise at Work Regulations 1989 — this regulation requires that the employee's hearing is protected. The first action level begins at 80 dBa. As a rule of thumb, if an ordinary conversation cannot be carried out at 1 m distance apart then it is considered to be too noisy.

The prime objective is to reduce the noise level at source. If this is not reasonable because of cost or production implications, then hearing protectors of the correct attenuation must be provided at no cost to the employee. Supervision must be effective, i.e. there is a duty to ensure that protection requirements are met, backed if necessary by Disciplinary Procedures. Failure to ensure protection of hearing may, in later years, result in a successful compensation claim for loss of hearing, which is now classed as an Industrial Disease.

BS 4434:1:1980 — requirements for refrigeration safety: an overall specification of the topic and often regarded in the refrigeration industry as 'the bible'.

BS 6164:1988 — safety in tunnelling in the construction industry: contains a very brief description of ground freezing, and precautions to be taken when the method is used.

BS 8004:1986 — §6.5.2 reference is made to excluding water from excavations by freezing the surrounding ground.

HS/G 30 — Storage of anhydrous ammonia under pressure in the UK.

The Cryogenics Safety Manual (3rd edition) — produced by the British Cryogenics Council and published by Butterworth-Heinemann in 1992. This manual constitutes an authoritative guide, *inter alia*, to the storage and handling of liquid nitrogen.

The Code of Practice *Mechanical integrity of vapour compression refrigerating systems for plant and equipment supplied and used in the UK* — Code of Practice published by the Institute of Refrigeration in 1979. Part 1 relates to the design and construction of systems using ammonia as the refrigerant. It includes a listing of standards and titles appropriate to refrigeration equipment.

The BCGA have published a series of documents, the most relevant of which are listed in the references.

Health and Safety Aspects of Ground Treatment Materials (Skipp and Hall) — was published by CIRIA in 1982 as report 95. This document is mainly concerned with toxicity effects associated with the use of chemicals, and section 7 includes some notes relating to ground freezing.

Control of groundwater for temporary works (Somerville) — published by CIRIA in 1986 as report 113. Some errors in the original text were corrected later in an errata sheet; the following comments should be noted by users.

- P10, Fig. 6 : the soil range over which freezing is applicable is not made clear: refer to Fig. 1.3 in this book.
- P11, Table 2, §b13 and 14: LN is unsuitable for long freeze-tubes; a long time is not always required.
- P12, Table 3, §13 and 14: comment should relate to a water seepage rate of less than 2 m/day, rather than to permeability.
- P17, Table 10: Freezing and compressed air are both expensive, but the freezing method has been shown to compare favourably with other methods on many occasions regardless of depth, and is much more versatile than electro-osmosis.

Code of Safe Drilling Practice — published by the British Drilling Association.

First Aid — it is strongly recommended that all personnel receive basic first-aid training.

Selected refrigerants

Ammonia

The term ammonia, a natural compound of hydrogen and nitrogen, refers to anhydrous ammonia (NH_3) in gaseous or liquid form. Ammonia has a characteristic pungent odour and can be detected by smell at concentrations of less than 10 ppm, i.e. well below the long term exposure limit of 25 ppm. Whilst it is the most toxic of all common fluids used as a primary refrigerant, it can be safely handled by competent and fully trained personnel using the correct equipment and procedures; the refrigeration industry in both the UK and the USA is convinced of its safety for use in industrial applications. It is cheap, has attractive thermodynamic properties, and does not contribute to the greenhouse effect.

Above 200 ppm concentration, the substance will cause irritation and discomfort of the mucus membrane and the eyes, but with no lasting consequences. The short term exposure limit is 35 ppm but for brief periods concentrations up to 500 ppm can be tolerated thus permitting safe evacuation from contaminated areas; above 1500 ppm tissues will be damaged or destroyed whilst exposures to 2500 ppm and above increase the risk of fatality. Liquid splashes of ammonia on the skin will cause burns, the severity of which depends on the amount of ammonia; even the smallest amount of liquid ammonia reaching the eyes may result in permanent injury.

Ammonia is flammable in air at concentrations of between 16% and 27% by volume, but ignition is difficult to achieve; it is toxic and inhalations may be lethal at the higher concentrations; breathing is impossible at such concentrations. Explosive and unstable compounds can be formed by reactions to mercury, halogens, hypochlorites, nitric acid, and some organic compounds.

Ammonia readily dissolves in water, liberating heat, and forming a strongly alkaline solution. Moist ammonia attacks copper, zinc, calcium and most of their alloys; also many rubbers and plastics; moisture in the system contaminates the compressor oil and reduces the operating efficiency of the plant, but does not lead to icing of valves as with halogen refrigerants. Oil and ammonia do not readily mix together, so the task of oil recovery from the bottom of flooded evaporators is easily carried out.

Tube-type cylinders of ammonia should be stored on wooden runners in the horizontal position but with the bottom raised slightly higher than the top-outlet. A protective cap should be fitted, and the storage space should be dry, well ventilated, and out of direct sunlight. 10 kg of anhydrous ammonia is equivalent to 14.05 m^3 gas at normal temperature and pressure.

It is hazardous to discharge ammonia directly to atmosphere unless purged slowly via a stack pipe above roof level when, due to its being lighteer than air, it will disperse fairly readily. As the substance is soluble with water, disposal is best effected by purging through a water filled purge pot.

Gas masks and ammonia proof respirators should be on site, and outside any location where leakage can occur, where they can readily be reached. Compressed air breathing apparatus is suitable for use in any concentration of ammonia; training in its use is essential, and records when used must be kept. Cold water should be hosed onto leaks until they can be dealt with, and onto cylinders close to sources of heat until they can be moved to a safe location. Deluge-type water showers should be provided close to plant exits for dowsing persons leaving ammonia atmosphere.

Note: The testing and design of components should be in accord with Part 1 of the Ammonia Code of Practice issued by the Institute of Refrigeration, see above.

Liquid nitrogen   Nitrogen is inert. Liquid nitrogen is produced by air separation techniques, and has a boiling point of $-196\,°C$ at ambient pressure. The resultant gas is therefore extremely cold and occupies over 600 times the volume of the liquid from which it derives. Thus the main hazards associated with the use of LN are cold burns or frostbite through contact with cold metal, breathing problems or asphyxiation through oxygen deficiency, and fire or explosion through oxygen enrichment adjacent to any sections of uninsulated metal LN conductor pipes. Calibrated gas/oxygen detection instruments must be used when working within plant.

The normal atmosphere contains 21% oxygen by volume; at 17% oxygen a naked flame extinguishes; at 14% the breathing rate increases and people appear intoxicated; at 10% pain goes unnoticed and people become ill tempered and less capable; at 6% muscular control is negligible, vomiting commences and consciousness is doubtful; below 6% breathing stops, the heart beats for only a few minutes and irreparable brain damage results.

It is therefore important, particularly in confined spaces — tunnels and shafts — that leakage of nitrogen is prevented; any nitrogen which does escape will form a 'slug' of negligible oxygen content which will move with the ventilation stream. All persons working or entering confined areas served by LN/gas systems must be instructed in and understand the safety procedures *before* work commences, be equipped with compressed air breathing apparatus, and always be within sight/sound of oxygen deficiency activated alarms. Other recommended safety precautions include in/out registers (as with mining), telephone link to the surface, pressure relief valves at the surface with associated valved bypasses to enable isolation of the circuit, and automatic safety valves along long liquid delivery lines with direct connection to the exhaust system.

References

APV BAKER LTD 1990. Ammonia, the negected alternative? *Refrigeration and Air Conditioning*, July.

BCC 1991. *Cryogenics safety manual*. Butterworth-Heinemann, Oxford, 3rd edn.

BCGA 1992. *CP21: Bulk liquid argon or nitrogen storage at customers premises*. 25pp.

BCGA 1992. *CP24: Application of the pressure systems and transportable gas containers regulations 1989 to operational process plant*. 31pp.

BCGA 1992. *CP25: Revalidation of bulk liquid oxygen, nitrogen, argon and hydrogen storage tannks*. 14pp.

BRITISH STANDARDS INSTITUTION 1989. BS 4434 *Safety aspects in the design, construction and installation of refrigerating appliances and systems*. BSI, 31pp.

ICI LTD. *Anhydrous ammonia industrial hazards bulletin*.

ICI LTD. *Anhydrous ammonia — safe handling of bulk liquid supplies*.

INSTITUTE OF REFRIGERATION 1979. *Mechanical integrity of vapour compression refrigerating systems for plant and equipment supplied and used in the United Kingdom*. Part 1 — design and construction of systems using ammonia as the refrigerant. 21pp.

JENKINS D. 1985. Survey of Ammonia refrigeration systems. *Proc. IoRef*, 10pp.

9. Contractual, cost and risk evaluation considerations

with contribution by David Jackson

Contractual arrangements

Artificial Ground Freezing (AGF) is a sequence of specialized operations which will influence the contractual relationship set up to cover its execution.

The particular set of circumstances generating the requirement for AGF may also suggest the appropriate form of contract. In this context 'circumstances' would mean physical circumstances, since any prerequisite for a competitive tender against likely alternatives (at quotation stage) could realistically prove unattainable (but see discussion on risk later in this chapter). Other considerations, e.g. successful track record, could overcome this disadvantage.

The decision to effect certain temporary works by the AGF method will be made either as part of the pre-contract design of the proposed permanent works, or to deal with difficulties that have arisen or may arise during execution of the permanent works, usually within the contract period. Selection at design stage will accompany designation within the full contract estabishment, i.e. as Main Contractor, or as a Nominated Sub-contractor, or as a domestic sub-contractor. In the latter case, an indication of specialist contractors acceptable to or pre-qualified by the Engineer may be included in the documentation.

The inter-relationship between temporary and permanent works operations imposes severe constraints when the role of the AGF specialist is Main Contractor, with the permanent works carried out under a sub-contract. The AGF Contractor would be fully liable to the Employer for his own direct work and that of his subcontractors. The ramifications of this are considerable, particularly for performance, permanent works completion, and maintenance, because the difficulties of control are compounded by having to oversee activity outside the rigid specialist specification associated with AGF.

Although it is important to avoid the impression that the AGF Contractor is strictly confined to the role of Sub-contractor (and there are several cases on record where the main contractor role was successfully assumed), he does usually emerge as such within the overall contract structure. Reasons of expediency, in contract terms, encourage this.

Also the Employer or his agent (Engineer/Architect) will wish to avoid possible contractual pitfalls that can arise with formal nomination of a specialist sub-contractor; these might include anomalies between sets of documentation, limitations on remedy against the Main Contractor for Sub-contractor defaults, and the effects of sub-contractor delays outwith main contractor control.

Nevertheless the Specialist Contractor may well be selected, approved or recommended by the Engineer at or before the final documentation stage. Such early 'approval' may well be dictated by the necessity to ensure effective lead time in advance of permanent construction operations. During this period the capacity preparations and technical proposals (feasibility, method statements and calculations) of the Specialist Contractor will be progressed, and areas of advantageous co-ordination between Main and Sub-contractor activities identified and developed.

When the need for ground freezing arises after the main contract has commenced it will usually be undertaken on a domestic sub-contract basis to rectify problems unanticipated (unforeseen) at commencement date. Exceptionally, an Employer/Specialist Sub-contractor Agreement might be created when an alternative stabilization method has previously been unsuccessfully attempted; as a consequence a new method, e.g. AGF, is ordered by the Engineer to recover the situation and achieve progression of the permanent works.

Unlike the Main Contractor, whose activities tend to be confined to a specific sector of industry, e.g. building, mining or civil engineering, the AGF contractor can be called upon to conduct his operations within many sectors as occasion demands. The contractual variations, arising from a plethora of 'standard' forms of contract, tend to impose differing obligations and liabilities, e.g. with regard to insurances, design, programme, indemnities and status, which the Specialist Sub-contractor has to absorb into his tendering philosophy. In the UK the most common form of sub-contract adopted is that associated with the most recent edition of the ICE Conditions of Contract, commonly known as the 'Blue Form', notwithstanding that such sub-contract 'standard forms' are often revised, adjusted or supplemented for compatibility with other main contract forms.

When 'physical conditions ... or artificial obstructions' are encountered, which the contractor contends will involve him in probable additional costs, the main contractor will inevitably be required to serve condition precedent Notices upon the Engineer/Architect [ICE Cl 12, see Jackson (1989)]. Assuming that the proposal for overcoming the problem necessitates the introduction of AGF, the response of the Engineer may well influence the qualification of the AGF tender with regard to payment. If the Main Contractor's contention is accepted, a formal Variation Order (VO) would probably be issued, with clear obligations for subsequent valuations and payment for the AGF works. However, if the Engineer is of the opinion that the difficulties could have been foreseen, and refuses to issue a VO for change of design ahead of AGF being mobilized, it will be necessary for the AGF Contractor to establish clearly his position with respect to payments for his work, i.e. a reimbursement procedure outside any 'pay-when-paid' clauses which might exist elsewhere in the documentation (see below).

AGF, in common with other specialist temporary works operations, is called upon to a significant degree to alleviate adverse conditions, often at relatively short notice. In entering into his sub-contract, the AGF contractor would be wise to promote acceptance of his own set of conditions supplementary to any 'standard' form of sub-contract, particularly as these forms are generally designed for use in conjunction with a 'standard' main contract form. Obviously such main contract forms correctly seek to bind the sub-contractor to their terms despite (in the case of AGF) apparent anomalies which, if accepted, could place the AGF Sub-contractor in breach even when his performance is unimpeded. To be effective such supplementary conditions must take precedence over all other terms and conditions in the full contract documentation where in conflict, if at all.

The AGF Contractor's 'offer' will usually be presented in the form of a schedule of items or rates outside the jurisdiction of any standard method of measurement, and which by the classification of its items is not a Bill of Quantities. The work is mostly variable by time or volume application, and is inconsistent with critical path principles embodied in most programmes. The ice-creation period (PFP) is indeterminate in specific terms, since the below-ground state (strata, groundwater flow, etc) is imprecise in nature, although very close approximations are usually achieved. Additionally some of the activities are entirely dependent on the productive capacity of others, e.g excavation and lining whilst the frozen state is maintained. A programme in itself is very often not contractual, e.g. ICE Cl 14 programme is not required until some time after the contract is formalized. It can therefore be seen that a qualification against contractual guarantee of programme is often considered necessary. It follows that the pass-on provisions of, *inter alia*, liquidated damages cannot be equitably apportioned, and a waiver with limitation of liquidated damages should receive recognition within the agreed terms.

The AGF design, which must be submitted to the Engineer (via Main Contractor) on request, will utilize the information offered with the contract documents and must take account of any 'guidance' therein with regard to inspection of the site and satisfaction as to the nature of the sub-soil. Qualification may be necessary, particularly when there is insufficient opportunity to inspect and satisfy '... so far as is practicable ...' [ICE 5th edition, Cl 11(1)], with respect, e.g. to the existence of a stratum which will perform as a cut-off; nevertheless any disclaimer which might exist against the accuracy of (borehole) information, and its supply 'for guidance only', must be observed. Any facilities and attendances required of the Main contractor by the AGF Contractor need to be clearly stated if conflict with possible countercharges for resources, e.g. cranes, scaffolding, etc, required as and when necessary, is to be avoided.

Payment provisions vary among the various forms of contract, and the AGF Sub-contractor has to beware of pay-when-paid clauses if he is to avoid risk-financing. Such clauses are prevalent in civil engineering standard sub-contract documentation, are generally absent from Building Forms, and are not subject to any period of payment in the Government Form GC/Works/1 Edition 3 whereas its GW/S sub-contract includes a 24 day stipulation.

Emergency responses can lead to contractual difficulties prejudiced to the serious detriment of the AGF contractor if he, in his understandable desire to respond to the urgency, does not clearly establish correct terms and conditions of contract prior to commencement. In common with other standard forms of sub-contract the standard (ICE) sub-contract seeks to establish that the Sub-contractor is deemed to have full knowledge of the main contract documentation, which includes specifications, geological data, site restrictions, ingress/egress, limitations on power and water services, drawings, programmes, etc. He will be bound by these provisions unless appropriate qualifications have been made prior to commencement.

Bond requirements may impose unrealistic proportional values and unfair justification for call. In-depth examination is necessary before embarking on a bonding arrangement, especially with International Forms where contractual pressure by abuse of performance bonds is a matter of record.

Cost

Although the general concepts of most ground freezing applications are similar, no two situations will be identical. Cost and price, too, will vary between situations, which results in very few contractors publishing price guidelines from which potential users could prepare reliable estimates.

This reluctance originates, perhaps, from the attitude that in-house technical know-how derived from many years experience, which is largely unsuited to protection by patents, is too valuable an asset to 'release'. The authors believe it is an outdated attitude, as evidenced by the much increased volume of technical material which has been published over the last few decades, coupled with the emergence of regular dedicated international symposia.

Those indicative costings which have been published necessarily differentiate between the two basic refrigerative methods, brine and LN. As with all financial considerations the date of publication is all important, and it is left to the user to update in line with changes in the value of money. No-one undertakes the publication of relevant indices; likewise there is no authoritative source of data to measure the cost effects of technical advances in AGF, or competitive changes. Only occasionally do authors of contract reports include details of costs.

Apart from commercial/management of resources criteria, there are many factors which can affect costs, and they include

- location and accessibility of the site
- availability of water and electric power
- depth and size of volume to be protected
- nature of the strata to be penetrated
- level and quality of the groundwater
- ambient conditions
- seasonal/time of year
- time (programme) limitations set by the client.

Remoteness and poor accessibility will increase costs, lack of power may favour the use of LN, large scale may preclude LN, hard or bouldery (intermediate) strata will increase drilling costs, significant depth may require expensive directional drilling, contaminated or flowing groundwater will lengthen the primary freeze period, high air temperatures will increase the power demand, while time constraints may dictate refrigerative capacity above the optimum. These topics are discussed in appropriate chapters.

In general it has been found that the cost variations for brine schemes, which usually result from a longer or shorter PFP than estimated, are relatively small, whereas for LN schemes if the quantity of LN consumed differs from the estimate, the effect on final cost can be significant. Many AGF specialist contractors are therefore willing, with brine schemes, to 'carry' the cost variables for the section of the work which is under their control by submitting a lump sum price for all the freezing activities up to and including creation of the ice-wall; thereafter, while the main or other sub-contractor is executing the construction stage within the protection of the ice-wall, the cost of ice-wall maintenance will be time based. With LN schemes few contractors will offer a lump sum price: all work will be on an as executed/consumed basis and the eventual account, therefore, somewhat 'open-ended' as with some other processes.

In Chapter 7, published cost data has been resourced for the contracts at *Aberdeen, Boulby, Selby Wistow,* and *Iver.* In each case the reliability and cost-effectiveness of ground freezing is apparent. At *Aberdeen* freezing cost less than had been spent on grouting, even though the bid price for grouting was lowest; at *Boulby* the expected time advantage of freezing versus grouting was exceeded sufficiently that, coupled with economies of scale, two freezings would have been the most cost-effective method; at *Wistow* the frozen shaft significantly outperformed the part-frozen, part-grouted shaft to the same depth; and at *Iver* freezing was successful where compressed air and grouting were inadequate.

Evaluation and selection of technique

Geotechnical works are sometimes regarded as 'particularly high risk activities' which are very much dependent on the quality and adequacy of the site investigation. Selection of a construction method and any necessary aids, within the context of 'temporary works', has often been based on a comparison of the apparent actual cost of the alternatives without real regard to the level of adverse risk or the advantage of beneficial effects on other elements of the works. This is particularly true with competitive tendering for contracts (the norm until recently), which has placed responsibility for temporary works with the contractor. Site investigation has often been restricted by financial limits, e.g. to 1% or less of the contract value, and is itself subject to competitive factors.

It is not surprising that the effect on a contract is severe when 'unforeseen' ground conditions arise during early stages of the works, and enhancements of or changes to the geotechnical process are needed to progress the works. Their cost often exceeds the contingency sum, typically 10% of the contract value, which has been the conventional way of making financial provision for risk.

Risk Analysis is a modern project management tool which is generally applied to the whole scheme; the techniques can also be applied equally effectively to individual elements of the works, e.g. site investigation and geotechnical processes, in fact such individual assessments form inter-relating parts of the whole risk management exercise. Individual analyses will examine the adverse and beneficial effects of technical uncertainties, programme changes and safety measures on security, time and cost, while the project analysis will also include consideration of the effects of politics, *force majeure* (flood, earthquakes), supply of materials, finance and taxes, industrial relations, etc in the particular country.

Objective

The purpose of what we may call a temporary works assessment is to identify and balance the advantages and shortcomings of each method in a rational manner, and evaluate their potential for use in the particular circumstances under review. Further consideration of the significance of each method on the project as a whole may be required before final decisions can be made, but this is outside the scope of this text.

Decisions taken at project appraisal stage are recognized as having a major impact on final cost; but changes after the initial commitments have been entered into can also significantly affect programme and other events and trades, as well as costs. For all these reasons, it is desirable that geotechnical decisions should be included in an in-depth pre-works appraisal, not directly influenced by lowest-bid tendering considerations.

Definitions

- *Hazard* — an uncertain event whose consequence is detrimental, having the potential for damage to the works or the environment, or injury to people.

- *Opportunity* — an uncertain event whose consequence is beneficial, usually in cost or time savings.
- *Risk* — an uncertain event, which may have a detrimental or a beneficial effect.
- *Uncertainty* — an event whose outcome cannot accurately be predicted.
- *Risk factor* — the likelihood that the perceived risk will materialize, combined with the comparative impact such would have on any aspect(s) of the works.
- *Risk value* — the integrated quantitative assessment of the probabalistic measure of the occurrence of the hazards, with a measure of the consequences of those events.

The above definitions, given in BGFS (1994), are based upon, but are not necessarily identical to, general usage.

Assessment

Subjective information based on experience, and objective information based on historical data and dedicated investigation, can be used to conduct qualitative and quantitative appraisals. Two simple models have been used for selection decisions and audits.

- *Model 'A'* — described by Harvey (1993), and based upon three headings: control of ground conditions, groundwater control, and safety during excavation and lining. Each aspect is assessed according to the perceived quality of the contractor's presentation of his design, execution, and safety proposals; its use by a client was instrumental in choosing AGF.
- *Model 'B'* — tabled by the author and based on five headings: installation, strata changes, adverse groundwater, deformation, and equipment breakdown, each being assessed on technical attributes (or lack thereof).

Both models allocate risk factors which, when aggregated, are applied to the tender prices to arrive at their relative values. Probability factors for model A range from 0 to 1 (although expressed as 0 to 100) (Table D.1); for model B two scoring ranges are given, from 0 to 1 (Table D.2), and from $0 \cdot 8$ to $1 \cdot 5$ (Table D.3), where $1 \cdot 5$ is the most adverse, 1 is 'no effect', and $0 \cdot 8$ represents a benefit or opportunity, a feature which is not possible with the conventional 0 to 1 range. The details of these models are given in Appendix D.

A model (C) based on Decision Trees [see Chapman *et al.* (1985), Chapman (1992)] has been developed and is outlined below.

- Identify the temporary works methods to be assessed and compared.
- Prepare a list of principal activities or events associated with each method.
- Tabulate the risks involved with each activity, and the responses to those risks.

- Allocate probability/certainty (of achievement) factors in the range 0 to 1, where 0 = maximum adverse effect (i.e. abandon), and 1 = total success.
- Apply to a Decision Tree, and evaluate.

Such a procedure can compare alternative strategies based on events, security, durations, or cost. The following strategy−cost assessments are based on an actual project. Although subject to the 'with hindsight' criticism, an attempt has been made to prepare 'uninfluenced' assessments.

Example
Remit. A shaft was to be sunk through clays to 20 m, clay with sand lenses to 23 m, sands and gravels to 30 m, and the remainder through competent strata.
Award. The tenders received included offers for dealing with the water-bearing strata by dewatering, grouting and ground freezing. The relative prices for ground treatment were ground freezing 100, grouting 70−125, and dewatering 105−115; the lowest grouting offer was adopted.
Event. The shaft flooded when the excavation reached the grouted zone; further grouting only permitted a small advance. Ground freezing was then adopted, and shaft sinking was completed without further incident. The relative actual expenditure on grouting was 180, and the actual expenditure on ground freezing, including associated main contractor attendances and delay costs was 150.
Analyses. See Tables 9.1−9.3, Fig. 9.1 and Appendix D.

The conclusion can be drawn from models A and B assessments that ground freezing offers the best value for money, both at bid stage and after the event.

In model C, the upper branch assesses the outcome if ground freezing had been selected from the outset, based upon the submitted price. The lower branch assumes that, of the alternatives, injection was apparently cheaper than AGF; it was selected and implemented but was unsuccessful, then requiring a second decision to adopt ground freezing to progress the works.

Table 9.1. Assessments using model A

Method	Tender price	Control of ground conditions: weight factor	Groundwater control: weight factor × 1·5	Safety during excavation and lining: weight factor	Total points	Points: £k*	Result
Using tender prices							
Freezing	100	45	60 × 1·5 = 90	55	190	1·9	1
Grouting	70	30	43 × 1·5 = 65	25	120	1·7	2
Dewatering	105	35	25 × 1·5 = 38	25	98	0·93	3
Using actual costs							
Grouting	180	30	65	25	120	0·67	2
Freezing	150	45	90	55	190	1·27	1

* Highest is best.

Table 9.2. Generation of risk value for model B assessment

Method	Risk	Probability	Impact			Risk factor	Risk value
			Security	Time	Cost		
Groundwater lowering	Installation	0·9	0·7	0·8	0·8	0·403	
	Strata changes	0·6	0·8	0·8	0·6	0·23	
	Deformation	0·9	0·8	0·8	0·7	0·403	
	Equipment breakdown	0·9	0·4	0·9	0·6	0·194	0·31
Injection	Strata changes	0·6	0·5	0·6	0·4	0·072	
	Deformation	0·9	0·8	0·8	0·8	0·461	
	Preferred channels	0·6	0·6	0·5	0·4	0·072	
	Equipment breakdown	0·9	1·0	0·9	0·9	0·729	0·33
Ground freezing	Installation	0·9	0·9	1·0	1·0	0·81	
	Strata chages	1·0	1·0	1·0	1·0	1·0	
	Adverse groundwater	0·8	0·8	0·7	1·0	0·448	
	Deformation	0·9	0·9	0·9	0·9	0·656	
	Inaccurate monitoring	1·0	0·8	1·0	1·0	0·8	
	Equipment breakdown	0·9	1·0	0·9	0·9	0·729	0·74

Table 9.3. Assessments using model B

	LTV	RV	LTV÷RV*	Order of merit by value
Using tender prices				
Dewatering	105	0·31	341	3
Grouting	70	0·33	210	2
Ground freezing	100	0·74	135	1
Using actual costs				
Grouting	180	0·33	540	2
Ground freezing	150†	0·74	203	1

LTV = Lowest tender value; RV = risk value
* Lowest is best
† Includes associated Main Contractor attendances and delay costs

The upper branches forecast that the final cost using AGF from the outset would have been 101 units compared with the 100 bid; the lower branches suggest that (a) the final cost using injection only, had it been successful, would have been 121 units compared with the 70 bid, whereas 180 units were actually expended before the decision was made to change from grouting to freezing; (b) the final cost of ground treatment using the injection approach would be 267·8 units compared with 330 actual expenditure, of which the AGF element was 146·8 units forecast compared with 150 units actual expenditure.

Whichever model is used, the judgement of the assessor for each case will play a crucial part in the accuracy of the result; he may need to refer to specialists in ground treatment processes if he has limited experience on which to base his assessment.

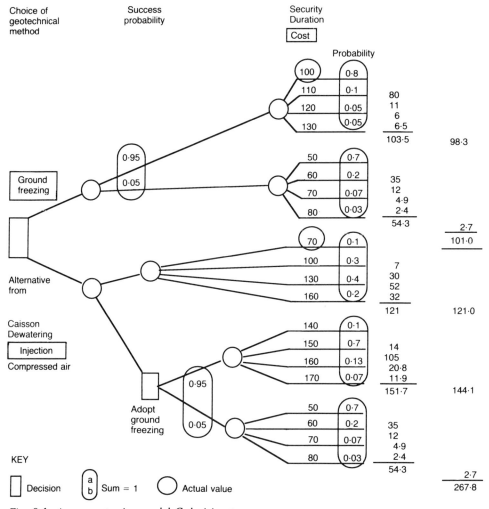

Fig. 9.1. Assessment using model C decision tree

Some historical events in AGF practice which might influence such assessments, with response actions in brackets, include

- settlement and loss of a tunnel following thawing of a foundation stratum which was sensitive to the freeze–thaw cycle (geotechnical appraisal/design assumptions)
- loss of the frozen body due to flooding, following probing to verify the thickness of the frozen ground (experience)
- flooding of a shaft when thawing preceded completion of back-wall grouting (programming)
- flooding of a shaft with a non-frozen core due to fissures in the supposed competent bedrock (geotechnical appraisal)
- flooding of the excavation due to a gap in the ice-wall associated with diverging freeze-tubes (borehole surveys, improved monitoring)

- excessive reduction of the excavated diameter due to creep, dictating a change in excavation and lining techniques and/or application of a lower refrigeration temperature (geotechnical appraisal/design assumptions)
- the presence of air voids when saturated voids were expected, negating the benefits of freezing (geotechnical appraisal/design assumptions)
- the discovery of obstructions after commencement of the ground freezing programme, dictating a change of procedure (desk study/geotechnical appraisal)
- under-allowance for low thermal conductivity of clay and the depressed freezing point due to salinity, requiring a much longer primary freeze-period than programmed (geotechnical appraisal/design assumptions)
- over-optimistic interpretation of the actual growth of the ice-wall leading to a partial collapse during the excavation stage (improved monitoring/supervision).

Reaction

The response to an assessment may be to retain, transfer, avoid or reduce the risk.

- Reduction or prevention of the probability and/or financial severity of the identified risk may be achieved by redesign or by a change of (the enabling) method.
- Avoidance, usually necessitating major reappraisal of the project, will be chosen when the impact is deemed too serious; the impact may be of a political, safety or financial nature. Such a decision often implies inadequate appraisal at the outset.
- Contractual, or non-insurance, transfer is sometimes possible by negotiation between the various parties involved, usually as a financial safeguard for the contractor by way of an indemnity.
- Planned risk retention by a contractor is a conscious act following recognition and assessment of the probability and effect of a risk, and may 'win' the contract; unplanned risk retention, on the other hand, arises when an event occurs from an unidentified risk, the costs of which accrue to the contractor.

Clearly careful appraisal of the hazards, the probability of their occurrence, and the cost and time effects attendant on their occurrence are very important activities; their omission or poor administration can lead, and has led, to many 'fire brigade' ground freeezing contracts.

Choosing ground freezing may appear to be expensive and too cumbersome, i.e. difficult to fit into preconceived programme times, but in many circumstances it provides a low-risk solution. This can be illustrated by reference to several contracts — fuller descriptions will be found in Chapter 7.

Ely Ouse A judgment was made by the designers, and ground freezing was selected as the right way to proceed; the designers had to argue their case against the doubters before awarding and progressing the works. All conditions which they set were met.

Aberdeen A proposal by the contractor to freeze instead of grout was declined; later, after two attempts by grouting were unsuccessful, freezing was adopted. The expenditure on freezing was less than had been spent on grouting. See also *Three Valleys*.

Boulby Two shafts were intentionally sunk by alternative methods and, as expected, the frozen shaft was completed more quickly than the grouted shaft even allowing for an unforeseen delay to the frozen shaft; however, the grouted shaft took so much longer than programmed that the supposed cost effectiveness of choosing two methods was lost.

Rogers Pass Ground freezing was considered but discarded at design/bid stage in favour of dewatering. In the event the deep well pumps could not lower the groundwater sufficiently, and the works were completed with the aid of ground freezing.

Du Toits Kloof Ground freezing was required to recover a collapse situation while an experimental/pilot tunnel was being driven. This led the designers to specify ground freezing for the main drive through weathered rock.

Runcorn The potential risk to the security of the neighbouring water courses dictated using the most certain geotechnical technique for the works, and the designer specified ground freezing. Similarly an ancient bridge at *Bromham* was protected against settlement while a trench was excavated between its abutments.

'What if?' conjectures can be made with respect to works which, while being completed without recourse to ground freezing, incurred high cost, long delays or adverse side-effects. For example, damage due to settlement following free air horizontal or inclined tunnelling at shallow depth, might have been minimized (perhaps avoided) had the overburden strata been frozen.

References

AL-BAHAR J.F. and CRANDALL K.C. 1990. Systematic risk management approach for construction projects. *J. Construction Engng and Man.*, Sept., **116**–3, 533–546.

BGFS 1994. Value, risk and cost effectiveness in AGF works. *Technical Memorandum* TM5, BGFS, 8pp.

BRITISH STANDARDS INSTITUTION 1991. BS 4778 *Quality vocabulary.* Section 3.1, Guide to concepts and related defitions.

CHAPMAN C.B. 1992. Risk management: Predicting and dealing with an uncertain future. *Rpt Independent Power Producers Society of Ontario*, 153pp.

CHAPMAN C.B. *et al.*, 1985. Problem-solving methodology design on the run. *J. Operational Research Soc.*, **36**–9, 769–778.

CHAPMAN C.B *et al.*, 1987. *Management for engineers*, Wiley, 654pp.

HARVEY S.J. 1993. Effective control of groundwater by use of freezing. *Proc. IC on Groundwater problems in urban areas*, ICE, London.

HELLARD R.B. 1993. *Achieving profitability with customer satisfaction.* Thomas Telford, London, 188pp.

JACKSON D. 1989, Contractual aspects of AGF. *Proc. 5th NSGF*, BGFS, Nottingham, 63–69.
PEACOCK W.S. and WHYTE I.L. 1992. Site investigation and risk analysis. *Proc. ICE*, **92**, 74–81.
PERRY J.G. and HAYES R.W. 1985. Risk and its management in construction projects. *Proc. ICE*, June, **78**, 499–521.
THOMPSON P. and PERRY J. 1993. *Engineering construction risks*. Thomas Telford, London, 55pp.

Appendix A: Symposia/Society proceedings

International Symposia on Ground Freezing (ISGF)

1st ISGF, 1978, Bochum, Germany
Proceedings in two volumes (unnumbered and 2 but referred to as 1 and 2 in the Bibliography) edited by H. L. Jessberger and published by Ruhr University, Bochum; out of print. All papers were reprinted (with different page numbers) in *Engineering Geology* **13** (1979) and *Developments in Geotechnical Engineering* **26**, both edited by H. L. Jessberger and published by Elsevier.

2nd ISGF, 1980, Trondheim, Norway
Proceedings in two volumes (Preprints and Proceedings but referred to as volumes 1 and 2 in the Bibliography) published by the Norwegian Institute of Technology; out of print. Most papers were reprinted (with different page numbers) in *Engineering Geology* **18** (1981) and/or *Geotechnical Engineering* **28**, both edited by P. E. Frivik, N. Janbu, R. Saetersdal and L. I. Finborud and published by Elsevier (the latter is not given in the Bibliography).

3rd ISGF, 1982, Hanover NH, USA
Proceedings in one volume, published by US Army Corps of Engineers, Hanover NH, and reprinted as Special Report 82–16 by USA CRREL. Many papers issued but not included in said volume (1), have been collected and are referred to as volume 2 in the Bibliography.

4th ISGF; 1985, Sapporo, Japan
Proceedings in two volumes: most papers appeared in volume 1 (entitled '*Ground Freezing*') published by Balkema, and the remainder plus poster summaries in volume 2 '*Ground Freezing 1985*' published by Hokkaido University Press, both edited by S. Kinosita and M. Fukuda.

5th ISGF, 1988, Nottingham, England
Proceedings in two volumes: most papers appeared in volume 1 (entitled '*Ground Freezing 88*'), and the remainder plus poster summaries in volume 2, both edited by R. H. Jones and J. T. Holden and published by Balkema.

6th ISGF, 1991, Beijing, China
Proceedings in two volumes: most papers appeared in volume 1

'*Ground Freezing 91*', and the remainder plus poster summaries in volume 2, both edited by Yu X. and Wang C., and published by Balkema.

State of the Art papers appeared in volume 1 in 1980 and 1988.

7th ISGF, 1994, Nancy, France
Proceedings in two volumes: most papers appeared in volume 1, published by Balkema, edited by M. Frémond, (volume 2 in press).

National Symposia on Ground Freezing (NSGF)

UK National annual Meetings or Courses organized by the British Ground Freezing Society were held at the University of Nottingham since 1982. Those held in

1983: 1st: Progress in AGF
1984: 2nd: Relating theory to practice in AGF
1985: 3rd: Symposium
1986: 4th: (collected papers only)
1989: 5th: Ground freezing in construction
1993: 6th: Frozen security — safety and cost effectiveness of the ground freezing method (collected papers only)
1994: 7th: (collected papers only)

are recorded in single volumes; they are referred to in this volume as *Proc'n th' NSGF*, and are available through the British Ground Freezing Society.

International Conferences on Permafrost (ICP)

1st ICP, 1963, Lafayette, Indiana, USA
Proceedings published as NRC Publication 1287 (1966), National Academy of Sciences, Washington DC. Available on microfilm US Library of Congress, #25−3138 Cold Regions Science & Technology Bibliography.

2nd ICP, 1973, Yakutsk, Russia
Proceedings, Vol 1 (1973) comprising N American Contribution, Vol 2 (1978) comprising USSR contribution, published by National Academy of Sciences, Washington DC.

3rd ICP, 1978, Edmonton, Alberta, Canada
Proceedings published in two volumes (1978), plus English translations of Soviet papers in two parts (1980), published by National Research Council of Canada.

4th ICP, 1983, Fairbanks, Alaska, USA
Proceedings in three volumes: Abstracts (1983) University of Alaska; Proceedings (1983) and Final Proceedings (1984) — published by National Academy Press, Washington DC.

5th ICP, 1988, Trondheim, Norway
Proceedings in three volumes, published by Tapir Publishers, Trondheim, Norway.

6th ICP, 1993, Beijing, China
Details not available to Author

In addition many papers on or relevant to ground freezing have been presented at Canadian Permafrost Conferences in 1962, 1964, 1969 and 1981, and US annual Rapid Excavation and Tunneling Conferences in 1972 (Chicago), 1979 (Atlanta), 1981 (San Francisco), 1983 (Chicago), 1985 (New York), 1987 (New Orleans), 1989 (Los Angeles), and 1991 (Seattle).

Appendix B: Conversion factors

General

in	× 25·4	= mm
ft	× 0·330 48	= m
yards	× 0·9144	= m
in^2	× 654·6	= mm^2
ft^2	× 0·0929	= m^2
$yards^2$	× 0·8361	= m^2
ft^3	× 0·028 32	= m^3
$yards^3$	× 0·7646	= m^3
in^3	× 16·387	= ml
ft^3	× 28·317	= 1 (litre)
pint	× 0·568	= 1
gall	× 4·546	= 1
lbs	× 0·4536	= kg
tons	× 1·016	= Mg = t (tonne)
lb/ft^3	× 16·02	= kg/m^3
lbf	× 4·448	= N
tonf	× 9·964	= kN
lbf/ft	× 14·59	= N/m
lbf/in^2	× 6·895	= kN/m^2 = kPa (Pa = pascal)
lbf/ft^2	× 47·88	= MN/m^2 = MPa (= N/mm^7)
$tonf/ft^2$	× 107·3	= kN/m^2 = kPa
ft/s	× 0·3048	= m/s
mph	× 1·609	= km/h
ft/s	× 0·3048	= m/s^2

Heat

(°F − 32) × 0·56 = °C
BTU × 1·055 = kJ
BTU/h × 0·293 = W
TR × 3519 = W

Thermal capacity BTU/lb°F × 4·187 = kJ/kg°C
Thermal conductivity BTU in/ft^2°F × 0·1442 = W/m°C
Intensity of heat flow rate BTU/ft^2h × 3·155 = W/m^2
Thermal transmittance BTU/ft^2h°F × 5·678 = W/m^2°C

Appendix C: Summary of ground freezing projects

Table C.1. Summary of ground freezing projects

Country	Site	Year	Application
Algeria	Arzew	1963	Shaft
Australia	Belmont John Darling Coll, BHP	1981	Shaft
Austria	Vienna	1980	Underpin
	Vienna	1987	Tunnel
	Vienna	1987	Tunnel
Belgium	Anderlues	1975	Shaft seals
	Andre Dumont	†	Two shafts
	Andre Dumont	1959	Shaft repair
	Antwerp	1932–33	Two shafts
	Antwerp	1932–33	Two shafts
	Antwerp	1932–33	Two shafts
	Antwerp	1967	Tunnel
	Antwerp	1967	Two shafts
	Antwerp, Ebes	1968	Two shafts
	Antwerp, Schelde	1969	Two shafts
	Antwerp, Kruisschans	1976	Two shafts
	Antwerp, Opera-Astrid tunnel	1980s	Tunnel
	Antwerp, Wesenbeke tunnel	1980s	Tunnel
	Antwerp	1982	Tunnel
	Beeringen	†	Two shafts
	Brussels, Anspach	1968	Two tunnels
	Brussels, Mintplein	1968	Tunnel
	Brussels, metro	1970	Tunnel
	Brussels, Banque	1971	Underpin
	Brussels, Assubel	1973	Underpin
	Brussels, Palace Hotel	1973	Underpin
	Brussels, Rogier Sq	1976	†
	Brussels, Senne	1977	Tunnel
	Brussels, CGER building	1980s	Pit
	Brussels, Cote d'Or	1980s	Metro
	Brussels, Gare Louise	1980s	Underpin
	Brussels, river Senne	1982	Tunnel
	Brussels, South Station	1980s	Underpin
	Dumont, Campine	1923	†
	Hainaut	1975–76	Shaft seal
	Haine	1885	Shaft
	Helchteren Zolder	1913–20	Two shafts
	Hensies	1921–23	Two shafts
	Houssu	1886?	Shaft
	Houthlaelen	1927–32	Two shafts
	Kumtich	1972	Tunnel
	Laura	1921–23	Two shafts
	Les Liegeois	†	Two shafts

Table C.1. Continued.

	Leuven	1973	Two drifts
	Liege	1976	Shaft
	Limbourg Meuse	1911–14	Two shafts
	Mol	1980s	Shaft
	Mol	1980s	Tunnel
	Thieu	1908	Shaft
	Vicq, Anzin Co	1894	Shaft
	Vicq, Anzin Co	1894	Shaft
	Winterslag	†	Two shafts
Brazil	São Paulo	1941	Underpin
Canada	Allan Potash Mines	1964	Two shafts
	Alwinsal Potash	1964	Shaft
	Brantford, Ontario	1973	Tunnel
	Esterhazy, IMC	1964–65	Shaft
	Goderich, Domtar Salt	1980	Shaft
	Lake Patience, PCA	1967	Two shafts
	Madison Is, Les Mine Salin	1980	†
	Montreal, C N Railways	1959	Tunnel
	Noranda	1965–66	Two shafts
	Ottawa	†	Tunnel
	Quebec, Selco Mining Corp	1979	Shaft
	Rogers Pass, CP Railway	1986	Shaft
	Timmins, Texas Gulf	1974	Pit
	Timmons, Ont, Asarco Exploration	1981	Shaft
	Unity	1953	Shaft
	Vansco, Cominco	1965	Two shafts
	Vansco, Duval	1966	Two shafts
	Windsor Ojibway, Ont	1953	Shaft
	Yarbo #2, IMC	1962	Shaft
China	Baodian (main)	1977	Shaft
(not a	Baodian (S air)	1978	Shaft
comprehensive	Beisu (auxiliary)	1971	Shaft
list)	Chianjiayen (E air)	1976	Shaft
	Dong Huan Tuo	1986–88	Four shafts
	Fangezhang (air)	1959	Shaft
	Haizi (cent air)	1979	shaft
	Haizi (main)	1979	Shaft
	Haizi (W air)	1979	Shaft
	Kongji (W air)	1983	Shaft
	Linxi (air)	1950	Shaft
	Liuqiao (auxiliary)	1971–75	Shaft
	Lüjiatuo (main)	1950	Shaft
	Lüjiatuo (auxiliary)	1960	Shaft
	Luling (auxiliary)	1961–63	Shaft
	Nantun (auxiliary)	1966	Shaft
	Nantun (main)	1967	Shaft
	Panji 1 (E air)	1974	Shaft
	Panji 1 (winding)	1978–81	Shaft
	Panji 2 (auxiliary)	1978	Shaft
	Panji 2 (main)	1975	Shaft
	Panji 2 (S air)	1979	Shaft
	Panji 2 (W air)	1978	Shaft
	Panji 3 (air)	1979	Shaft
	Panji 3 (E air)	1981	Shaft
	Pingtingshan (air)	1968	Shaft

Table C.1. Continued.

	Suoli (auxiliary)	1966	Shaft
	Tangshan (air)	1950	Shaft
	Weicun (auxiliary)	1986	Shaft
	Xiegiao (auxiliary)	1984–85	Shaft
	Xingtai (main)	1950	Shaft
	Zhangi (main and auxiliary)	1971–72	Two shafts
Congo	Pointe Noire	1965	Shaft
Egypt	Alexandria, City	1986	Tunnel repair
	Cairo, City	1990	Two tunnels
Finland	Helsinki, KluuviCleft	1981	Tunnel
France	F1	1886–96	Eight shafts
	F2	1886–1905	26 shafts
	F3	1886–1905	Ten shafts
	F4	1886–1905	Four shafts
	Lyon	†	Shaft
	Nancy	1970	Sewer
	Nice	1986	Tunnel
	Paris, Les Marines	1971	Shaft
	Paris, Place St Michel	1907	Tunnel
	Paris, RAPT	1967	Metro
	Pompey	†	Pit
	Schönensteinbach	1970	Shaft
	Staffelden	1968	Shaft
	Vernejoul	†	Shaft
	Wattrelos	1964	Gallery
Germany	G1	1898–1904	Eight shafts
	G2	1906	Twelve shafts
	G3	1905	Shaft
	G4	1908	Shaft
	Aller Harmonia	1912–14	Shaft
	Altendorf	†	Shaft
	Anhaltische	1898?	Shaft
	Archibald #IX	1883	Shaft
	Arenberg	1911	Shaft
	Conow Lubteen	1911–13	Shaft
	Dortmund	1971	Tunnel
	Duisburg	1978	Tunnel
	Dusseldorf	1978	Tunnel
	Essen	†	Tunnel
	Fallers-Lebon	1911–13	Shaft
	Finsterwalde	1884	Two shafts
	Frankfurt, Frankfurt City	1978	Tunnel
	Friedrichroda	1911–12	Shaft
	Friedrichroda	1911–12	Shaft
	Friedrichroda	1911–12	Shaft
	Haltern	1979–82	Two shafts
	Hamberg Prince Adalbert	1913–14	Shaft
	Hamberg	1967	Tunnel
	Hamborn, Thyssen V	1905	Shaft
	Hansa	1911–13	Shaft
	Hansa	1938	Shaft
	Herne-ost	†	†
	Homberg	1951	Shaft

Table C.1. Continued.

	Project	Date	Type
	Ilsenburg	†	Shaft
	Jessenitz	1886	Shaft
	Königswusterhausen	1884	Shaft
	Leverkusen Auguste Victoria	1928	Shaft
	Lohberg	1905	Two shafts
	Mariagluck	1911–13	Shaft
	Marie	†	Shaft
	Niedersachsen	1906	Shaft
	Nuremberg	†	Tunnel
	Prinz Adalbert, Ovelgonne	1909	Shaft
	Prosper, Ruhrkohle	1980	Shaft
	Sopjie	†	Shaft
	Stuttgart	1976	Tunnel
	Voerde	1979–83	Two shafts
	Wallach #2	1924	Shaft
	Walsam	1978	Shaft
	Weser	1910	Shaft
Gibraltar	Gibraltar 1, PSA	1976	Two piers
	Gibraltar 2, PSA	1977	Two piers
	Gibraltar 3, PSA	1979	Two piers
	Gibraltar 4, PSA	1981	Two piers
Holland	Hendrik #IV	1953–56	Shaft
	Langveld	1951	Shaft
	Laura en Vereeniging	1924	†
	Limburg	1905	Two shafts
	Maurits #III	1954–57	Shaft
	Osterleek	1973	Pipe repair
Italy	Agri Sauro	1986	Tunnel
	Calabria	1966	Tunnel
	Como Lake	1972	Foundation
	Genova	†	†
	Milan	†	†
	Naples	1979	Tunnel
	Salerno	1973	Tunnel
	Santo Marco	1977	Tunnel
Japan	Chiba Inba-numa River	1976–77	Shaft connxn
	Chiba, canal	1980	†
	Chiba, Nemeto	1983–84	Tunnel
	Hirano River	1967–68	Tunnel
	Horokiri	1972–74	Shaft
	Inba-Numa river basin	1976–77	T/S intersection
	Keihin	1985	Tunnel
	Nishiosaki, Meguro River	1967–68	Tunnel
	Nunobiki	1984	Tunnel
	Osaka, Joto river	1965–68	Tunnel
	Samezo, Megyro River	1980	Tunnel
	Tatsumi, Hirano River	1970	Tunnel
	Tatsumi, Joto Canal	1972	Tunnel
	Teito, TR 6 and 9	1967–68	Two tunnels
	Tokyo, Ayasegawa	1972–74	Shaft
	Tokyo, Furukawa River	1965–68	Tunnel
	Tokyo, Furukawa River	1970–74	Tunnel
	Tokyo, Furukawa	1977–78	Shaft
	Tokyo, Kanda River	1969–72	Tunnel

Table C.1. Continued.

	Tokyo, Kodo	1979	Tunnel
	Tokyo, Megro River	1966–68	Tunnel
	Tokyo, Nihonbashi River	1969–72	Tunnel
	Tokyo, Nihonbashi River	1977–80	Tunnel
	Tokyo, Okinomiya	1981–83	Manhole
	Tokyo, Osaki, Negro River	1979–80	Tunnel
	Tokyo, RT9	1967–71	Tunnel
	Tokyo	1976–77	Tunnel
	Tokyo	1985	Tunnel
	and over 100 more		
Norway	Liland	1970	Tunnel
	Oslo	1980	Tunnel
	Tverrfjellet	1967	Tunnel
Poland	Lubin	†	Eleven shafts
	Solno	1924–27	Shaft
	Solno	1956	Shaft
	Wapno, Solvay	1912–14	Shaft
	unstated	pre 1980	20 shafts
Russia	Leningrad (now St Petersburg)		Two Tunnels
	Moscow	1932–35	Three drifts
	Moscow	1932–35	Two galleries
	(central)	1980	Shaft
Silesia	Frohlich	1913	Shaft
	Tarnowitz	1889	Shaft
South Africa	Du Toits Kloof pilot, NTC	1974	Tunnel
	Du Toits Kloof, National Transport Comm.	1981	Tunnel
	Durban	†	Foundation pit
Spain	Barcelona	1980s	Pit
	Burgos	1986	Underpin
	Llausett Dam	†	Foundation
	Seville	1983	Tunnel
	Secuita	1982	Tunnel
	Valencia	1987	Foundation
Sweden	Stockholm	1884–86	Tunnel
Switzerland	Born, Swiss Railways	1969	Tunnel
	Geneva, CERN	†	†
	Lhongrin, Lausanne	1968	Tunnel
	Zurich, Limmat	1987	Tunnel
	Zurich, Milchbuck	1978	Tunnel
United Kingdom	Aberdeen, City	1980	Shaft
	Asfordby, BC	1985	Two shafts
	Bevercotes, NCB	1952–55	Two shafts
	'B'	1924–26	Two shafts
	Billingham, ICI	1991	Shaft
	Birmingham, Stirchley, STWA	1974	Tunnel
	Birmingham, Sutton Coldfield, STWA	1974	Pit
	Blackpool, Hawtin Industries	1971	Pits
	Blackwall Tunnel, GLC	1967	Two tunnel eyes
	Boulby, B. Potash	1974	Shaft
	Bromham Bridge	1991	†
	Bullcroft Colliery	1910	Two shafts

Table C.1. Continued.

Calverton #2, Bestwood Clly Co	1947–49	Shaft
Canvey Island, B Gas	1967	Four pits
Carcroft, NCB	1979	Trench
Cheltenham, Littlewoods	1971	Sheet piling
Cotgrave, NCB	1955	Two shafts
Crewe, Midland Rollmakers	1960	Shaft
Driffield, Barratt Homes	1974	Pit
Dundee, Tayside Regional Council	1981	Tunnel
Durham, Dawdon, Castlereagh	pre 1941	Shaft
Durham, Dawdon, Theresa	pre 1941	Shaft
Durham, Seaham Harbour	1926	Two shafts
Durham, South Hetton	1955	Shaft
Durham, Wearmouth	pre 1909	Shaft
Durham, Wearmouth, NCB	1956–59	Shaft
Edinburgh, Craigentinney	1974	Tunnel
Edinburgh City, McDonald Rd	1974	Tunnel
Ely Ouse, Essex River Authority	1968	Shaft 3
Ely Ouse, Essex River Authority	1968	Shaft 1
Ely Ouse, Essex River Authority	1968	Shaft 2
Ely Ouse, Essex River Authority	1968	Shaft 7
Ely Ouse, Essex River Authority	1968	Shaft 8
Ely Ouse, Essex River Authority	1968	Shaft 10
Farnworth, Town	1960	Shaft
Farnworth, Town	1966	Two shafts
Fleetwood, Town	1954–55	Two pits
Folkestone 2, Transmanche	1990	Shaft
Folkestone 3, Transmanche	1990	Shaft
Folkestone 4, Transmanche	1990	Shaft and tunnel
Fosdyke, British Gas	1971	Sheet piling
Glasgow Kinning, G. Passenger Transport	1978	Pit
Glasgow, Potter St, Strathclyde R C	1982	Two pits
Glasgow, Renfrew	1974	Tunnel
Handen Hold, NCB	1961	Pit
Hawthorn, NCB	1956–57	Shaft
Isle of Grain, CEGB	1974	Tunnel
Kellingley, NCB	1958–60	Two shafts
Lea Hall, NCB	1952–55	Two shafts
London, Chelsea, Thames Water	1987	Drift
London, Hornchurch	1966	Tunnel
London, Lower Thames St	1974	Piles
London, Pimloco, LTE	1970	Drift
London, St Pauls, GPO	1966	Shaft
London, Thamesmead, PSA	1989	Tunnel
London, Tottenham Hale, LTE	1966	Drift
London, Vauxhall Cross, LTE	1968	Drift
Lowestoft, Docks Bd	1967	Shaft
Lowestoft, Docks Bd	1968	Tunnel stub
Lynemouth, Alcan	1969	Tunnel
Margam	1950	Skip pit
Millom Moorbank, Hodbarrow	1928	Shaft
Millom	1933	Dam
Millom, M and Askham	1957	Shaft
Mouldsworth	1940	Shaft
Newcastle, Northumbrian W	1974	Shaft
Norwich, City	1961–62	Trench

Table C.1. Continued.

	Port Talbot, Wales Steel	1949–50	Skip pit
	Radcliffe, Town	1969	Shaft
	Rhos Ddu, Denbigh CC	1971	Shaft seal
	River Medway, CEGB	1976	Tunnel
	Rugely	1953	Two shafts
	Runcorn, NWWA	1978–79	Two tunnels
	Selby Gascoigne Wd, NCB/BC	1978–80	Two drifts
	Selby North Selby, NCB/BC	1978	Two shafts
	Selby Riccall, NCB/BC	1978	Two shafts
	Selby Stillingfleet #1, NCB/BC	1978	Shaft
	Selby Stillingfleet #2, NCB/BC	1979	Shaft
	Selby Whitemoor, NCB/BC	1979	Two shafts
	Selby Wistow #1, NCB/BC	1977	Shaft
	Selby Wistow #2, NCB/BC	1977	Shaft
	Sheffield, City	1981	Tunnel
	Sherburn, BPB	1963	Shaft
	Sherburn, BPB	1966	Drift
	Solway, Workington Iron	1937	Two shafts
	Stoke-on-Trent, City	1963	Tunnel
	Stonehouse, Strathclyde RC	1986	Tunnel
	Sunderland	1975	Tunnel
	Swansea	1862	Shaft
	Swansea, Tir John PS	1933	Shaft
	Swansea, Tir John PS	1933	Shaft
	Three Valleys	1986	Tunnel
	Tilbury, Essex Water	1990	Shaft
	Tooting Bec, Thames Water	1990	Shaft/tunnel
	Widnes, United Sulphuric	1954	Trench
	Workington, United Steel	1937–38	Two shafts
	York, City	1951	Tunnel
USA	Arizona	†	Pit
	Belle Isle Rehab, Cargill Salt	1974	Shaft
	Belle Isle, La, Cargill Salt	1970–73	Two shafts
	Brooklyn NY	†	†
	California	†	Four shafts
	Carlsbad, NM	1952	Shaft
	Carlsbad, NM, Duval	1963	Two shafts
	Carlstadt	1966	Pit
	Chapin, Michigan	1888–91	Shaft
	Cleveland	†	†
	Colorado	†	Trench
	Cote Blanche, La, Carey Salt	1964	Shaft
	Detroit	1979	Tunnel
	George Dam	†	†
	Georgia	†	Tunnel
	Grand Coulee Dam	1936	Slide
	Illinois, Amax Coal	1974	Shaft
	Illinois, Peabody Coal	1978	Shaft
	Illinois, Turris Coal	1981	Three shafts
	Illinois, White County Coal	1979	Shaft
	Indiana	†	Trench
	Iron Mountain, MI	1988	Pit
	Island Creek Coal, Kentucky	1975	Shaft
	Jefferson Is, La, Morton Salt	1977–78	Shaft
	Lafayette, La, Diamond Crystal	1975	Shaft

Table C.1. Continued.

	Location	Year	Type
	Lake Charles, La	1964	Pit
	Maine	†	Pit
	Manhattan NY 1	1959	Shaft
	Manhattan NY 2	1971	Tunnel
	Michigan	†	†
	Michigan	†	Nine shafts
	Michigan	†	Shaft
	Milwaukee, MMWB	1985	Three shafts
	Milwaukee, MMWB	1985	Linked shafts
	Minnesota	†	Underpin
	Nebraska	†	Pit
	New York	†	Pit
	New York	†	Tunnel
	New York	1980	Tunnel
	New York	†	24 shafts
	New York	†	Shaft
	Ohio	†	Pit
	Paducah, Kentucky	†	Pit
	St Omer	†	Shaft
	Seattle, Renton	†	Shaft
	Skagit river, George Dam	†	†
	S Carolina	†	Tunnel
	Utah	†	Pit
	Utah — Salt Lake City	1963	Pit
	Washington DC	1978	Tunnel
	Week's Is, La; Morton Salt	1978	Shaft
	Wells, Maine	1977	Two pits
	Wyoming	†	Pit

† Information not available to Author
Note: These listings are not complete; the Author will be pleased to receive details of other ground freezing projects to update the foregoing and for inclusion in a database for possible future publication.

Appendix D: Evaluation Models

Model A Risk assessment based on comparison of contractors' proposals [after Harvey (1993)]

A set of suggested conditions is applied to each (contractor's) submission for scoring in the table below (Table D.1).

1 Ground conditions, weight factor × 1·0.
   100> Assessed all data and understands varying soil properties; method proposed is not affected by ground conditions as the excavation is continuously supported.
   75> Assessed all data and understands varying soil properties and taken this into account in the method; includes plans to deal with changes if they occur.
   50> Assessed all data and aware of variable ground conditions; has made provision for change of excavation method but no detailed plan.
   25> Has fair understanding of ground conditions and referenced possible method changes; methods to be finalized as the excavation progresses.
   0> Has taken no account of varying ground conditions.

2 Control of groundwater, weight factor × 1·5.
   100> Method proposed is not affected by water and can operate under any soil or groundwater conditions.
   75> Understands groundwater problem and details control of water flowing prior to excavation commencing.
   50> Has assessed water flows at various horizons; no plan to control water before excavation but has plan in event of water flows being encountered.
   25> Has understanding of groundwater regime but no specific action plan to deal with water during excavation.
   0> Has taken no account of site hydrology.

Table D.1. Model A: Evaluation of contractors proposals*

Contractor (method)	Tender price: £k	Control of ground conditions	Groundwater control	Safety during excavation and lining	Total points	Points: £k	Result
A (freezing)		$f \times 1$	$f \times 1 \cdot 5$	$f \times 1$	(sum)		(highest is best)
B (grouting)		$f \times 1$	$f \times 1 \cdot 5$	$f \times 1$	(sum)		
C (dewatering)		$f \times 1$	$f \times 1 \cdot 5$	$f \times 1$	(sum)		

* f = weight factor

EVALUATION MODELS

3 Safety, weight factor × 1·0.
- 100> Has provided precise, detailed safety plan; ventilation and hoisting systems duplicated; need to test for gas; HSE approved methods.
- 75> Using conventional excavation methods and detailed safety arrangements; has detailed equipment to be used and critical equipment duplicated.
- 50> Has provided outline details on safety in respect of hoisting, ventilation, gas testing; has not detailed critical equipment which is duplicated.
- 25> Has stated that safety plan will be drawn up later; very little mention of safety in presentation.
- 0> Has provided little or no information about safety standards; cavalier attitude to safety at presentation (trust us, we know what we are doing attitude).

Model B Risk assessment based on comparison of geotechnical techniques (Harris)

Table D.2. Model B: evaluation of geotechnical methods, factor range 0–1

Method	Risk	Probability	Impact			Risk Factor	Risk value (average)
			Security	Time	Cost		
Sheet/bored piling	Installation	0·4–1	0·3–0·9	0·4–1	0·4–0·9		
	Strata changes	0·6–1	0·4–0·9	0·4–1	0·6–0·9		
	Deformation	0·4–1	0·3–0·8	0·3–0·8	0·3–0·7		
	Equipment breakdown	0·9–1	1	0·4–0·9	0·9–1		
Caisson sinking	Installation	0·5–1	0·4–1	0·4–0·9	0·6–0·9		
	Strata changes	0·5–1	0·4–1	0·4–0·9	0·4–1		
	Deformation	0·4–1	0·3–0·7	0·3–0·7	0·3–0·7		
	Equipment breakdown	0·9–1	1	0·5–0·9	0·7–1		
Groundwater lowering	Installation	0·5–1	0·4–1	0·7–1	0·9–1		
	Strata changes	0·4–1	0·4–1	0·5–1	0·6–1		
	Deformation	0·4–1	0·4–1	0·7–1	0·6–0·9		
	Equipment breakdown	0·9–1	0·3–0·8	0·4–0·9	0·4–0·7		
Injection	Strata changes	0·6–1	0·4–0·9	0·6–0·9	0·4–0·9		
	Deformation	0·7–1	0·4–0·9	0·6–0·9	0·4–0·9		
	Preferred channels	0·4–1	0·3–0·9	0·4–0·9	0·3–0·8		
	Equipment breakdown	0·9–1	1	0·7–0·9	0·9–1		
Compressed air	Strata changes	0·4–1	0·4–1	0·4–0·9	0·4–1		
	Adverse groundwater	0·6–1	0·6–1	0·8–1	0·5–0·9		
	Equipment breakdown	0·9–1	0·3–0·9	0·5–0·9	0·4–0·9		
Electro-osmosis	Strata changes	0·3–1	0·4–0·9	0·4–1	0·5–0·9		
	Equipment breakdown	0·9–1	0·4–0·9	0·7–0·9	0·5–0·9		
Ground freezing	Installation	0·7–1	0·7–1	0·7–1	0·8–1		
	Strata changes	0·7–1	0·8–1	0·6–1	0·7–1		
	Adverse groundwater	0·6–1	0·4–1	0·4–1	0·4–1		
	Deformation	0·7–1	0·6–1	0·7–1	0·6–1		
	Innaccurate monitoring	0·8–1	0·5–1	0·7–1	0·6–1		
	Equipment breakdown	0·9–1	0·9–1	0·9–1	0·9–1		

Preliminaries.

- List all practical construction methods and associated geotechnical processes.
- Tabulate the *hazards* and *opportunities* for each method.
- Describe the *effect/outcome* should the hazard occur: could be detrimental or beneficial leading to a loss or a gain.
- Carry out the assessment procedure.

The assessment procedure suggested is as follows.

- Consider each METHOD and eliminate those which are considered to be inappropriate or unsuitable.
- Add any further hazards relevant to each method being assessed that become apparent, assess their effect and impact and allocate a range of Probability Values.

Table D.3. *Model B: Evaluation of geotechnical methods, factor range 0·8–1·5*

| Method | Opportunity/hazard | Probability | Impact | | | Risk |
			Security	Time	Cost	

Sheet/ bored piling	Installation
	  incomplete drivage
	Strata changes
	  parted clutches
	Deformation
	  uplift of base
	Equipment breakdown
		1·0–1·4
		
		1·0–1·3
		
		1·0–1·4
		1·0–1·2
			1·1–1·5
			
			1·1–1·4
			
			1·1–1·5
			1
				1·1–1·4
				
				1·1–1·4
				
				1·1–1·5
				1·1–1·4
					1·1–1·4
					
					1·1–1·3
					
					1·1–1·4
					1·0–1·1	

Caisson sinking	Installation
	  shallow obstruction
	Strata changes
	  excessive friction
	Deformation
	  uplift of base
	Equipment breakdown
		1·0–1·4
		
		1·0–1·4
		
		1·0–1·4
		1·0–1·2
			1·0–1·3
			
			1·0–1·4
			
			1·1–1·5
			1
				1·1–1·4
				
				1·1–1·4
				
				1·1–1·5
				1·1–1·4
					1·1–1·3
					
					1·0–1·5
					
					1·1–1·4
					1·0–1·2	

Ground- water lowering	Installation
	Strata changes
	  less permeable
	  more permeable
	Deformation
	  loss of fines
	  uplift of base
	Equipment breakdown
		1·0–1·3
		
		1·0–1·4
		1·0–1·4
		
		1·0–1·4
		1·0–1·4
		1·0–1·2
			1·0–1·4
			
			1·0–1·4
			1·0–1·3
			
			1·0–1·4
			1·1–1·5
			1·0–1·5
				1·0–1·2
				
				1·0–1·2
				1·1–1·3
				
				1·0–1·2
				1·1–1·5
				1·2–1·4
					1·1–1·3
					
					0·9–1·2
					0·9–1·3
					
					1·1–1·3
					1·1–1·4
					1·2–1·4	

Injection	Installation
	Strata changes
	  smaller voids
	  larger voids
	  more extensive voids
	  preferred channels
	Deformation
	Equipment breakdown
		1·0–1·4
		
		1·0–1·2
		1·0–1·2
		1·0–1·2
		1·0–1·4
		1·0–1·1
		1·0–1·2
			1·1–1·4
			
			1·0–1·2
			1·1–1·3
			1·1–1·4
			1·1–1·4
			1·1–1·3
			1
				1·1–1·2
				
				0·9–1·1
				1·1–1·2
				1·1–1·3
				1·1–1·4
				1·1–1·3
				1·1–1·2
					1·1–1·4
					
					0·8–1·2
					1·1–1·4
					1·1–1·4
					1·1–1·4
					1·1–1·4
					1·1–1·2	

For each HAZARD the procedure is as follows.

- Allocate a value to the Probability Factor scoring from 0 (most adverse) to 1·0 (totally successful).
- Allocate Impact Factors for security, time and cost, taking account of all likely effects, and scoring on a similar basis.
- At this stage there is the opportunity, if one of the above four headings is considered to be of lesser or greater importance in the overall assessment, to 'weight' that value up or down as appropriate.
- Multiply all four Probability/Impact factors to obtain the Risk Factor.
- For each METHOD it is then necessary to select a unique figure, which may be the average or lowest of the Risk Factors obtained under the various headings: this is the Risk Value.

Table D.3. Continued

Method	Opportunity/hazard	Probability	Impact			Risk
			Security	Time	Cost	
Compressed air	Strata changes					
	more voided strata	1·0–1·2	1·0–1·3	1·0–1·1	1·0–1·2	
	Adverse groundwater					
	increased water head	1·0–1·3	1·0–1·2	1·0–1·1	1·1–1·3	
	Equipment breakdown	1·0–1·2	1·1–1·5	1·1–1·3	1·1–1·2	
Electro-osmosis	Installation	1·0–1·1	1·1–1·4	1·1–1·4	1·1–1·3	
	Strata changes					
	change in permeability	1·0–1·4	1·1–1·4	1·1–1·5	1·1–1·3	
	Equipment breakdown	1·0–1·2	1·1–1·4	1·1–1·3	1·1–1·3	
Ground freezing	Installation					
	converging FTs	1·0–1·2	1·0–1·2	1·0–1·1	1·0–1·1	
	diverging FTs	1·0–1·2	1·0–1·3	1·0–1·3	1·0–1·4	
	broken FT	1·0–1·2	1·0–1·2	1·0–1·1	1·0–1·1	
	Strata changes					
	strata more clayey	1·0–1·2	1·1–1·2	1·1–1·3	1·1–1·2	
	strata more granular	1·0–1·2	0·9–1·0	0·8–1·1	0·9–1·0	
	large voids above WT	1·0–1·2	1·1–1·3	1·1–1·4	1·0–1·3	
	large voids below WT	1·0–1·2	1·1–1·4	1·1–1·3	1·1–1·3	
	poor cut-off stratum	1·0–1·2	1·1–1·4	1·1–1·3	1·1–1·3	
	Adverse groundwater					
	fluctuating WL	1·0–1·2	1·1–1·3	1·1–1·3	1·1–1·3	
	reduced WL	1·0–1·2	1·1–1·2	1·1–1·2	1·1–1·2	
	multiple aquifers	1·0–1·2	1·0–1·2	1·0–1·2	1·0–1·2	
	flowing water	1·0–1·3	1·1–1·4	1·1–1·4	1·1–1·4	
	saline soil/water	1·0–1·4	1·1–1·3	1·1–1·3	1·1–1·3	
	Deformation					
	ground heave	1·0–1·2	1·0–1·2	1	1·0–1·3	
	creep deformation	1·0–1·2	1·0–1·2	1·0–1·3	1·0–1·2	
	Inaccurate monitoring	1·0–1·2	1·0–1·4	1·0–1·3	1·0–1·3	
	Equipment breakdown	1·0–1·2	1·0–1·2	1·0–1·2	1·0–1·1	

Table D.4. Some risk effects and their mitigation for different geotechnical methods

Method	Opportunity/hazard	Effect	Mitigation
Sheet/ bored piling	Installation incomplete drivage	Poor/no seal; influx of soil/water; base heave	More comprehensive geotechnical appraisal
	Strata changes parted clutches	Influx of soil/water	Change pile section
	Deformation uplift of base	Flooding and instability	More comprehensive geotechnical appraisal, control
	Equipment breakdown	Delay	Maintenance, provide backup
Caisson sinking	Installation shallow obstruction	Tilting; loss of seal	More comprehensive geotechnical appraisal, control
	Strata changes excessive friction	Reduced/no advance	More comprehensive geotechnical appraisal, control
	Deformation uplift of base	Flooding and instability	More comprehensive geotechnical appraisal, control
	Equipment breakdown	Delay	Maintenance, provide backup
Ground- water lowering	Installation Strata changes less permeable more permeable	Poor drawdown Larger yield	More comprehensive geotechnical appraisal More comprehensive geotechnical appraisal
	Deformation loss of fines uplift of base	Subsidence of nearby property Flooding and instability	Improve the filters More comprehensive geotechnical appraisal, control
	Equipment breakdown	Flooding; collapse	Maintenance, provide backup
Injection	Installation Strata changes smaller voids larger voids more extensive voids preferred channels	Less grout; equal/less effective More grout and/or voids unfilled; less effective More grout; less effective Grout travels away	More comprehensive geotechnical appraisal More comprehensive geotechnical appraisal More comprehensive geotechnical appraisal More comprehensive geotechnical appraisal
	Deformation Equipment breakdown	Delay	Maintenance, provide backup

Table D.4. Continued

Compressed air	Strata changes 　more voided strata 　Adverse groundwater 　　increased water head 　Equipment breakdown	Loss of air Increased pressure; medical complications Reduction in pressure	More comprehensive geotechnical appraisal More comprehensive geotechnical appraisal Maintenance, provide backup
Electro-osmosis	Installation Strata changes 　change in permeability Equipment breakdown	Ineffective Delay	More comprehensive geotechnical appraisal Maintenance, provide backup
Ground freezing	Installation 　converging FTs 　diverging FTs 　broken FT Strata changes 　strata more clayey 　strata more granular 　large voids above WT 　large voids below WT 　poor cut-off stratum Adverse groundwater 　fluctuating WL 　reduced WL 　multiple aquifers 　flowing water 　saline soil/water Deformation 　ground heave 　creep deformation 　Inaccurate monitoring 　Equipment breakdown	 Damage to neighbouring FT; thicker ice-wall Longer PFP. 'window' in ice-wall; flooding/soil influx Loss of brine to strata; depressed freezing point Longer PFP Shorter PFP Chilled air contributes nothing Ice is weaker than frozen ground Flooding of excavation Longer PFP; less effective Low mc yields weaker frozen strength Interconnection via drill-holes Longer PFP; load may exceed plant capacity Depressed freezing point; longer PFP Distress to neighbouring property Damage to FTs and/or to structural lining Misinterpretation Longer PFP; (eventual) loss of ice-wall support	 Monitoring control/redrill or install extra FT Monitoring control/redrill or install extra FT Monitoring control/install salvage FT More comprehensive geotechnical appraisal More comprehensive geotechnical appraisal More comprehensive geotechnical appraisal More comprehensive geotechnical appraisal More comprehensive geotechnical appraisal More comprehensive geotechnical appraisal, monitoring control Increase the mc by irrigation Seal annuli Reduce flow-rate by filling voids or stopping pumping Dilute by irrigation Monitoring control Monitoring control More frequent calibration checks, education Maintenance, provide backup

- 'Normalize' the Tender prices by dividing each by the lowest (i.e. the lowest then becomes 1).
- Divide the normalized tender price by the Risk Value for each method to obtain the Notional Value of that method.
- Sort the Notional Values in 'Lowest-is-Best' ORDER OF VALUE.

A generalized comparison for various methods of temporary works (see Chapter 1) is given in Tables D.2 and D.3, which include suggested ranges of Factors under the several headings.

Characteristics which may lead to elimination of particular methods from consideration prior to the Risk Assessment stage include

- the excess area needed with groundwater lowering compared with the area to be protected
- the very limited applicability of electro-osmosis with respect to particle size
- the medical constraints imposed with compressed air
- noise pollution associated with driven piles.

Headings as similar as possible have been used for the principal hazards for each method; inevitably the sub-headings for each method will differ in number and description. Some sub-headings will not be applicable in the circumstances being assessed, and so will be omitted; this of itself is not significant: what matters is that the assessment of each method is as thorough as possible. When the Probability Values are averaged (see below), comparison of the resultant Risk Values can be used in an assessment.

Bibliography

AAS G. Laboratory determination of strength properties of frozen salty marine clay. *Proc. 2nd ISGF*, Trondheim, 1, 144–156, and in *Engng Geol.*, 18, 67–78.

ABZHALIMOV R. S. 1987. Additional loads developed as a result of freezing backfill soil on walls of underground crossing. *SM & FE*, 24–6, 258–261.

ADAM D. 1924. Shaft sinking by the freezing process. *Colliery Engng*, 1, Nov., 409–424.

AERNI K. and METTIER K. 1980 Ground freezing for the construction of the three-lane Milchbuck road tunnel, Zurich. *Proc. 2nd ISGF*, Trondheim, 1, 889–895, 2, 115–120.

AGUIRRE-PUENTE J. et al. 1985. Experimental measurements and a numerical method for ice sublimation. *Proc. 4th ISGF*, Sapporo, 2, 1–8.

AGUIRRE-PUENTE J. et al. 1994. Interactions 'structures/sols gelés': gazoducs enterrés dans le peréglisol; une collaboration franco-canadienne (in French) *Proc. 7th ISGF*, Nancy, France, 1, 407–408.

AIREY E. M. 1983. A method of predicting temperatures during artificial ground freezing. *Proc. 1st NSGF Progress in AGF*, BGFS, Nottingham, 15–22.

AKAGAWA S. 1980. Poisson's ratio of sandy soil under long term stress by creep tests. *Proc. 2nd ISGF*, Trondheim, 1, 235–246.

AKAGAWA S. 1988. Evaluation of the X-ray radiography efficiency for heaving and consolidation observation. *Proc. 5th ISGF*, Nottingham, 1, 23–28.

AKAGAWA S. et al. 1982. Review and findings of laboratory tests on the mechanical properties of artificially frozen soils. *Shimizu Tech. Research Bulletin*, Mar., 1, 7–17.

AKAGAWA S. et al. 1985. Frost heave characteristics and scale effect of stationary frost heave. *Proc. 4th ISGF*, Sapporo, 1, 137–146.

AKILI W. 1971. Stress–strain behaviour of frozen fine-grained soils. *US Highway Research Record*, 360, 8pp.

AKIYAMA T. 1980. A model tank test using artificial ground freezing method. *Proc. 2nd ISGF*, Trondheim, 1, 1025–1036.

ALBERT M. R. 1983. Computer models for two-dimensional transient heat conduction. *CRREL Rpt CR 83–12*, 66pp.

ALBERT M. R. 1984. Modelling two-dimensional freezing using transfinite mappings and a moving mesh FE technique. *CRREL Rpt CR 84–10*, 45pp.

ALKIRE B. D. 1977. Generalised thaw settlement of soil. *Proc. 15th NS Engng Geol. and Soils Engng*, Idaho, Apr., 20pp.

ALKIRE B. D. 1978. Some factors that affect thaw strain. *Transport Research Record*, 675, 6–10.

ALKIRE B. D. 1980. A mechanism for predicting the effect of cyclic freeze–thaw on soil behaviour. *Proc. 2nd ISGF*, Trondheim, 1, 285–296.

ALKIRE B. D. and ANDERSLAND O. B. 1973. Effect of confining pressure on the mechanical properties of sand-ice materials. *J. Glaciology*, 12–16, 469–481.

ALKIRE B. D. and MORRISON J. A. 1982. Comparative response of soils to freeze–thaw and repeated loading. *Proc. 3rd ISGF*, Hanover NH, 1, 89–96.

ALMQVIST E. 1988. Liquid nitrogen ground freezing study. *Aga AB Innovation Rpt* (U).

ALNOURI I. 1969. *Time-dependent strength behaviour of two soil types at lowered temperatures.* Michigan State University, *PhD thesis (U)*, 111pp.

ALTOUNYAN P. F. R. *et al.* 1982. Temperature, stress and strain measurements during and after construction of concrete linings in frozen sandstone. *Proc. 3rd ISGF*, Hanover NH, **1**, 343–348.

ALTOUNYAN P. F. and FARMER I. W. 1981. Tunnel lining pressures during ground water freezing and thawing. *Proc. 5th RETC*, San Francisco, May, **1**, 784–800.

AMPE A. 1935. Creusement de deux puits par le procede de la congelation, *Revue Independent Mining 3360*, 15 Dec., 603–609. Abstract 'Sinking of two shafts by freezing process' in *Colliery Guardian*, **152**, 9 Apr., 1936, 687–688.

ANDERSLAND O. B. and AL-MOUSSAWI M. 1988. Cyclic thermal strain and crack formation in frozen soils. *Proc. 5th ISGF*, Nottingham, **1**, 167–172.

ANDERSLAND O. B. and LADANYI B. 1994. *An introduction to frozen ground engineering.* Chapman Hall.

ANDERSLAND O. B. and ALNOURI I. 1970. Time dependent strength behaviour of frozen soils. *J. ASCE SM & F Div*, **96**, SM4, 1249–1265.

ANDERSLAND O. B. and ALWAHHAB M. R. M. 1982. Bond and slip of steel bars in frozen sand. *Proc. 3rd ISGF*, Hanover NH, **1**, 27–34.

ANDERSLAND O. B. and ANDERSON D. M. (eds) 1978. *Geotechnical engineering for cold regions.* McGraw-Hill, 566pp.

ANDERSLAND O. B. and DOUGLAS A. G. 1970. Soil deformation rates and activation energies. *Géotechnique*, **20**–1, 1–16.

ANDERSLAND O. B. *et al.* 1978. Mechanical properties of frozen ground. In Andersland O. B. and Anderson D. M. (eds), *Geotechnical engineering for cold regions.* McGraw-Hill, 216–275.

ANDERSLAND O. B. *et al.* 1994. Ground water remediation by controlled soil freezing. *Proc. 7th ISGF*, Nancy, France, **1**, 57–63.

ANDERSON D. M. and MORGENSTERN N. R. 1973. Physics, chemistry and mechanics of frozen gound: a review. *Proc. 2nd IC on Permafrost*, Yakutsk, 257–288.

ANDERSON D. M. and TICE A. R. 1972. Predicting unfrozen water contents in frozen soils from surface area measurements. *US Highway Research Record*, 393, 12–18.

ANDERSON D. M. and WILLIAMS P. J. 1985. Freezing and thawing of soil–water systems — State of the practice report. *ASCE*.

ANNAN A. P. and DAVIS J. L. 1978. High frequency electrical method for the detection of freeze–thaw interfaces. *Proc. 3rd IC on Permafrost*, Edmonton, **1**, 496–500.

ANON 1937. Freezing arch across toe of east forebay slide, Grand Coulee dam. *Reclamation Era*, Jan, **12**.

AOYAMA K, OGAWA S. and FUKUDA M. 1985. Temperature dependencies of mechanical properties of soils subjected to freezing and thawing. *Proc. 4th ISGF*, Sapporo, **1**, 217–222.

ARCONE S. A. and DELANEY A. J. 1989. Investigations of dielectric properties of some frozen materials using cross borehole radiowave pulse transmissions. *CRREL Rpt 89–4*, Mar.

ASSUR A. 1963. Discussion on creep of frozen soils. *Proc. 1st IC on Permafrost*, Indiana, NAS-NRCC Publication 1287, 339–340.

ASTM *D. 5520-94. Standard test method for laboratory determination of creep properties of frozen soil samples in uniaxial compression.*

ATKINS R. T. 1983. In-situ thermal conductivity measurements. *Rpt Alaksa Department Trnsptn*, 38pp.

AUGHENBAUGH N. B. and HUANG S. L. 1987. Sublimation of pore ice in frozen silt. *J. Cold Regions Engng*, Dec., **1**–4, 171.

AULD F. A. 1983. Design of temporary support for frozen shaft construction. *Proc. 1st NSGF Progress in AGF*, BGFS, Nottingham, 1–8.

AULD F. A. 1984. Freeze-wall strength and stability in deep shaft sinking — whose theory do you believe? *Proc. 2nd NSGF Relating theory to practice in AGF*, BGFS, Nottingham, 35–42.

AULD F. A. 1985a. Freeze-wall strength and stability design problems in deep shaft sinking — is current theory realistic? *Proc. 4th ISGF*, Sapporo, **1**, 343–350.

AULD F. A. 1985b. Freeze-wall structural design and case histories. *Proc. 3rd NSGF*, BGFS, Nottingham, 35–44.

AULD F.A. 1986. The effects of temporary works frost heave and thaw settlement on the construction of the Selby Wistow mine shaft air and fan drift permanent works. *Proc. 4th NSGF*, BGFS, Nottingham, 48–63.

AULD F. A. 1988a. Design and installation of deep shaft linings in ground temporarily stabilised by freezing: Part 1: shaft lining deformation characteristics. *Proc. 5th ISGF*, Nottingham, **1**, 255–262.

AULD F. A. 1988b. Design and installation of deep shaft linings in ground temporarily stabilised by freezing: Part 2: shaft lining and freeze-wall deformation compatibility. *Proc. 5th ISGF*, Nottingham, **1**, 263–277.

AULD F. A. 1988c. The application of AGF to the construction of the ventilation drifts at Selby Wistow mine. *Proc. 5th ISGF*, Nottingham, **2**, 513–524.

AULD F. A. 1989. High strength, superior durability, concrete shaft linings. *Proc. IC on Shaft Engng*, IMM, Harrogate, **25–37**.

AULD F. A. 1993a. Concreting in cold environments. *Proc. 6th NSGF Frozen security: Safety and cost effectiveness of the ground freezing method*, BGFS, Nottingham, 15–18.

AULD F. A. 1993b. Casting concrete against frozen ground, *Technical Memorandum TM8, BGFS*, 3pp.

AZIZ K. A. and LABA J. T. 1976. Rheological model of laterally stressed frozen soil. *J. Geotech. Engng*, ASCE, Aug., **102**–8, 825–839.

BABA H. U. 1993. *Factors influencing the frost heave of soils*. University of Nottingham, *PhD thesis*, 453pp.

BABENDERENDE S. 1980. Application of the NATM for metro construction in Germany. *Proc. Eurotunnel80*, Basle, 54–58.

BAERLOCKER J. R. 1982. Horizontal ground freezing for large tunnels. *Proc. 3rd ISGF*, Hanover NH, **2***, 11–29.

BAKER A. C. J. and JAMES A. N. 1990. Tunnel connection to Thames reservoirs. *Proc. ICE*, 1–88, 929–954.

BAKER G. C. and OSTERKAMP T. E. 1988. Salt redistribution during laboratory freezing of saline sand columns. *Proc. 5th ISGF*, Nottingham, **1**, 29–33.

BAKER T. H. W. 1976a. Transportation, preparation and storage of frozen soil samples for laboratory testing. *ASTM Publication 599*, 88–112.

BAKER T. H. W. 1976b. Preparation of artificially frozen sand specimens. *NRCC, Paper NRCC 15349 DBR 682*. 16pp.

BAKER T. H. W. 1978a. Strain rate effect on the compressive strength of frozen sand. *Proc. 1st ISGF*, Bochum, **2**, 73–80, and in *Engng Geol.*, **13**, 223–231.

BAKER T. H. W. 1978b. Effect of end conditions on the uniaxial strength of frozen sand. *Proc. 3rd IC on Permafrost*, Edmonton, 6pp.

BAKER T. H. W. and DAVIS J. L. 1982. Application of time-domain reflectometry to determine the thickness of the frozen zone in soils. *Proc. 3rd ISGF*, Hanover NH, **2***, 3–10.

BAKER T. H. W. *et al.* 1982. Guidelines for classification and laboratory testing of artificially frozen ground. *Proc. 3rd ISGF*, Hanover NH, 2*, 141–167.

BAKER T. H. W. *et al.* 1982. Confined and unconfined compression tests on frozen sands. *Proc. 4th Cndn C. on Permafrost (1981) (R. J. E. Brown Mem. Vol.)*, Alberta, Mar., 387–393.

BAKER T. H. W. and KONRAD J. M. 1985. Effect of sample preparation on the strength of artificially frozen sand. *Proc. 4th ISGF*, Sapporo, **2**, 171–176.

BAKER T. H. W. and KURFURST P. J. 1985. Acoustic and mechanical properties of frozen sand. *Proc. 4th ISGF*, Sapporo, **1**, 227–234.

BALL E. F. 1967. Simple transient flow method of measuring thermal conducitivity and diffusivity. *Proc. IoRef.* †

BAOLAI W. and FRENCH H. M. 1994. Determination of permafrost creep parameters from *in situ* measurements. *Proc. 7th ISGF*, Nancy, France, **1**, 117–120.

BARRY and JACOBOVICS 1913. Freezing method and cementation, applications in sinking the potassium shaft, Wendland. *Glückauf*, 1885.

BATTELINO D. and SKRABL S. 1989. Computer design of tunnelling with soil freezing. *Proc. 12th IC SM & FE*, Brazil, **2**, 1441–1444.

BAZANT Z. J. 1976. Dewatering and hydraulic problems, soil freezing and chemical techniques. *Proc. 6th IC SM & FE*, Vienna, **1–3**, 33–76.

BEBI P. C. and METTIER K. R. 1979. Ground freezing for the construction of the 3-lane Milchbuck road tunnel, Zurich. *Proc. Tunnelling 79*, IMM, London, Ch. 16, 245–255.

BELL F. G. (ed) 1975. *Methods of treatment of unstable ground*. Butterworth, 159–171.

BELL F. G. 1992. Ground freezing. In Bell F. G. (ed), *Engineering in rock masses*, 321–333.

BELL M. J. 1982. The design of shaft linings in coal measure strata. *Proc. NS on Strata Mechanics*, Univ. of Newcastle, 160–166.

BELL M. J. 1983. Ice-walls — freezing and thawing. *Proc. 1st NSGF Progress in AGF*, Nottingham, 75–82.

BELL M. J. 1984. Concrete in temporarily frozen ground. *Proc. Concrete in the Ground*, Concrete Soc., May, 11pp.

BELTON J. 1993. Environmental benefits of AGF: a case history: Billingham. *Proc. 6th NSGF Frozen security: Safety and Cost Effectiveness of the Ground Freezing Method*, BGFS, Nottingham, 19–28.

BERGGREN A. L. 1991. Recent ground freezing in Scandinavia. *Proc. 6th ISGF*, Beijing, **2**, 517–520.

BERGGREN A. L. and FINBORUD L. I. 1980. Deformation properties of frozen soils and their time dependency. *Proc. 2nd ISGF*, Trondheim, **1**, 262–271.

BERGGREN A. L. and FURUBERG T. A new Norwegian creep model and creep equipment. *Proc. 4th ISGF*, Sapporo, **1**, 181–186.

BIQUET M. 1953. Sining shafts more than 1000 m deep through quicksand. *Proc. NC on Mining*, Brussels, July, **52**, 558–574.

BIRCH F. and CLARK H. 1940. The thermal conductivity of rocks and its dependence upon temperature and composition. *American J. Sci.*, **238**–8, 529–558, and **238**–9, 613–635.

BITTNER F. 1985. Completion of freeze shaft at Voerde. *Proc. Coll. Shaft sinking and Tunnelling, Glückauf*, **121**, 1423–1479.

BLACK P. B. 1991. Interpreting unconfined water content. *Proc. 6th ISGF*, Beijing, **1**, 3–6.

BLANCHARD D. and FRÉMOND M. 1982. Cryogenic suction in soils. *Proc. 3rd ISGF*, Hanover NH, **1**, 233–237.

BLANCHARD D. and FRÉMOND M. 1985. Soils frost heaving and thaw settlement. *Proc. 4th ISGF*, Sapporo, **1**, 209–216.

BONDARENKO G. I. 1994. Regulation of temperature as a method to control deformation of spoil heaps in polar regions. *Proc. 7th ISGF*, Nancy, France, **1**, 75–78.

BONDARENKO G. I. and SADOVSKY A. V. 1975. Strength and deformability of frozen soil in contact with rock. *SM & FE*, **12**–3, 174–178.

BONDARENKO G. I. and SADOVSKY A. V. 1991. Water content effect of the thawing clay soils on shear strength. *Proc. 6th ISGF*, Beijing, **1**, 123–128.

BOOTITE J. and MEYER J. 1981. Freezing with liquid nitrogen (in French). In G. Filliat (ed), *La pratique des sols et fondations*, 734–785.

BOSCH H. J. 1978. Construction of a sewer in artificially frozen ground.

Proc. 1st ISGF, Bochum, **2**, 127–132, and in *Engng Geol.*, 1979, **13**, 537–550.

BOURBONNAIS J. and LADANYI B. 1985a. The mechanical behaviour of frozen sand down to cryogenic temperatures. *Proc. 4th ISGF*, Sapporo, **1**, 235–244.

BOURBONNAIS J. and LADANYI B. 1985b. The mechanical behaviour of a frozen clay down to cryogenic temperatures. *Proc. 4th ISGF*, Sapporo, **2**, 237–244.

BOUTONNET M. 1994. Le dimensionnement au gel des chaussés françaises et la sauvegarde du patrimoine routier pendant l'hiver (in French). *Proc. 7th ISGF*, Nancy, France, **1**, 255–263.

BOZOZUK M. and BAKER T. W. H. 1978. Measuring total volumetric strains during triaxial tests on frozen soils: a new approach. *Cndn Geotech. J.*, **15**–4, 620–621.

BRAGG R. A. and ANDERSLAND O. B. 1980. Strain rate, temperature and sample size effects on compression and tensile properties of frozen sand. *Proc. 2nd ISGF*, Trondheim, **1**, 34–47.

BRAGG R. A. and ANDERSLAND O. B. 1982. Strain dependence of Poisson's ratio for frozen sand. *Proc. 4th Cndn C. on Permafrost (1981) (R. J. E. Brown Mem. Vol.)*, Alberta, 365–372.

BRAITHWAITE T. R. 1968. Freezing techniques for shaft support. *Proc. NC on Tunnels and Shafts*, Ch. 27, 356–358.

BRANDTL H. 1980. The influence of mineral composition on frost susceptibility of soils. *Proc. 2nd ISGF*, Trondheim, **1**, 815–823.

BRAUN B. 1972. Ground freezing for tunnelling in water bearing soil at Dortmund. *Tunnels and Tunnelling*, **4**, 27–32.

BRAUN B. 1985. German and Swiss experiences with ground freezing. *Tunnels and Tunnelling*, Dec., 47–50.

BRAUN B. and MACCHI A. 1974. Ground freezing techniques at Salerno. *Tunnels and Tunnelling*, Mar., 81–89.

BRAUN B. and NASH W. R. 1982. Ground freezing applications in underground mining construction. *Proc. 3rd ISGF*, Hanover NH, **1**, 319–326.

BRAUN B. and SHUSTER J. A. 1979. Ground freezing — Technology opens new applications. *World Construction*, Nov., 26–31.

BRAUN B. *et al.* 1978. Ground freezing for support of open excavations. *Proc. 1st ISGF*, Bochum, **2**, 137–156.

BRAUN W. M. 1968a. Cryogenic underground tanks for LNG. *Ground Engng*, Sept., 21–23.

BRAUN W. M. 1968b. Liquid nitrogen for instant ground freezing. *Ground Engng*, Sept.,

BRENDENG E. Early experiences with ground freezing in Norway. *Proc. 2nd ISGF*, Trondheim, **1**, 896–906.

BRIGHENTI G. 1970. Influence of cryogenic temperatures on the mechanical characteristics of rocks. *Proc. 2nd IC of Rock Mechanics*, 473–477.

BRODSKARA A. G. 1965. Compressibility of frozen ground. *Transl by CRREL*, NTIS AD 715087, 2–83.

BROKENSTEIN D. *et al.* 1991. Construction of a shallow tunnel under protection of a frozen soil structure: Fahrlachtunnel at Mannheim, Germany. *Proc. 6th ISGF*, Beijing, **2**, 481–489.

BROMS B. B. and YAO L. Y. C. 1964. Shear strength of a soil after freezing and thawing. *J ASCE SM & FE Div*, SM4, 1–25.

BROWN A. C. J. and JAMES A. N. 1990. The Three Valleys Water Committee: tunnel connection to Thames water reservoirs. *Proc. ICE*, **1**, Dec., 929–954.

BRUSHKOV A. and CHEKHOVSKY A. 1994. Adfreezing shear strength peculiarities of different soils. *Proc. 7th ISGF*, Nancy, France, **1**, 207–213.

BUDKOWSKA B. B. and FU Q. 1989. Analysis of creep effects in frozen soils. *Proc. 3rd IS Numerical Models in Geomechanics*, Niagara Falls, Elsevier.†

BURGHARDT G. *et al.* 1982. Crack formation during shaft sinking in salt rock due to freezing temperature. *Kali and Steinsalz J*, **9**, 294–315.

BURT T. P. and WILLIAMS P. J. 1976 Hydraulic conductivity in frozen soils. *Earth surface processes*, **1**–3, 349–360.

BUTKOVICH T. R. 1954. Ultimate strength of ice. *US Army Research Rpt 11*.

BUTTIENS E. 1978. Conversion of abandoned collieries in south Belgium to LPG storge with description of special plugging of various shafts. *Proc. 1st ISGF*, **1**, 349–356, and in *Proc. Rockstore*, Sweden, **1**, 101–105.

CAMES-PINTEAUX A. M. and AGUIRRE-PENTE J. 1988. Cylindrical Stefan problem sensitiveness to thermal and geometrical parameters. *Proc. 5th ISGF*, Nottingham, **2**, 547–550.

CAMES-PINTEAUX A. M. *et al.* 1985. Underground cryogenic cavities — field measurements and numerical methods. *Proc. 4th ISGF*, Sapporo, **1**, 55–64.

CAPITAINE P. and REBHAN D. 1988. Ground freezing with liquid nitrogen. *Proc. 5th ISGF*, Nottingham, **1**, 273–277.

CHAICHANAVONG T. 1976. *Dynamic properties of ice and frozen clay under cyclic triaxial loading conditions.* University of Michigan, PhD thesis, 355pp.

CHALMERS B. and JACKSON K. A. 1970. Experimental and theoretical studies of the mechanism of frost heaving. *CRREL Research Rpt 199*.

CHAMBERLAIN E. J. 1973. Mechanical properties of frozen ground under high pressure. *Proc. 2nd IC on Permafrost*, Yakutsk, 295–305.

CHAMBERLAIN E. J. 1980. Overconsolidation effects of ground freezing. *Proc. 2nd ISGF*, Trondheim, **1**, 325–337, and in *Engng Geol.*, **18**, 97–110.

CHAMBERLAIN E. J. 1981. Frost susceptibility of soil: review of index tests. *CRREL Monograph 81–2*, 110pp.

CHAMBERLAIN E. J. 1982. Comparative evaluation of frost susceptibility. *Research Record Trnsptn*, 809, 42–52.

CHAMBERLAIN E. J. 1983. Frost heave of saline soils. *Proc. 4th IC on Permafrost*, Fairbanks, 121–126.

CHAMBERLAIN E. J. 1985. Shear strength anisotropy in frozen saline and freshwater soils. *Proc. 4th ISGF*, Sapporo, **2**, 189–194.

CHAMBERLAIN E. J. 1986. Evaluation of selected frost susceptibility test methods. *CRREL, Rpt 86–14*, 51pp.

CHAMBERLAIN E. J. 1987. A freeze–thaw test to determine the frost susceptibility of soils. *CRREL Rpt 87–1*, 90pp.

CHAMBERLAIN E. J. 1991. Freeze-thaw effects on clay covers and liners. *Proc. 6th IC on Cold Regions Engineering*, CRREL.†

CHAMBERLAIN E. J. 1993. Physical changes in clays due to frost action and their effect on engineering structures. *Frost in Geotech. Engng*, **2**, 863–886.

CHAMBERLAIN E. J. and BLOUIN S. E. 1978. Densification by freezing and thawing of fine material dredged from waterways. *Proc. 3rd IC on Permafrost*, Edmonton, 623–628.

CHAMBERLAIN E. J. and GOW A. 1978. Effect of freezing and thawing on the permeability and structure of soils. *Proc. 1st ISGF*, Bochum, **1**, 31–44, and in *Engng Geol.*, 1979, **13**, 73–92.

CHAMBERLAIN E. J. and HOEKSTRA 1970. The isothermal compressibility of frozen soil and ice to 30 kbars at $-10°C$. *CRREL Rpt 225*, 33pp.

CHAMBERLAIN E. J. *et al.* 1977. Resilient modulus and Poisson's ratio for frozen and thawed silt and clay subgrade materials. *Proc. NS on Applications of Soil Dynamics in Cold Regions*, ASCE, 229–281.

CHAMBERLAIN E. J. *et al.* 1979. Resilient response of two frozen and thawed soils. *ASCE J. GED*, **105**−2, 257−271.

CHAMBERLAIN E. J. *et al.* 1984. Survey of methods for classifying frost susceptibility in frost action and its control. *ASCE Tech. Comm. Cold Regions Engineering Monograph*, 102−142.

CHANGJIAN T. and ZONGYAN S. 1980. Horizontal frost heave thrust acted on buttress constructions. *Proc. 2nd ISGF*, Trondheim, **1**, 725−734.

CHEN H. P. and WU T. W. 1978. Experimental research on the principal mechanical properties of freezing and frozen soils in China. *Proc. 3rd IC on Permafrost*, Edmonton, **2**, 159−177.

CHEN R. *et al.* 1994. Analysis on the frozen fringe and the ice segregation temperature. *Proc. 7th ISGF*, Nancy, France, **1**, 9−12.

CHEN W., TANG Z. and LI G. 1991. Causes and countermeasures of flooding in shafts constructed by freezing method. *Proc. 6th ISGF*, Beijing, **1**, 313−318.

CHEN X. 1988. Mechanical characteristics of artificially frozen clays under triaxial stress conditions. *Proc. 5th ISGF*, Nottingham, **1**, 173−180.

CHEN X, LI, K. and ZHANG Y. 1991. Temperature effects on mechanical characteristics of an artificially frozen typical clay under triaxial stress conditions. *Proc. 6th ISGF*, Beijing, **1**, 129−134.

CHEN X. and SU L. 1991. Mechanical properties: general report. *Proc. 6th ISGF*, Beijing, **2**, 429−436.

CHEN X. and WANG Y. 1988. Frost heave prediction for clayey soils. *Cold Regions Sci. and Tech.*, **15**−3, 233−238.

CHEN X. and WANG Y. 1991. Prediction and control of frost damage to engineering projects in seasonally frost regions. *Proc. 6th ISGF*, Beijing, **1**, 7−10.

CHEN X. B. *et al.* 1988. On salt heave of saline soil. *Proc. 5th ISGF*, Nottingham, **1**, 35−39.

CHEN X. *et al.* 1994. Centrifuge modelling of frost heave of pipelines. *Proc. 7th ISGF*, Nancy, France, **1**, 91−96.

CHEN X. B. *et al.* 1994. Some behaviours of saline soils mixed with Na_2SO_4 and NaCl during cooling. *Proc. 7th ISGF*, Nancy, France, **1**, 385−389.

CHOU W. 1985. Stress on reinforcing ribs and concrete strain from in situ measurement during shaft sinking by freezing process. *Proc. 4th ISGF*, Sapporo, **1**, 147−152.

CHOU W. 1988a. A case study on freezing pressure in a deep shaft. *Proc. 5th ISGF*, Nottingham, **2**, 481−486.

CHOU W. 1988b. Design problems of shaft lining and freezing wall in deep shaft sinking by the freezing process. *Proc. 5th ISGF*, Nottingham, **2**, 487−490.

CHRISTOTINOV L. V. 1980a. A cryoscopic method of measuring the unfrozen water content in soils. *Proc. 2nd ISGF*, Trondheim, **1**, 374−382.

CHRISTOTINOV L. V. 1980b. Pore water migration studies at a freezing boundary. *Proc. 2nd ISGF*, Trondheim, **1**, 647−655.

CHUBAROVA N. P. *et al.* 1980. Hot plate field tests on frozen soils. *SM & FE*, **16**−4, 228−230.

CHUNICHEV B. D. 1972. Phase composition of water in frozen ground under pressure. *CRREL Transl 319*, 9pp.

CHUVILIN YE M. and YAZYNIN O. M. 1988. Frozen soil macro- and micro-texture formation. *Proc. 5th IC on Permafrost*, Trondheim, **1**, 320−328.

CLARK *et al.* 1968. The Victoria Line: escalator tunnel at Tottenham Hale Station. *Proc. ICE*, 7270S, 423−429, 340−341.

CLARK M. A. *et al.* 1986. Developments in the use of thermal neutron radiography for studying mass transfer in a partially frozen soil. *Proc. 4th NSGF*, Nottingham, 31−47.

CLARKE A. P. J. and MACKENZIE C. N. P. 1994. Overcoming ground difficulties at Tooting Bec. *Proc. ICE*, **102**−2, 60−75.

CLEASBY J. V. *et al.* 1975. Shaft sinking at Boulby Mine, Cleveland Potash Ltd. *Proc. IMM*, **84**, 7–28, and 147–8.

CLOUD G. 1976. Air shaft will move with the soil around it. *Engng News Record*, 10 Aug., 27.

COCKCROFT T. N. *et al.* 1982. Construction of a section of the Du Toits Kloof tunnel by use of ground freezing. *Tunnelling 82*, IMM, London, 105–116.

COLE D. M. *et al.* 1985. Repeated load triaxial testing of frozen and thawed soils. *Geotech. Testing J*, Dec., **8**–4, 166–170.

COLE D. M. *et al.* 1986. Resilient modulus of freeze–thaw affected granular soils for pavement design and evaluation. *CRREL Rpt*, **86**–4.

COLLINS S. P. and DEACON W. G. 1972. Shaft sinking by ground freezing in Ely-Ouse Essex scheme. *Proc. ICE*, Supp. vii & xx, May, 129–156, 319–336.

COMINI G. *et al.* 1974. Finite element solution of non-linear heat conduction problems with special reference to phase change. *Intl J for Numerical Methods in Engng*, Jan., **8**, 613–624.

COMPARINI E. 1988. On a model for quasi-steady freezing processes of saturated porous media. *Proc. 5th ISGF*, Nottingham, **1**, 41–50.

CONNOLLY W. 1976. The big Kentucky freeze. *Engng News Record*, 19 Aug., 27.

CORAPCIOGLU M. Y. and PANDAY S. 1985. A mathematical model of ground movement due to thaw action in unsaturated soils. *Proc. 4th ISGF*, Sapporo, **2**, 115–119.

CORBOULEIX S. and MOUROUX B. 1994. Les phénomènes périglaciaires en France: inventaire et intérêt (in French). *Proc. 7th ISGF*, Nancy, France, **1**, 413–414.

COUTTS R. I. and KONRAD J. M. 1994. Finite element modelling of transient non-linear heat flow using the node state method. *Proc. 7th ISGF*, Nancy, France, **1**, 39–47.

CUI G. and YANG W. 1991. Stress analysis on freezing pipes by modelling test. *Proc. 6th ISGF*, Beijing, **1**, 219–224.

CUMMINGS J. B. and HAYWARTH R. G. 1958. Freezing for shaft sinking, *Mining Congress J*, Washington, **44**, Dec., 52–56.

CZAJKOWSKI R. L. and VINSON T. S. 1980. Dynamic properties of frozen silt under cyclic loading. *ASCE J Geotech. Engng*, **106**–9, 963–980.

CZURDA K. A. and SCHABABERLE R. 1988. Influence of freezing and thawing on the physical and chemical properties of swelling clays. *Proc. 5th ISGF*, Nottingham, **1**, 51–58.

CZURDA K. A. and WAGNER J. F. 1985. Frost heave and clay expansion in freshwater clays. *Proc. 4th ISGF*, Sapporo, **1**, 123–128.

DAICHAO S. *et al.* Estimation of frost heave for stratified soil profile. *Proc. 7th ISGF*, Nancy, France, **1**, 129–141.

DANIELYAN Y. S. and YANITSKY P. A. 1991. Heat and moisture exchange for freezing and thawing grounds. *Proc. 6th ISGF*, Beijing, **1**, 105–112.

DASH J. G. 1994. Ground freezing for management of hazardous wastes. *Proc. 7th ISGF*, Nancy, France, **1**, 351–354.

DAW G. P. and WILLIS A. J. 1983. Location of ice-wall in Gascoigne Wood drifts — Selby project. *Proc. 1st NSGF Progress in AGF*, BGFS, Nottingham, 59–66.

DAWSON, A. S. 1954. Quicksand frozen for shaft sinking. *Cndn Mining J*, **75**, 68–70.

DE BEER E., BUTTIENS E. and MAERTENS J. 1980. Research of the behaviour of non-cohesive soils when treated by artificial ground freezing. *Proc. 2nd ISGF*, Trondheim, **1**, 338–353.

DE VRIES 1963. Thermal properties of soils. *Physics and plant environment*, North Holland Publishing Co, 210–235.

DE VRIES 1966. Thermal properties of soils. In Van Wijk W. R. (Ed) *Physics of plant environment*. North Holland Publishing Co.

DEIX F. and BRAUN B. 1988. Vienna subway construction — use of brine freezing in combination with NATM under compressed air. *Proc. 5th ISGF*, Nottingham, **1**, 321–330.

DEMARS K. R. and VANOVER E. A. 1980. Strength of partially saturated sand–clay mixtures. *Proc. 2nd ISGF*, Trondheim, **1**, 132–143.

DESCHARTRES M. H. *et al.* 1988. Acoustic measurements on soaked soils during thaw and unfrozen water content determination. *Proc. 5th ISGF*, Nottingham, **2**, 551–554.

DIEKMAN N. and JESSBERGER H. L. 1982. Creep behaviour and strength of an artificially frozen silt under triaxial state. *Proc. 3rd ISGF*, Hanover, **2***, 31–37.

DIRKSEN C. 1964. *Water movement and frost heaving in unsaturated soil without an external source of water*. Cornell University, PhD thesis.

DJABALLAH N. and AGUIRRE-PUENTE J. 1994. Perméabilité des sols gelés, approche théoretique, première confrontation avec des mesures expérimentales (in French). *Proc. 7th ISGF*, Nancy, France, **1**, 69–74.

DOIG P. J. 1985. Grouting and freezing for shaft water control in Milwaukee. *Proc. RETC*, New York, **2**, 1211–1224.

DOIG P. J. 1991. NS9 drop shaft and ancillary facilities (Milwaukee). *Proc. 10th RETC*, Seattle USA, 685–704.

DOMKE O. 1915. On the stresses in a frozen cylinder of ground used for shaft sinking (in German). *Glückauf*, **51–47**, 1129–1135.

DUBINI M. M. 1983. Interaction of frozen soils with wells and pipelines. *Proc. 3rd ISGF*, Hanover NH, **1**, 349–354.

DUDEK, S. J. M. 1980. *The experimental and theoretical prediction of frost heave in granular soils*. University of Nottingham, PhD thesis.

DUDEK S. J-M. and HOLDEN J. T. 1979. A theoretical model for frost heave. *Proc. 1st IC on Numerical Methods in Thermal Problems*, Swansea, July, 216–229.

DUMONT-VILLAIRES A. 1956. The underpinning of the 26 storey 'Companhia Panlista de Segmos' building, São Paulo, Brazil. *Géotechnique*, **6**, 1–14.

DYSLI M. 1991. Resilient modulus-of freeze-thaw or resilient frost heave. *Proc. 6th ISGF*, Beijing, **1**, 225–230.

EBEL W. 1985. Influence of specimen end conditions and slenderness ratio on the mechanical properties of frozen soils. *Proc. 4th ISGF*, Sapporo, **2**, 231–236.

EBELING V. 1954. The sinking, flooding and reopening of #3 Hansa potash works at Empelde. *Kali ni Steinsalz*, Feb., **4**, 3–19.

ECKARDT H. 1978. Bending behaviour of frozen soils (in German). *Vortrag bau grundtag*, Berlin, Sep., 20pp.

ECKARDT H. 1980. Bearing behaviour of frozen soil. *Proc. 2nd ISGF*, Trondheim, **1**, 272–284.

ECKARDT H. 1981. Laboratory borehole creep and relaxation tests in thick walled cylindrical samples of frozen sand. *Montreal Polytechnic Rpt 665–222*, 134pp.

ECKARDT H. 1982. Creep tests with frozen soils under uniaxial tension and compression. *Proc. 4th Cndn C. on Permafrost (1981) (R. J. E. Brown Mem. Vol.)*, Alberta, 394–405.

EDGERS L. *et al.* 1990. Field evaluation of criteria for frost susceptibility of soils. *Trnsptn Research Record*, **1190**, 73.

EFIMOV S. S. *et al.* 1980. The influence of cyclic freezing-thawing on heat and mass transfer characteristics of clay soil. *Proc. 2nd ISGF*, Trondheim, **1**, 462–469.

EINCK H. B. and WEILER A. 1982. Experiences and investigations using gap freezing to control ground water flow. *Proc. 3rd ISGF*. Hanover NH, **1**, 193–204.

ELLIS D. R. and McCONNELL J. 1959. Use of the freezing process in the construction of a pumping station at Fleetwood, Lancs. *Proc. ICE*, Feb., 175–186.

ENDO K. 1969. Artificial soil freezing method for subway construction. *Proc. Japan. Soc. of Engineers.* †

ERLINGHAGEN E. 1924. Development of shaft sinking according to the freezing process during the past 20 years. *Zeit, Verein D. Ingenieure*, **68**, 383–393.

ERSHOV E. D. et al. 1991. Water and ion migration of frozen soils in open system. *Proc. 6th ISGF*, Beijing, **1**, 11–16.

ERSHOV E. D. et al. 1994. Heavy metal ions transfer in frozen soils. *Proc. 7th ISGF*, Nancy, France, **1**, 103–108.

ERSOY T. and TOGROL E. 1976. Factors affecting the shear strength of frozen soil. *Proc. 2nd IS on Cold Regions Engineering*, Alaska, 40–43.

ERSOY T. and TOGROL E. 1978. Temperature and strain effects on the strength of compacted frozen silty-clay. *Proc. 3rd IC on Permafrost*, Edmonton, July, 5pp.

EVDOKIMOV P. D. and ZAUERBREY I. I. 1972. Effects of freezing on the mechanical properties of clay moraine. *CRREL Transl 323*, 6pp.

EVERITT D. H. and HAYNES J. M. 1965. Capillary properties of some model pore systems with special reference to frost damage. *Bulletin RILEM*, June, **27**, 31–38.

FAIRHURST C. 1970. Investigation of brittle fracture of frozen soil. *CRREL Rpt*, Aug., 88pp.

FAROUKI O. T. 1983. Thermal properties of soils relevant to ground freezing — design techniques for their estimation. *Proc. 3rd ISGF*, Hanover NH, **1**, 139–146.

FAROUKI O. T. 1986. Thermal properties of soils. *Transl Tech Publications*, 136pp. Reprint of *CRREL Monograph 81–1*, 1981.

FERGUSON G. A. and CRYSTAL L. 1984. Control of ground and water for shaft sinking. *Proc. 2nd NSGF Relating theory to practice in AGF*, BGFS, Nottingham, 7–12.

FETZ L. B. 1979. Short-cut frost heaving test for soils. *Frost-i-jord*, **22**, 41–48.

FINBORUD I. and BERGGREN A. L. 1980. Deformation properties of frozen soils. *Proc. 2nd ISGF*, Trondheim, 262–271.

FINN W. D. L. and YONG R. N. 1977. Dynamic response of frozen ground. *NS on Application of Soil Dynamics in Cold Regions (ASCE)*, 1–95.

FINN W. D. L. and YONG R. N. 1978. Seismic response of frozen ground. *J. ASCE Geotech. Div.*, **104**–10, 1225–1241.

FIRTH G. W. and GILL J. J. 1963. The sinking of Kellingley shafts. *Mining Engineer (Proc. IME)*, **123**, Dec., 147–166.

FISH A. M. 1980. Kinetic nature of the long term strength of frozen soils. *Proc. 2nd ISGF*, Trondheim, **1**, 95–108.

FISH A. M. 1982. Deformation and failure of frozen soils and ice at constant and steadily increasing stresses. *Proc. 4th Cndn C. on Permafrost (1980) (R. J. E. Brown Mem. Vol.)*, Alberta, Mar., 419–428.

FISH A. M. 1983. Thermodynamic model of creep at constant stresses and constant strain rates. *CRREL Report 83-33*, 18pp.

FISH A. M. 1985. Creep strength, strain rate, temperature and unfrozen water relationship in frozen soil. *Proc. 4th ISGF*, Sapporo, **2**, 29–36.

FISH A. M. 1987. Shape of creep curves in frozen soils and polycrystalline ice. *Cndn Geotech. J.*, Nov., **24**–4, 623–629.

FISH A. M. 1991. Strength of frozen soil under a combined stress state. *Proc. 6th ISGF*, Beijing, **1**, 135–145.

FISH A. M. 1994. Creep strength of Ottawa frozen sand under varying mean stress. *Proc. 7th ISGF*, Nancy, France, **1**, 103–108.

FISH A. M. and SAYLES F. H. 1981. Acoustic emissions during creep of frozen soils. *ASTM STP-750*, 194−205.

FLOESS C. H., LACY H. S. and GERKEN D. E. 1989. Artificially frozen ground tunnel — case history. *Proc. 12th IC SM & FE*, Brazil, **2**, 1445−1448.

FORIERO A. and LADANYI B. 1994. Pipe uplift resistance in frozen soil and comparison with measurements. *ASCE J. Cold Regions Engng*, **8**−3, 93−111.

FØRLAND T. and RATKJE S. K. 1980a. Irreversible thermodynamic treatment of frost heave. *Proc. 2nd ISGF*, Trondheim, **1**, 611−617.

FØRLAND T. and RATKJE S. K. 1980b. On the theory of frost heave. *Frost-i-jord*, **21**, 25−48.

FORREST W. 1982. Selby Drifts: ground treatment with particular reference to freezing technique. *Tunnelling 82*, IMM, London, 117−120.

FORSTER-BROWN E. O. 1927. Freezing in vertical shaft sinking. 154−183; 402−406.

FORTIER R. *et al.* 1994. Links between *in situ* determined mechanical and electrical properties of a silty frozen ground. *Proc. 7th ISGF*, Nancy, France, **1**, 143−152.

FOWLER A. C. and NOON C. G. 1993. A simplified numerical solution of the Miller model of secondary frost heave. *Cold Regions Sci. Tech.*, **21**, 327−336.

FREDEN S. 1985. Mechanism of frost heave and its relation to heat flow. *Proc. 6th IC SM & FE*, Montreal, **1**, 41−45.

FREDEN S. and STENBERG L. 1980. Frost heave tests on tills with an apparatus for constant heat flow. *Proc. 2nd ISGF*, Trondheim, **1**, 760−771.

FRÉMOND M. 1994. Supercooling: a macroscopic predictive theory. *Proc. 7th ISGF*, Nancy, France, **1**, 79−84.

FRÉMOND M. and MIKKOLA M. 1991. Thermomechanical modelling of freezing soil. *Proc. 6th ISGF*, Beijing, **1**, 17−24.

FRÉMOND M. and SAN-MARTIN H. J. 1994. Soil freezing in presence of solute. *Proc. 7th ISGF*, Nancy, France, **1**, 379−383.

FRIVIK P. E. 1980. State of the Art report: Ground Freezing: thermal properties, modelling of processes and thermal design. *Proc. 2nd ISGF*, Trondheim, **1**, 354−373, and in *Engng Geol.*, **18** (1981), 115−133.

FRIVIK P. E. and COMINI G. 1982. Seepage and heat flow in soil freezing. *Proc. ASME Heat Transfer*, **104**, 323−328.

FRIVIK P. E. and THORBERGSEN E. 1980. Thermal design of artificial soil freezing systems. *Proc. 2nd ISGF*, Trondheim, **1**, 556−567.

FROLOV A. D. 1973. Elastic and electrical properties of frozen ground. *Proc. 2nd IC on Permafrost*, Yakutsk, 307−311.

FROLOV A. D. 1978. Problems and possibilities of studying the process of dynamic relaxation in frozen earth materials. *Proc. 3rd IC on Permafrost*, Edmonton, **1**, 149−168.

FROST S. R. 1984. A model of frost heave and ice lensing including overburden pressure. *Proc. 2nd NSGF Relating theory to practice in AGF*, BGFS, Nottingham, 13−18.

FRYDMAN S. *et al.* 1979. Effect of pre-freezing on the strength and deformation properties of granular soils. *Soils Foundations*, **19**−4, 31−42.

FUKUDA M. 1982. Heat flow measurements in freezing soils with various freezing front advancing rates. *Proc. 4th Cndn C. on Permafrost (1981) (R. J. E. Brown Mem. Vol.)*, Alberta, 445−452.

FUKUDA M. and INOUE M. 1973. On the dynamic moduli of frozen soils. *Physical Science, Hokkaido Univ.*, Ser A, **31**, 245−259.

FUKUDA M. and ISHIZAKI T. 1991. Heat and mass transfer: general report. *Proc. 6th ISGF*, Beijing, **2**, 409−416.

FUKUDA M. and NAKAGAWA S. 1985. Numerical analysis of frost heaving based on the coupled heat and water flow model. *Proc. 4th ISGF*, Sapporo, **1**, 109−118.

FUKUDA M. and TOMATSU Y. 1994. Effects of frost depths of the ground to the degrees of liquefaction in Kushiro region under the seismic activities of 1993 Kushiro-Oki earthquake. *Proc. 7th ISGF*, Nancy, France, **1**, 295−302.

FUKUDA M. *et al.* 1985. Experimental study on prevention of frost heave using heat pipe. *Proc. 4th ISGF*, Sapporo, **2**, 341−346.

FUKUDA M. *et al.* 1991. The evaluation of reducing method of total heave amounts using soil cement mixtures. *Proc. ISGF91*, Beijing, **2**, 417−424.

FUKUO Y. 1966. On the rheological behaviour of frozen soil. *Bulletin Kyoto Univ.*, Mar., **15**−3, 11−7; Sept., **16**−1, 31−40.

FUKUO Y. 1974. Deformation rate of frozen soil. *Bulletin Nara Univ.*, Nov., **23**−2, 41−50.

FUKUO Y. and ARIGA Y. 1966. On the deformation of frozen soil. Geophysics Inst., Kyoto Univ., 6, 187−192.

FUKUO Y. and ARIGA Y. 1967. Unfrozen water content of artificially frozen soil. *Bulletin Kyoto Univ.*, Oct., **17**−2, 73−77.

FUKUO Y. *et al.* 1968. On the deformation of frozen soil due to uniaxial compression. *Rpt Kyoto Univ.*, Mar., 6pp.

FUNCKEN R. *et al.* 1983. Construction of an experimental laboratory in deep clay formation. *Proc. Eurotunnel 83*, Switzerland, 76−86.

FUNG S. I. and STEENHUIS T. S. 1986. Heat and moisture transfer in a partially frozen non-heaving soil. *Soil Sci. Soc. America*, **50**−5, 1114−1122.

GAIL C. P. 1954. Application of the freezing method to shaft sinking. *Cndn Mining and Metallurgy Bulletin*, Sept., **509**, 586−589.

GAIL C. P. 1961 Stabilisation of excavations by freezing. *Proc. ASCE*, **42**, 37−40.

GAIL C. P. 1972. Tunnel driven using subsurface freezing. *Civil Engng (ASCE)*, May, **42**−5, 37−40.

GALLAVRESI F. 1980. Ground freezing − application of the mixed method brine−liquid nitrogen. *Proc. 2nd ISGF*, Trondheim, **1**, 928−939.

GALLAVRESI F. 1982. Recent developments in ground freezing. *Proc. 3rd ISGF*, Hanover NH, **2***, 39−42.

GALLAVRESI F. 1991a. The ground freezing technique with experiences in Italy. *BGFS Newsletter*, **16**, May, 3pp.

GALLAVRESI F. 1991b. Considerations on the choice among the different freezing methods and competitiveness of the system as compared to alterntive solutions. *Proc. 6th ISGF*, Beijing, **1**, 319−324.

GAO W. and XU X. 1988. Field observations of water migration in unsaturated freezing soils with different ground water tables. *Proc. 5th ISGF*, Nottingham, **1**, 59−64.

GARAND R. and LADANYI B. 1983. Frost susceptibility testing of a compacted glacial till. *Proc. 3rd ISGF*, Hanover NH, **1**, 277−284.

GARDNER A. R. *et al.* 1982. Strength and creep testing of frozen soils. *Proc. 3rd ISGF*, Hanover NH, **1**, 53−60.

GARDNER A. R. *et al.* 1984. A new creep equation for frozen soils and ice. *Cold Regions Sci. and Tech.*, **9**, 271−275.

GARNETT G. A. 1981. Tunnelling on development of first crossing of Manchester Ship Canal. *Cooling Prize presentation*, BGS, 13pp.

GARUS B. *et al.* 1980. Selected problems of the freezing process in rock formations and the control of this process in the Polish copper field. *Proc. 2nd ISGF*, Trondheim, **1**, 1001−1013.

GEDDES J. D. 1958. The prediction and measurement of temperatures in concrete structures with particular reference to some recent construction in frozen ground. *Proc. ICE Northern Counties Assn*. In Bulletin **11**, Durham Kings College, 43pp.

GEMANT A. 1950. The thermal conductivity of soils. *J. Applied Physics*, **21**, 750−752.

GILBERT P. A. 1982. Freezing of triaxial compression test specimens of cohesionless soil to determine internal density variations. *Proc. 3rd ISGF*, Hanover NH, **2***, 43–57.

GILPIN R. R. 1980. A model for the prediction of ice lensing and frost heave in soils. *Water Resources Research*, **16**–5, 918–930.

GLOUGHNOUR R. R. and ANDERSLAND O. B. 1968. Mechanical properties of a sand–ice system. *Proc. ASCE SM4*, **94**, 923–950.

GLUDICE S. D. *et al.* 1977. Finite element simulation of freeze process in soils.†

GOLDBERG D. T. 1975. Design and construction of freeze techniques specific to cut and cover tunnelling.†

GONGCHUN Y. 1986. Prevention of cracking and water leakage in the concrete inner lining of frozen shafts. *J. Mining Geol. Engng, Oct.*, **4**–3, 247–251.

GONZE P. 1983a. Verification of the stability of excavations in artificially frozen ground. *Proc. IS on Field Measurements in Geomechanics*, Switzerland, 1021–1031.

GONZE P. 1983b. Computer based ground freeze design. *Tunnels and Tunnelling*, Mar., 49.

GONZE P. 1986. Artificial ground freezing in Belgium. *Proc. 4th NSGF*, BGFS, Nottingham, 8–30.

GONZE P. 1994. The computer code GEOCAL. *Proc. 7th ISGF*, Nancy, France, **1**, 13–16.

GONZE P. *et al.* 1985. Sand ground freezing for the construction of a subway station in Brussels. *Proc. 4th ISGF*, Sapporo, **1**, 277–284.

GONZE P. *et al.* 1985. Determination of rheological parameters of frozen soils by laboratory tests. *Proc. 4th ISGF*, Sapporo, **2**, 195–200.

GORI F. and UGHI M. 1983. Experimental results on the freezing of saturated sands. *Proc. 3rd ISGF*, Hanover NH, **1**, 185–192.

GORLE D. 1980. Frost susceptibility of soils — influence of thermal variables and depth of the water table. *Proc. 2nd ISGF*, Trondheim, **1**, 772–783.

GORODETSKII S. E. 1975. Creep and strength of frozen soils under combined stress (transl from Russian). *SM & FE*, **12**–3, 205–209.

GOTO S. and TAKAHASHI Y. 1982. Frost heave characteristics of soil under extremely low frost penetration rate. *Proc. 3rd ISGF*, Hanover NH, **1**, 261–268.

GOTO S. and TANAKA M. 1985. Field frost heaving test on diluvial clayey soil. *Proc. 4th ISGF*, Sapporo, **2**, 157–162.

GOTO S. and TANAKA M. 1988. Direct shear test at a frozen/unfrozen interface. *Proc. 5th ISGF*, Nottingham, **1**, 181–186.

GOTO S. *et al.* 1985. The measurement of frost heaving pressure on an LNG inground tank. *Proc 4th ISGF*, Sapporo, **1**, 337–356.

GOTO S. *et al.* 1985. Frozen earth pressure on the inground LNG tank wall. *Proc. 4th ISGF*, Sapporo, **1**, 327–336.

GOTO S. *et al.* 1988. The comparison between results of heat transfer analysis and the observed values on a refrigerated LPG inground tank. *Proc. 5th ISGF*, Nottingham, **1**, 331–338.

GOTO S. *et al.* 1991. Results of the heat transfer analysis compared to thermal measurements made in arcuate heater-LNG inground storage system. *Proc. 6th ISGF*, Beijing, **1**, 25–32.

GOUGHNOUR R. R. and ANDERSLAND O. B. 1968. Mechanical properties of a sand-ice system. *J. ASCE SM & FD Div.*, **94**–SM4, 923–950.

GRAHAM E. B. *et al.* 1983. The British Gas Canvey Island LNG Terminal: a review of developments. *Proc. IC LNG7*, Jakarta, May, 17pp.

GRAHAM J. and AU V. C. S. 1985. Effects of freeze–thaw and softening on a natural clay at low stresses. *Cndn Geotech. J.*, **22**, 69–78.

GRECHISHCHEV S. Y. 1973. Basic laws of thermorheology and temperature cracking of frozen ground. *Proc. 2nd IC on Permafrost*, Yakutsk, 228–233.

GRECHISHCHEV S. E. 1980. Thermorheological principles of heaving. *Proc. 2nd ISGF*, Trondheim, **1**, 618–625.

GRECHISHCHEV S. E. and PAVLOV A. V. 1994. Freezing–thawing kinetics in water and water-saturated soils. *Proc. 7th ISGF*, Nancy, France, **1**, 49–55.

GREENE D. P. et al. 1994. Investigation of frost/pipe heave associated with large diameter chilled gas pipeline operation using small-scale laboratory model. *Proc. 7th ISGF*, Nancy, France, **1**, 409–410.

GREGORY O. and MAISHMAN D. 1973. Motorway construction meets an unusual problem in old shaft treatment. *Trans. IME (Manchester)*, Feb. 24pp.

GRIFFIN A. R. 1987. Tunnelling through frozen ground: a case history at Iver, Bucks. *Proc. 23rd NC on Engineering geology of underground movements*, Engineering Group of Geological Soc., Sept., 63–76.

GROB H. 1984. Ground freezing as tunnelling support. *Proc. Beijing IC on Advanced Tunnelling and Subsurface Use*, **4**–4, 265–269.

GROENEVELT P. H. and KAY B. D. 1975. Hydrostatics of frozen soil. *Proc. NC Soil-water problems*, Canada, May, 192–199.

GROENEVELT P. H. and KAY B. D. 1980. Pressure distribution and effective stress in frozen soils. *Proc. 2nd ISGF*, Trondheim, **1**, 597–610.

GURYANOV I. E. 1985. Thermal-physical characteristics of frozen, thawing and unfrozen grounds. *Proc. 4th ISGF*, Sapporo, **2**, 225–230.

GUYMON G. L. et al. 1980. A one-dimensional frost heave model based on simultaneous heat and water flux. *Cold Regions Sci. and Tech.*, **3**, 122, 253–262.

GUYMON G. L. et al. 1977. FE model of transient heat conduction with isothermal phase change. *CRREL Special Rpt 77-38*, 167pp.

GYSI H. and MADER P. 1987. Ground freezing technique: its use in extreme hydrological conditions. *Proc. Int. Conf. on Groundwater Effects in Geotechnical Engineering*, Dublin, 165–168.

HAINES F. D. et al. 1975. Strain rate effect on the strength of frozen silt. *CRREL Research Rpt 350*, Hanover NH.

HAMPTON C. N. 1986. *Strength and creep testing for artificial ground freezing*. University of Nottingham, PhD thesis, 221pp.

HAMPTON C. N. et al. 1985. Modelling the creep behaviour of frozen sands. *Proc. 3rd NSGF*, BGFS, Nottingham, 27–33.

HAMPTON C. N. et al. 1988. The time dependent response of frozen soils subject to triaxial stress. *Proc. 5th ISGF*, Nottingham, **2**, 559–560.

HARANDA K. et al. 1994. Measurement of electrical resistivity of frozen soils. *Proc. 7th ISGF*, Nancy, France, **1**, 153–156.

HARLAN R. L. 1973. Analysis of coupled heat-fluid transport in partially frozen soil. *Water Resources Research*, **9**–5, 1314–1323.

HARLAN R. L. and NIXON J. F. 1978. Ground thermal regime. In *Andersland O. B. and Anderson D. M. (eds), Geotechnical engineering for cold regions*, McGraw-Hill, 103–163.

HARRIS J. S. 1972a. The control of groundwater by freezing. *Proc. ICE*, Feb., 405–407.

HARRIS J. S. 1972b. Ground freezing seals sheet-pile leaks. *Contract J.*, 10 Feb, 1pp.

HARRIS J. S. 1974. Cryogenic treatment of shafts and tunnels. *Tunnels and Tunnelling*. Sept., 67–70.

HARRIS J. S. 1975. Freezing relieves bad ground conditions. *Underground Services*. **3**–2, 22–23.

HARRIS J. S. 1976. Ground freezing — some uses of this geotechnical process in civil engineering applications during the last decade. *Proc. C. on the Next Decade*, Jan, 4pp.

HARRIS J. S. 1979. Ground freezing for shaft sinking: a review of the technique. *Proc. Tech S. on Coal Mining*, Anhui, Oct.
HARRIS J. S. 1984b. Some applications of artificial ground freezing to foundation engineering. *Proc. 2nd NSGF Relating Theory to Practice in AGF*, BGFS, Nottingham, 1–6.
HARRIS J. S. 1985a. Freeze-hole drilling. *Proc. NS on Drilling Methods*, BDA, Sept., 53–58.
HARRIS J. S. 1985b. Optimum ice-wall construction. *Proc. 3rd NSGF*, BGFS, Nottingham, 53–58.
HARRIS J. S. 1988. State of the Art — Tunnelling using artificially frozen ground. *Proc. 5th ISGF*, Nottingham, **1**, 245–253.
HARRIS J. S. 1989. Factors affecting the choice of a geotechnical process; construction and monitoring; Case histories. *Proc. 5th NSGF Ground Freezing in Construction*, BGFS, Nottingham, 8–13, 56–62.
HARRIS J. S. 1991. Case histories: general report. *Proc. 6th ISGF*, Beijing, **2**, 473–480.
HARRIS J. S. 1993a. Ground freezing. In Thorburn S. and Littlejohn G. S. (eds), *Underpinning and Retention*, Blackie, 220–241.
HARRIS J. S. 1993b. AGF processes. *Technical Memorandum TM2*, BGFS, 2pp.
HARRIS J. S. 1993c. Site investigation for AGF works. *Technical Memorandum TM3*, BGFS, 3pp.
HARRIS J. S. 1993d. Tunnel freezings. *Technical Memorandum TM7*, BGFS, 5pp.
HARRIS J. S. 1993e. Value for money: a case history: Aberdeen. *Proc. 6th NSGF Frozen Security: Safety and Cost Effectiveness of the Ground Freezing Method*, BGFS, Nottingham, 29–34.
HARRIS J. S. and BELL M. J. 1993. Pits and shallow shaft freezings. BGFS, *Technical Memorandum TM6*, 4pp.
HARRIS J. S. and FAIR A. R. 1986. Finite element analysis of artificially frozen ground. *Proc. NC on Computer Applications in Geotechnical Engineering*, MGS Birmingham, April, 167–174.
HARRIS J. S. and NORIE E. H. 1982. Construction of two short tunnels using artificial ground freezing. *Proc. 3rd ISGF*, Hanover NH, **1**, 383–388.
HARRIS J. S. and POLLARD C. A. 1985. Some aspects of groundwater control by the ground freezing and grouting methods. *Proc. NS on Groundwater in Engineering Geology*, Sheffield, Sept., 499–519.
HARRIS J. S. and REED R. J. 1975. Ground freezing at Stirchley. *Ground Engng*, Sept., 46–48.
HARRIS J. S. and WILLS A. J. 1993. Introduction to AGF. *Technical Memorandum TM1*, BGFS, 4pp.
HARRIS J. S. and WOODHEAD F. A. 1977. Ground freezing deals with tunnel instability. *Ground Engng*, Sept., 47–48.
HARRIS J. S. and WOODHEAD F. A. 1978. Ground freezing for large diameter foundation piers. *Consulting Engineer*, Jan.†
HARVEY S. J. 1983. Ground freezing successfully applied to the construction of the Du Toits Kloof tunnel. *Proc. 1st NSGF Progress in AGF*, BGFS, Nottingham, 51–58.
HARVEY S. J. 1993. Effective control of groundwater by use of freezing. *Proc. IC on Groundwater Problems in Urban Areas*, London.
HARVEY S. J. and BELTON J. A. 1994. Ground freezing to aid construction of an effluent treatment shaft within a chemical plant. *Proc. 7th ISGF*, Nancy, France, **1**, 317–325.
HARVEY S. J. and MARTIN C. J. 1988. Construction of the Asfordby mine shafts through Bunter Sandstone by use of ground freezing. *Proc. 5th ISGF*, Nottingham, **1**, 339–348, and in *Mining Engineer*, **148**–323, 51, 53–58.
HARVEY S. J. and WILLS A. J. 1994. Cost effectiveness and risk assessment for AGF works. *Technical Memorandum TM5*, BGFS, 8pp.

HASHEMI H. E. and SLIEPCEWICH C. M. 1973. Effect of seepage stream on artificial soil freezing. *J. ASCE SM & FE Div.*, 267–289.

HAYNES F. D. 1978. Strength and deformation of frozen silt. *Proc. 3rd IC on Permafrost*, Edmonton, **1**, 655–661.

HAYNES F. D., CARBEE D. and VANPELT D. 1980. Thermal diffusivity of frozen soil. *CRREL Rpt 80-38*, 30pp.

HAYNES F. D. and KARALIUS J. A. 1977. Effect of temperature on the strength of frozen silt. *CRREL Rpt 77–3*, Feb., 27pp.

HAYNES F. D. *et al*. 1975. Strain rate effect on the strength of frozen silt. *CRREL Rpt 350*.

HEGEMANN J. 1980. A new concept for sinking freeze shafts into great depths. *Proc. 2nd ISGF*, Trondheim, **1**, 989–1000.

HEGEMANN J. 1983. Sinking in Voerde shaft (in German). *Proc. C. on Shaft Sinking and Tunnelling*, Berlin, Glückauf, **119**–20, 972–978.

HEGEMANN J. and JESSBERGER H. L. 1985. Deep frozen shaft with gliding liner system. *Proc. 4th ISGF*, Sapporo, **1**, 357–373.

HEINER A. 1972. Strength and compaction properties of frozen soil. *Swedish National Building Research Document 11*, 71pp.

HEINRICH D. *et al*. 1978. Ground freezing monitoring techniques. *Proc. 1st ISGF*, Bochum, **1**, 271–288, and in *Engng Geol.*, **13** (1979), 455–471.

HEISE F. 1929. The behaviour of clay deposits in freezing shafts. *Bergbau*, **42**, 603–607.

HEISE F. 1930. New tests on the behaviour of clay deposits in freezing shafts. *Bergbau*, **43**, 693.

HELLSTROM W. 1954. Experiences with a modified freezing method employed in the sinking of Bruno shaft. *Bergbau Tech, Berlin*, June, **4**, 312–320.

HEMERIJCKX E. 1985. Antwerp pre-metro works. *Tunnel*, Nov., 221–227.

HENDERSON J. B. and MCNAIR P. R. 1987. Design and construction of the Aberdeen sea outfall. *Tunnels and Tunnelling*, Apr., 57–58.

HENRARD J. L. and WHETTON J. T. 1927. The sinking of Londonderry Colliery, Seaham Harbour, by the freezing process. *Proc. IME*, **75**, 358–391.

HENRY K. S. 1990. Laboratory investigation of the use of geotextiles to mitigate frost heave. *CRREL Rpt 90–6*, Aug., 28pp.

HERZOG P. and HOFER A. 1980. Uniaxial creep tests on morainic material from Switzerland. *Proc. 2nd ISGF*, Trondheim, **1**, 212–222.

HERZOG P. and HOFER A. 1983. Uniaxial creep tests on a morainic material from Switzerland. *Swiss Inst. SM*, **116**, 79–87.

HIEATT M. J. and DRAPER A. R. 1985. Three Valleys: the reality of a rolling freeze. *Proc. 3rd NSGF*, BGFS, Nottingham, 45–52, and in *Civil Engng*, Nov./Dec., 33–39.

HIGASHI A. 1953. On thermal conductivity of soil, with special reference to that of frozen soil. *Proc. American Geophysical Union*, **34**–5, 737–748.

HIGHTER W. E., ALTSCHAEFFL A. G. and LOVELL C. W. 1970. Low temperature effects on the compaction and strength of a sandy clay. *US Highway Research Record 304*, 45–51.

HILL A. 1950. Shaft sinking and equipment at Calverton Colliery. *Trans. IME*, **110**, Feb., 77–97.

HIROBE R. and IKARASHI T. 1979. Unconfined compression test of frozen sandy soil. *National Research Centre Disaster Rpt 22*, 167–173.

HOEKSTRA P. 1966. Moisture movement in soils under temperature gradients with the cold-side temperature below freezing. *Water Resources Research*, **2**, 241–250.

HOEKSTRA P. 1969a. The physics and chemistry of frozen soil. *US Highway Research Record 103*, 78–90.

HOEKSTRA P. 1969b. Water movement and freezing pressures. *Soil Sci. Soc. of America*, **33**–4, 512–518.

HOEKSTRA P. and KEUNE R. 1967. Pressure effects on the conductance of frozen monymorillonite suspensions. *Clays and Clay Minerals*, **15**, 215–225, Pergamon Press.
HOEKSTRA P. *et al.* 1959. Frost heaving pressure. *CRREL Rpt 51.*
HOFFMANN D. 1942. FH Poetsch and the freezing method. *Glückauf*, **78**, *713–715.*
HOLDEN J. T. 1983a. Approximate solutions for Millers theory of secondary heave. *Proc. 4th Int. Conf. on Permafrost*, **1**, 498–503.
HOLDEN J. T. 1983b. A brief review of methods of thermal analysis. *Proc. 1st NSGF Progress in AGF*, BGFS, Nottingam, 23–30.
HOLDEN J. T. 1985. Thermal aspects and analysis. *Proc. 3rd NSGF*, Nottingham, 1–6.
HOLDEN J. T. 1991. Towards a three dimensional model of frost heave. *Proc. 6th ISGF*, Beijing, **1**, 231–236.
HOLDEN J. T. *et al.* 1980. Heat and mass flow associated with a freezing front. *Proc. 2nd ISGF*, Trondheim, **1**, 502–514, and in *Engng Geol.*, 1981, **18**, 153–164.
HOLDEN J. T. *et al.* 1985. Some developments in rigid ice model frost heave. *Proc. 4th ISGF*, Sapporo, Aug., **1**, 93–99.
HOLZMANN P. 1979. Soil freezing in tunnel construction; experience gained on major road and rail tunnel projects. *Tech. Rpt*, Holzmann GmbH, Apr. [based on Krabbe (1978)].
HORIGUCHI K. 1978. Effects of the rates of heat removal on the rate of frost heaving. *Proc. 1st ISGF*, Bochum, **1**, 25–30.
HORIGUCHI K. 1985. Determination of unfrozen water content by differential scanning calorimetry (DSC). *Proc. 4th ISGF*, Sapporo, **1**, 33–38.
HORIUCHI Y. and MAENO N. 1985. Measurements of pressures developed in freezing water after breakdown of supercooling. *Proc. 4th ISGF*, Sapporo, **2**, 69–75.
HU D. and CAO J. 1991. Ground freezing in Yangzou mine area. *Proc. 6th ISGF*, Beijing, **1**, 325–330.
HUANG J. H. 1971. Effective thermal conductivity of porous rocks. *J. Geophysical Res.*, **76**–26, 6420–6427.
HUANG J. 1993. Research of the frostheaving project survey. *J. Glaciol. Geocryology* (Chinese), **15**–3.
HUANG S. L. *et al.* 1986. Stability study of CRREL permafrost tunnel. *J. Geotech. Engng*, ASCE **112**–8, 777–790.
HUANG S. L. and SPECK R. C. 1985. The deformational behaviour of a tunnel in permafrost. *Proc. 4th ISGF*, Sapporo, **2**, 277–282.
HUANG S. L. *et al.* 1991. Effects of temperature on swelling coal shale. *Proc. 6th ISGF*, Beijing, **1**, 41–48.
HUDER J. 1979(a). Technology of frozen soil (in German). *Proc. Swiss NC on Soil and Rock Mechanics*, Zurich, June, **100**, 1–8.
HUDER J. 1979(b). Geotechnical AGF analyses. (in German).
HUDER J. 1981. Milchbuck tunnel. *Proc. 10th IC SM & FE.*
HUNEAULT P. *et al.* 1988. Effect of frozen creep on stresses at the moving thaw front around a wellbore. *Proc. 5th ISGF*, Nottingham, **1**, 279–284.
HUNTER J. R. 1957. The shaft water problem at Thorne. *Trans. IME*, **116**, Jan., 970–987.
HUTCHINSON M. T. and AULD F. A. 1989. Design of an ice-wall. *Proc. 5th NSGF Ground Freezing in Construction*, BGFS, Nottingham, 30–55.

ILYINA N. P. and SHAFARENKO Y. M. 1978. Strength analysis of frozen ground enclosure of a low temperature storage reservoir for LNG. *Proc. 1st ISGF*, Bochum, **1**, 227–234, and in *Engng Geol.*, **13**–4, 367–374.
INOUE M. and KINOSITA S. 1975. Mechanical properties of frozen soil. *Low Temp. Science Series*, **33**, 243–253.

ISAEV O. N. et al. 1991. The progress of the method of static sounding in the investigation of geotechnical properties of frozen soils. *Proc. 6th ISGF*, Beijing, **1**, 147–154.

ISGF WORKING GROUP 1 1991. Testing methods for frozen soils. *Proc. 6th ISGF*, Beijing, **2**, 493–502.

ISGF WORKING GROUP 2 1991. Frozen ground structures — basic principles of design. *Proc. 6th ISGF*, Beijing, **2**, 503–516.

ISHIZAKI T. and FUKUDA M. 1991. Frost heaving of a saturated soil under various temperature conditions and overburden pressures. *Proc. 6th ISGF*, Beijing, **2**, 521–522.

ISHIZAKI T. and NISHIO N. 1985. Experimental study of final ice-lens growth in partially frozen saturated soil. *Proc. 4th ISGF*, Sapporo, **1**, 71–78.

ISHIZAKI T. and NISHIO N. 1988. Experimental study of frost heaving of a saturated soil. *Proc. 5th ISGF*, Nottingham, **1**, 65–72.

ISHIZAKI T. et al. 1994. Temperature dependence of unfrozen water film thickness of frozen soils. *Proc. 7th ISGF*, Nancy, France, **1**, 85–89.

ISSMFE 1989. Work report 1985–1989 of the Technical Committee on Frost TC8. *IS VTT94, Frost in Engineering*, H. Rathmeyer (ed), Espoo, Finland, **1**, 15–45.

ITO H. et al. 1991. Thermal analysis of frozen soil wall during partially stopping of cooling. *Proc. 6th ISGF*, Beijing, **2**, 523–526.

IZUTA H. et al. 1988. Mechanical properties of circular slabs of frozen sand and clay in bending tests *Proc. 5th ISGF*, Nottingham, **1**, 187–192.

JACKSON D. 1989. Contractual aspects of AGF. *Proc. 5th NSGF, Ground Freezing in Construction*, Nottingham, 63–79.

JACKSON K. A. and CHALMERS B. 1958. Freezing of liquids in porous media with special reference to frost heave in soils. *J. Applied Physics*, **29**–8, 1178–1181.

JAME Y. and NORUM D. I. 1976. Heat and mass transfer in freezing unsaturated soil in a closed system. *Proc. 2nd NC on Soil-water Problems in Cold Regions*, 46–56.

JAMSON L. E. 1964. Frost penetration in sandy soil. *Proc. Royal Institute of Technology, Stockholm*, Civ. Eng. **10**–31.

JANSEN F. and GLEBE E. 1959. Shaft sinking in the West German coal mining industry. *Proc. IS on Shaft Sinking and Tunnelling*, London, 6, 27pp.

JAPAN GAS ASSOCIATION. Recommended practice for LNG inground storage. *Japanese Society of Gas*, 1036–1059.

JEFFREY R. I. 1984. Ice-wall design in shaft sinking. *Proc. 2nd NSGF Relating Theory to Practice in AGF*, BGFS, Nottingham, 29–34.

JESSBERGER H. L. 1977. Strength and time dependent deformation of artificially frozen soil. *Proc. IS on Frost Action in Soils*, Lulea, Feb., 157–167.

JESSBERGER H. L. 1980a. Theory and application of ground freezing in civil engineering. *Cold Regions Sci. and Tech.*, **3**, 3–27.

JESSBERGER H. L. 1980b. State of the Art Report: Mechanical properties, process and design in ground freezing. *Proc. 2nd ISGF*, Trondheim, **1**, 1–33.

JESSBERGER H. L. 1982. Applications of ground freezing to soil improvement in engineering practice. *Proc. S. on Recent Improvements in Ground Improvement Techniques*, Bangkok, Nov., 469–482.

JESSBERGER H. L. 1987. Artificial freezing of the ground for construction purposes. In Bell F. G. (ed), *Ground Engineer's Reference Book*, 31/1–17.

JESSBERGER H. L. 1991. Opening address. *Proc. 6th ISGF*, Beijing, **1**, 399–403.

JESSBERGER H. L. and JAGOW R. 1989. Determination of frost susceptibility of soils. *Proc. Frost 89*, Finland, 449–469.

JESSBERGER H. L. and JORDAN P. 1982. Frozen saline sand subjected to dynamic loads. *Proc. 3rd ISGF*, Hanover NH, **1**, 19−26.

JESSBERGER H. L. and KOTTHAUS M. 1991. Mechanical properties of synthetic cometary material. *Proc. 6th ISGF*, Beijing, **1**, 155−162.

JESSBERGER H. L. and MAKOWSKI E. 1980. Optimisation of the freeze-pipe arrangement and the necessary refrigeration plant capacity by a FEM program. *Proc. 2nd ISGF*, Trondheim, **2**, 43−78, and in *Engng Geol.*, **18** (1981) Elsevier, 175−188.

JESSBERGER H. L. *et al.* 1976. Study of the stress behaviour of frozen solids (in German). *Proc. 6th IC SM & FE*, Vienna, **1**, 233−239.

JESSBERGER H. L. *et al.* 1979. Shaft sinking in oil sand formations. *Proc. 7th EC SM & FE*, Brighton, Sept., **1**, 189−194.

JESSBERGER H. L. *et al.* 1981. Calculation of frozen soil structures considering the temperature dependent strength and creep behaviour of frozen soils. *Proc. IC on Numerical Methods for Coupled Problems*, Swansea.

JESSBERGER H. L. *et al.* 1985. Thermal calculation for ground freezing with liquid nitrogen. *Proc. 4th ISGF*, Sapporo, **2**, 95−101.

JESSBERGER H. L. *et al.* 1987. Frozen soil as a temporary cutoff system for underpassing the Limmat river. *Proc. NC on Groundwater Effects in Geotechnical Engineering*, Germany, 169−173.

JESSBERGER H. L. *et al.* 1988. Thermal design of a frozen soil structure for stabilisation of the soil on top of two parallel metro tunnels. *Proc. 5th ISGF*, Nottingham, **1**, 349−356.

JIAJIE W. *et al.* 1994. Thermal and mechanical characteristics of clay soil artificially frozen by liquid nitrogen. *Proc. 7th ISGF*, Nancy, France, **1**, 199−203.

JIAN G. W. 1985. Research for frost heave behaviour of Planosol. *Proc. 4th ISGF*, Sapporo, **2**, 59−62.

JILIANG S. 1980. Shaft sinking through water-bearing strata. *Proc. IS on Mine Planning and Development*. Beidine, Sept.†

JIU Y. Y. *et al.* 1991. Heat conduction analysis around cooled underground opening using the finite element method. *Proc. 6th ISGF*, Beijing, **1**, 99−104.

JOHANSEN N. I. *et al.* 1980. Sublimation and sublimation control in the CRREL tunnel. *Proc. 2nd ISGF*, Trondheim, **1**, 952−968.

JOHANSEN N. I. and RYER J. W. 1982. Permafrost creep measurements in the CRREL tunnel. *Proc. 3rd ISGF*, Hanover NH, **1**, 61−64.

JOHANSEN Ø. and FRIVIK P. E. 1980. Thermal properties of soils and rock minerals. *Proc. 2nd ISGF*, Trondheim, **1**, 427−453.

JOHANSEN P. M. *et al.* 1982. The performance of a high pressure propane storage cavern in unlined rock, Norway. *Proc. Norwegian Hard Rock Tunnelling*.

JOHNSON B. D. and KETTLE R. J. 1980. Frost heaving and hydraulic conductivity. *Proc. 2nd ISGF*, Trondheim, **1**, 735−747.

JOHNSON J. B. and ESCH D. C. 1985. Frost jacking forces on H and pipe piles embedded in Fairbanks silt. *Proc. 4th ISGF*, Sapporo, **2**, 125−133.

JOHNSON T. C. *et al.* 1986. Frost action predictive techniques: an overview of research results. *Research Record*, **1089**, 147−161.

JOHNSON T. C. *et al.* 1979. Effect of freeze−thaw cycles on resilient properties of fine grained soils. *Engng Geol.*, **13**−4, 247−276.

JOHNSTON G. H. 1981. (ed.) Engineering characteristics of freezing and thawing soils. *Permafrost Engineering Design and Construction*, John Wiley & Sons, 73−147.

JONAS A. 1941. The mechanical strength of the frost wall in shaft sinking by the freezing method and its influence on the strength of the shaft lining. *Glückauf*, **77**, 365−377, 384−388.

JONES J. S. 1980. State of the Art Report: Engineering practice in artificial ground freezing. *Proc. 2nd ISGF*, Trondheim, **1**, 837−856.

JONES J. S. and BROWN R. E. 1978a. Soft ground tunnelling by ground freezing. *Tunnelling and Underground Structures*, US Trnsptn Research Board, **684**, 28–36.

JONES J. S. and BROWN R. E. 1978b. Design of tunnel support systems using ground freezing. *Proc. 1st ISGF*, Bochum, **1**, 235–254, and in *Engng Geol.*, 1979, **13**, 375–395.

JONES J. S. and BROWN R. E. 1979. *Artificial ground freezing for tunnel construction: a case history.* US Trnsptn Research Board,

JONES J. S. and VAN ALLER H. W. 1982. Development of a methodology for predicting ground surface movement above tunnels in soft ground supported by freezing. *Proc. 3rd ISGF*, Hanover NH, **2***, 59–67.

JONES R. H. 1977 (ed). Frost heave testing and research. *Proc. C. on Frost Heave*, Univ. of Nottingham, 141pp.

JONES R. H. 1978. Frost heave damage and its prevention. In Peel P. S. (ed.), *Developments in Highway Engineering*, Applied Science Publishers, 43–77.

JONES R. H. 1980a. Discussion on design parameters for special soil conditions. *Proc. 7th EC SM & FE*, **4**, 194, 196–7, 206–7.

JONES R. H. 1980b. Developments and applications of frost susceptibility testing. *Proc. 2nd ISGF*, Trondheim, **1**, 748–759, and in *Engng Geol*, **18** (1981), 269–280.

JONES R. H. 1981. The prediction of frost heave. *Univ. of Nottingham.*†

JONES R. H. 1982. Ground movements associated with artificial ground feezing. *Proc. 3rd ISGF*, Hanover NH, **1**, 295–304.

JONES R. H. 1983 (ed). Progress in artificial ground freezing. *Proc. 1st NSGF Progress in AGF*, BGFS, Nottingham, 82pp.

JONES R. H. 1984a (ed.). Relating theory to practice in artificial ground freezing. *Proc. 2nd NSGF Relating theory to practice in AGF*, BGFS, Nottingham, 54pp.

JONES R. H. 1984b. Role of field observations in bridging the gap between theory and practice in AGF. *Proc. 2nd NSGF Relating theory to practice in AGF*, BGFS, Nottingham, 49–54.

JONES R. H. 1985. Mechanical properties of frozen ground. *Proc. 3rd NSGF*, BGFS, Nottingham, 21–26.

JONES R. H. 1989. Background, history and literature (on ground freezing); introduction to the basic properties of freezing, frozen and thawing soils. *Proc. 5th NSGF Ground Freezing in Construction*, BGFS, Nottingham, 7–13, 14–29.

JONES R. H. 1993a. Control of ground movements. *Proc. 6th NSGF Frozen security: Safety and Cost Effectiveness of the Ground Freezing Method*, BGFS, Nottingham, 3–14.

JONES R. H. 1993b. Control of ground movement in AGF works. *Technical Memorandum TMS*, BGFS, 8pp.

JONES R. H. and BABA H. U. 1994. Modelling heave-time curves from laboratory tests. *Proc. 7th ISGF*, Nancy, France, **1**, 169–174.

JONES R. H and DUDEK S. J-M. 1979. Comparison of the precise freezing cell with other facilities for frost heave testing. *Transport Research Record*, 705, 63–71.

JONES R. H. and HURT J. T. 1975. Improving the repeatability of frost heave tests. *Highways and Roads Construction*, July/Aug., **43** (1787–8), 8–13, and **43** (1791), 37 and 52.

JONES R. H. and HURT K. G. 1978. An osmotic method for determining rock and aggregate suction characteristics with applications to frost heave studies. *QJEG*, **11**–3, 245–252.

JONUSCHEIT G. P. 1978. Subway construction in Stuttgart under protection of frozen soil roof. *Proc. 1st ISGF*, Bochum, **2**, 133–136.

JOOSTEN H. 1906a. Use of shaft sinking by freezing for 2 shafts at the Government Mine B (Wilhelmina) at Limburg, Netherlands. *Glückauf*.

JOOSTEN H. 1927. Deep freezing process for shaft sinking. *Glückauf*, 577.

JORDAN P. and JAGOW-KLAFF R. 1991. Control and prediction of the temperature in a clay formation during construction of a deep frozen shaft in Germany. *Proc. 6th ISGF*, Beijing, **2**, 455–462.

JORDAN P. et al. 1994. Use of artificial ground freezing in sections of the Dusseldorf subway. *Proc. 7th ISGF*, Nancy, France, **1**, 327–339.

JØSANG T. 1983. Ground freezing techniques used for tunnelling in Oslo city centre. *Proc. Norwegian Soil and Rock Engng Assn*, 39–44.

JUMIKIS A. R. 1957. The effect of freezing on a capillary meniscus. *US Highway Research Record*, 168.

JUMIKIS A. R. 1966. *Thermal soil mechanics*. Rutgers University Press.

JUMIKIS A. R. 1978a. Cryogenic texture and strength aspects of artificially frozen soils. *Proc. 1st ISGF*, Bochum, **1**, 75–86.

JUMIKIS A. R. 1978b. Some aspects of artificial thawing of frozen soils. *Proc. 1st ISGF*, **1**, 183–192.

JUMIKIS A. R. 1980. Thermal modelling of freezing soil systems. *Proc. 2nd ISGF*, Trondheim, **1**, 470–483.

JUMIKIS A. R. 1982a. Thermal modelling of freeze–thaw depths in soils. *Proc. 3rd ISGF*, Hanover NH, **1**, 213–216.

JUMIKIS A. R. 1982b. Modelling of influx of groundwater into excavation supported by an artificially frozen soil wall. *Proc. 3rd ISGF*, Hanover NH, **2***, 69–76.

JUNGST 1907. Project of the Klein Rossein Mine near Stieringen (Lorraine). *Glückauf*, 623.

KANG S. et al. 1994. Field observation of solute migration in freezing and thawing soils. *Proc. 7th ISGF*, Nancy, France, **1**, 397–398.

KAPLAR C. W. 1969. Laboratory determination of dynamic moduli of frozen soils and ice. *CRREL Research Rpt 163*, Jan., 20pp.

KAPLAR C. W. 1971. Some strength properties of frozen soil and effect of loading rate. *CRREL Rpt 159*, 20pp.

KARLOV W. D. 1978. The research of the frost heave on non-water-saturated loamy soil in field conditions. *Proc. 1st ISGF*, Bochum, **1**, 13–24.

KARLOV V. D. and ARAFYEV S. V. 1988. On the regularities of the change of shear strength of soils with thawing and dynamic loadings. *Proc. 5th ISGF*, Nottingham, **1**, 193–196.

KARLSSON E. G. and WINGBRO N. T. 1972. A numerical method for calculation of basic data for earth stabilisation by freezing with liquid nitrogen. *Proc. ICEC4, 4pp*.

KARPOV V. M. and VELLI Y. Y. 1968. Displacement resistance of frozen saline soils. *SM & FE*, **4**, 277–279.

KATAOKA T. et al. 1980. Mechanical properties of frozen soils (in Japanese). *Proc. 15th Jap. NC SM & FE*, 661–664.

KATSUYUKI S. and TAKASHI M. 1985. Frost heaving of volcanic ash soils. *Proc. 4th ISGF*, Sapporo, **2**, 163–169.

KAY B. D. and GROENEVELT P. H. 1974. On the interaction of water and heat transport in frozen and unfrozen soils. *Proc. Soil Sci. Soc. of America*, **38**, 395–400.

KAY B. D. and GROENEVELT P. H. 1983. The redistribution of solutes in freezing soil: exclusion of solutes. *Proc. 4th IC on Permafrost*, Fairbanks, 584–588.

KAY B. D. and PERFECT E. 1988. State of the Art: Heat and mass transfer in freezing soils. *Proc. 5th ISGF*, Nottingham, **1**, 3–21.

KELLAND J. D. and BLACK J. C. 1969. Cominco's Saskatchewan potash shafts. *Proc. 9th Commonwealth C. Mining and Metallurgy* (IMM) London, May, **35**, 20pp.

KELSH D. J. and TAYLOR S. 1988. Measurement and interpretation of electrical freezing potential of soil. *CRREL Rpt 88–10*.

KENEDY G. F. and LIELMEZS J. 1973. Heat and mass transfer of freezing water-soil system. *Water Resources Research*, **9**-2, 395–400.

KENT D. D. *et al.* 1975. Variables controlling behaviour of a partly frozen saturated soil. *Proc. C. Soil-water Problems in Cold Regions*, Calgary, May, 70–88.

KEROLA P. *et al.* 1981. Freezing in the mining industry (in Finnish). *Bergshant*, **39**–2, 94–96.

KERSTEN M. S. 1949. Laboratory research for the determination of the thermal properties of soils. *ACFEL Tech Rpt 23*, and in *Univ. of Minnesota Engineering Exptl Stn*, Bulletin 28.

KERSTEN M. S. 1963. Thermal properties of frozen ground. *Proc. 1st IC on Permafrost*, Lafayette, 301–305.

KERSTEN M. S. and COX A. E. 1951. The effect of temperature on the bearing value of frozen soils. *US Highway Research Board Bulletin*, **40**, 32–38.

KESSURU Z. *et al.* 1987. Simultaneous modelling of rock freezing and water seepage and its practical applications. *I. J. Mine Water*, Hungary, **6**, 1–32.

KESTLER M. A. and BERG R. L. 1991. Use of insulation for frost prevention. *CRREL Rpt 91–1*, Jan.

KETTLE R. J. and MCCABE E. Y. 1985. Mechanical stabilisation for the control of frost heave. *Cndn J. Civil Engng*, **12**–3, 899–905.

KETTLE R. J. and WILLAMS R. T. 1976. Frost heave and heaving pressure measurements in colliery shales. *Cndn Geotech. J.*, **14**, 127–138.

KHAKIMOV K. R. 1957. Problems in the theory and practice of artificial freezing of soil. *Academy of Sciences, Moscow*, 121pp.

KHAKIMOV K. R. 1966. Artificial freezing of soils — theory and practice. *Transl from Russian by A Barouch*, US Dept of Interior.†

KHAZIN B. G. and GONCHAROV B. V. 1974. Use of ultrasound to estimate the strength of frozen soils during working. *SM & FE*, **11**–2, 122–125.

KING M. S. 1977. Acoustic velocities and electrical properties of frozen sandstones and shales. *Cndn J. Earth Sci.*, **14**–5, 1004–1013.

KING M. S. and GARG O.P. 1982. Compressive strengths and dynamic elastic properties of frozen and unfrozen ore from Northern Quebec. *Proc. 4th Cndn NC on Permafrost* (1981) (R. J. E. Brown Mem. Vol.), Calgary, 374–381.

KING M. S. *et al.* 1974. Ultrasonic velocity measurements on frozen rocks and soils. *Proc. NS Permafrost Hydrology and Geophysics*, Calgary, 8pp.

KINOSITA S. 1973. Creep property of frozen soil. *Low Temp. Sci.*, **31**, 261–269.

KINOSITA S. 1978. Effects of initial water conditions on frost heaving characters. *Proc. 1st ISGF*, Bochum, **1**, 3–12.

KINOSITA S. *et al.* 1982. Frost action freezing ground surrounding underground storage of a cold liquid. *Proc. 3rd ISGF*, Hanover NH, **1**, 305–310.

KINOSITA S. and ISHIZAKI T. 1980. Freezing point depression in moist soil. *Proc. 2nd ISGF*, Trondheim, **1**, 640–646.

KIRIBAYASHI E. *et al.* 1985. Stress–strain characteristics of an artificially frozen sand in uniaxially compressive tests. *Proc. 4th ISGF*, Sapporo, **2**, 177–182.

KIRIYAMA S. *et al.* 1980. Artificial ground freezing in shield work. *Proc. 2nd ISGF*, Tronheim, **1**, 940–951.

KIVEKAS L. *et al.* 1986. Brittleness of reinforced concrete structures under arctic conditions. *CRREL Rpt 86–2*, 42pp.

KLECZEK Z. 1971. Stress and strain of frozen rocks in shaft surroundings as a function of time. *Univ. of Krakow Mining Bulletin*, **37**.

KLEIN J. 1978. Time and temperature dependent stress–strain behaviour for frozen Emscher-marl (in German). *Ruhr University Rpt*, Dec., Series G.

KLEIN J. 1979. The application of finite elements to creep problems in ground freezing. *Proc. 3rd Int. Conf. on Numerical Methods in Geomechanics*, Aachen, Apr., 493–502.

KLEIN J. 1980a. Calculating the strength of frost walls in freezing shaft construction (in German). *Shaft construction study group*, Bochum, 14pp.
KLEIN J. 1980b. Influence of temperature gradient on steady state creep in freeze shaft design. *Proc. 21st S. on Rock Mechanics*, May, 192–196.
KLEIN J. 1980c. Structural design of freeze shafts in frictionless clay formations taking account of the time factor (in German), *Glückauf*, **41**, 51–56.
KLEIN J. 1980d. Compromose cone — a useful form of isotropic yield surface for freeze shaft design. *Proc. 2nd ISGF*, Trondheim, **1**, 1014–1024.
KLEIN J. 1981a. Dimensioning of ice-walls around deep-freeze shafts sunk through sand formations of type B=2, under consideration of time (in German). *Glückauf Forschungsh*, **42**–3, June, 112–120.
KLEIN J. 1981b. Finite element method for time-dependent problems of frozen soils. *I. J. for Numerical Methods in Geomechanics*, **5**, 263–283.
KLEIN J. 1982. Present state of freeze shaft design in mining. *Proc. S. on Strata Mechanics*, Newcastle upon Tyne, 147–153.
KLEIN J. 1985a. *Handbuch des Gefrierschachbaus im Bergbau*, Glückauf.†
KLEIN J. 1985b. Influence of friction angle on stress distribution and deformational behaviour of freeze shafts in nonlinear creeping strata. *Proc. 4th ISGF*, Sapporo, **2**, 307–315.
KLEIN J. 1988. State of the Art: Engineering design of shafts. *Proc. 5th ISGF*, Nottingham, **1**, 235–244.
KLEIN J. 1991. Minimum ice-wall thickness in freeze-shaft design. *Proc. 6th ISGF*, Beijing, **2**, 527–528.
KLEIN J. and JESSBERGER H. L. 1978. Creep stress analysis of frozen soils under multiaxial states of stress. *Proc. 1st ISGF*, Bochum, **1**, 217–226, and in *Engng Geol.* (1979), **13**, 353–365.
KLEMENTOVICH V. V. and MINENKO V. I. 1950. Use of expanding cement for waterproof cast-iron linings when sinking frozen shafts (in Russian). *Ugol*, Moscow, Nov., 30–31.
KLUBER T. 1981. Potential applications for soil freezing in tunnel construction. *Proc. Tunnel 81*, Essen, **1**, 105–124.
KNUTSSON S. 1981. Shear strength of frozen soils. *Proc. 10th IC SM & FE*, Stockholm, **3**, 731–732.
KNUTSSON S. *et al.* 1985. Analysis of large scale laboratory and in situ frost heave tests. *Proc. 4th ISGF*, Sapporo, **1**, 65–70.
KONRAD J. M. 1987. Procedure for determining the segregation potential of freezing soils. *Geotech. Testing J.*, June, **10**–2, 51–58.
KONRAD J. M. and MCCAMMON A. W. 1990. Soluble partitioning in freezing soils. *Cndn Geotech. J.*, **27**–6, 726–736.
KONRAD J. M. and MORGENSTERN N. R. 1980. A mechanistic theory of ice lens formation in fine grained soils. *Cndn Geotech. J.*, **17**–4, 473–486.
KONRAD J. M. and MORGENSTERN N. R. 1981. The segregation potential of a freezing soil. *Cndn Geotech. J.*, **18**, 482–491.
KONRAD J. M. and MORGENSTERN N. R. 1982. Effects of applied pressure on freezing soils. *Cndn Geotech. J.*, **19**, 494–505.
KONRAD J. M. and MORGENSTERN N. R. 1984. Frost heave prediction of chilled pipelines buried in unfrozen soils. *Cndn Geotech. J.*, **21**, 100–115.
KONRAD J.-M. and SHEN M. Simulation of retaining wall displacement by frost action using the segregation potential approach. *Proc. 7th ISGF*, Nancy, France, **1**, 265–270.
KONZ P. *et al.* 1978. Railway tunnel, Born (Switzerland). *Consulting Eng.*, Sept., 42–43.
KOOPMANS D. and MILLER R. D. 1966. Soil freezing and soil water characteristic curves. *Proc. Soil Sci. Soc. of America*, **30**, 680–685.
KOVARI K. *et al.* 1979. New developments in the instrumentation of underground openings. *Proc. 4th RETC*, Atlanta, June, 817–837.

KRABBE W. 1978. Practical experience using ground freezing process in large traffic tunnels (in German). *Vortrag baugrundtag*, Berlin, Sept.†

KREKLER H. 1919. Value of low temperature process for shaft sinking. *Glückauf*, **55**, 589–597.

KRONIK Y. A. 1982. Thermomechanical enthalpy model for ground freezing design. *Proc. 3rd ISGF*, Hanover NH, **1**, 167–176.

KRZEWINSKI T. G. and TART R. G. Jr 1985. Thermal design considerations in frozen ground engineering — State of the practice report. *Proc. ASCE.*†

KUBO H. et al. 1985. Estimating method in freezing index. *Proc. 4th ISGF*, Sapporo, **2**, 103–108.

KUJALA K. 1989. Frost action and the mechanical properties of an artificially frozen test plot. *Proc. 12th IC SM & FE*, Brazil, **2**, 1449–1454.

KUJALA K. 1991. Assessment of frost susceptibility of soils. *Proc. 6th ISGF*, Beijing, **1**, 49–54.

KUNIEDA T. et al. 1991. Numerical case studies of ground freezing for the construction of drain pump chambers. *Proc. 6th ISGF*, Beijing, **1**, 237–244.

KURFURST P. J. and KING M. S. 1972. Static and dynamic elastic properties of two sandstones at permafrost temperatures. *J. Petroleum Tech.*, Apr., 495–504.

KURFURST P. J. and PULLAN S. 1985. Field and laboratory measurements of seismic and mechanical properties of frozen ground. *Proc. 4th ISGF*, Sapporo, **1**, 255–264.

KURFURST P. J. and PULLAN S. E. 1988. Acoustic properties of frozen near-shore sediments, Southern Beaufort Sea. *Proc. 5th ISGF*, Nottingham, **1**, 197–204.

KURODA T. 1985. Theoretical study of frost heaving — kinetic process at water layer between ice lens and soil particles. *Proc. 4th ISGF*, Sapporo, **1**, 39–46.

KUTTER, W. 1965. The sinking of shaft #3, Baden division potash mines. *Kali Steinsalz, Hanover*, Feb., **4**–4, 117–126.

KUTVITSKAYA N. B. 1994. Ground freezing for using foundation structures. *Proc. 7th ISGF*, Nancy, France, **1**, 245–248.

LABA J. T. 1971. Viscoelastic properties of a laterally confined sand-ice system subjected to temperature increase. *US Highway Research Record*, 360, 26–36.

LABA J. T. 1974. Adfreezing of sands to concrete. *US Trnsptn Research Board*, **497**, 31–39.

LABA J. T. 1975. Forces exerted on rigid retaining walls by a confined frozen soil layer. *Proc. 4th SE Asian C. Soil Engineering*, Malaya, Apr., 5–18.

LABA J. T. and AZIZ K. A. 1972. Pressure–time relationship in laterally stressed frozen granular soils. *US Highway Research Board*, **393**, 79–87.

LACY H. S. and FLOESS C. H. 1990. Minimum requirements for temporary support with artificially frozen ground. *Cndn Trnsptn Research Record*, **1190**, 46–56.

LACY H. S. et al. 1982. A case history of a tunnel constructed by ground freezing. *Proc. 3rd ISGF*, Hanover NH, **1**, 389–396.

LADANYI B. 1972. An engineering theory of creep of frozen soils. *Cndn Geotech. J.*, **9**–1, 63–80.

LADANYI B. 1973. Evaluation of in situ creep properties of frozen soils with the pressuremeter. *Proc. 2nd IC on Permafrost*, Yakutsk, July 310–318.

LADANYI B. 1975. Bearing capacity of strip footings in frozen soil. *Cndn Geotech. J.*, **12**–3, 393–407.

LADANYI B. 1977. Field pressuremeter and penetrometer testing in frozen soils. *Proc. S. Permafrost Field Methods and Permafrost Geophysics*, Saskatoon, NRRC Tech memo **124**, 31–42.

LADANYI B. 1980. Stress and strain rate controlled borehole dilatometer tests in permafrost. *Proc. Wksp on Permafrost Engineering*, Quebec, NRCC Tech. memo ACGR 130, 57−69.

LADANYI B. 1981a. Mechanical behaviour of frozen soils. *Proc. IS on Mechanical Behaviour of Structured Media*, Ottawa, **B**, 205−245.

LADANYI B. 1981b. State of the Art: Determination of creep settlement of shallow foundations in permafrost. *Proc. NC ASCE*, St Louis MI.

LADANYI B. 1982a. Determination of geotechnical parameters of frozen soils by means of the cone penetration test. *Proc. 2nd EurS on Penetration Testing*, Amsterdam, May, 671−678.

LADANYI B. 1982b. Ground pressure development on artificially frozen soil cylinder in shaft sinking. *Special Vol: Prof De Beer: Inst Geotech Brussels*, 21pp.

LADANYI B. 1982c. Borehole creep and relaxation tests in ice-rich permafrost. *Proc. 4th Cndn Conf. on Permafrost* (1981) (R. J. E. Brown Mem. Vol.), 406−415.

LADANYI B. 1983. Shallow foundations on frozen soil: Creep settlement. *Proc. ASCE*, **109**−11, 1434−1448.

LADANYI B. 1984. Tunnel lining design in a creeping rock. *Proc. ISRM/BGS C. on Design and Performance of Underground Excavations*, 18−26.

LADANYI B. 1991. Pressure variation on a wellbore casing during permafrost thawing. *Proc. 6th ISGF*, Beijing, **1**, 245−250.

LADANYI B. 1992a. Theory and practice of ground freezing. *Proc.* 7−22, Balkema.

LADANYI B. 1992b. Design of shaft linings in frozen ground. *Proc. 2nd IS Mining in the Arctic*, Fairbanks, Alaska, 51−60.

LADANYI B. 1993. Remoulded test specimen preparation. (U)

LADANYI B. 1994 (in press). Frozen soil-structure interfaces. Chapter in *Mechanics of material interfaces*, Elsevier.

LADANYI B. and ARTEAU J. 1978. Effect of specimen shape on creep response of a frozen sand. *Proc. 1st ISGF*, Bochum, **1**, 141−154, and in *Engng Geol.*, **13** (1979), 207−222.

LADANYI B. and BENYAMINA M. B. 1993. Triaxial relaxation testing of a frozen sand. *46th Cndn Geotech Conference*, preprints, 319−332.

LADANYI B. and JOHNSTON G. H. 1972. In situ testing of frozen soils. *Northern Engineer*, **4**−1, 6−8.

LADANYI B. and JOHNSTON G. H. 1973. Evaluation of in situ creep properties of frozen soils with the pressuremeter. *Proc. 2nd Int. Conf. on Permafrost*, Yakutsk, 310−318.

LADANYI B. and JOHNSTON G. H. 1978. Field investigations of frozen ground. In Andersland O. B. and Anderson D. M. (eds), *Geotechnical Engineering for Cold Regions*, 459−504.

LADANYI B. and MU SHEN 1989. Mechanics of freezing and thawing in soils. *Frost in Geotechnical Engineering* (ed. Rathmeyer H.), IS VTT94, Espoo, **1**, 73−103.

LADANYI B. and MU SHEN 1993. Freezing pressure development on a buried chilled pipeline. *Proc. 2nd IS on Frost and Geotech Engng*, Anchorage, 23−33.

LADANYI B. and PAQUIN J. 1978. Creep behaviour of frozen sand under a deep circular load. *Proc. 3rd IC on Permafrost*, Edmonton, July, 7pp.

LADANYI B. and SGAOULA J. 1992. Sharp cone testing of creep properties of frozen sand. *Cndn Geotech Jn.*, **29**, 757−764.

LADE P. V. *et al.* 1980. Stress−strain and volumetric behaviour of frozen soil. *Proc. 2nd ISGF*, Trondheim, **1**, 48−64.

LAHAV N. and ANDERSON D. M. 1973. Montmorillonite-benzidine reactions in the frozen and dry states. *Clay and Clay Minerals*, **21**, 137−139.

LAKE L. M. and NORIE E. H. 1982. Application of horizontal freezing in tunnel construction — two case histories. *Proc. Tunnelling 82*, IMM, London, 283−289.

LAMÉ and CLAPEYRON 1833. Mémoire sur l'equibre intérieur des corps solides homogènes. *Mém divers savans*, **4**.

LANDGRAEBER W. 1926. Progress in the use of refrigeration in mining. *Angew Chemie*, **39**, 816–822.

LANDGRAEBER W. 1949. Innovations in the application of artificial freezing in mining. *Bergbau Rdsch*, Oct., 205–207.

LARSON D. B. et al. 1973. Shock wave studies of ice and two frozen soils. *Proc. 2nd IC on Permafrost*, Yakutsk, 318–325.

LATZ J. E. 1952. Freezing method solves problem in Carlsbad shaft, New Mexico. *Mining Engng*, Oct., **4**–10, 942–947.

LAW G. J. 1960. Soil freezing to construct a railway tunnel. *Proc. ASCE*, 2639.†

LAW K. T. 1977. Design of a loading platen for testing ice and frozen soil. *Cndn Geotech. J.*, **14**–2, 266–271.

LAW K. T. 1978. Analysis of uniaxial loading on frozen soil and ice. *NRCC Tech Rpt*.

LEBRET P. et al. 1994. Modélisation de la profondeur du pergélisol au cours du dernier cycle glaciaire en France (in French) *Proc. 7th ISGF*, Nancy, France, **1**, 415–416.

LEBRETON F. 1885. Mémoire sur la méthode de congélation de M. Poetsch. *Annals des Minrs (Paris)*, **7**–8, 111pp.

LECLAIR P. et al. 1994. Ondes élastiques dans les milieu poreux soumis au gel — application à l'étude des sols et des roches en conditions froides (in French). *Proc. 7th ISGF*, Nancy, France, **1**, 175–179.

LEE J. 1969. Escalator tunnel at Tottenham Hale Station, Victoria Line. *Proc. ICE*, 7270S, 423–429.

LENK K. 1944. Simplification of the freezing process in foundation construction. *Die Bautechnik*, **33**–36, 418–420.

LENZINI P. A. and BRISS B. 1975. Ground stabilisation: Review of grouting and freezing for underground openings. *Dept of Trnsptn*, Washington, FRA-ORD, D-75-95, 86pp.

LEROUEIL S. et al. 1991. Effects of frost on the mechanical behaviour of Champlain Sea clays. *Cndn Geotech. J.*, **28**, 690–697.

LEVALLOIS J. 1989. Freezing by liquid nitrogen for alluvium consolidation in tunnels. *Proc. 12th IC SM & FE*, Brazil, **2**, 1455–1456.

LEWIS R. W. and BASS B. R. 1976. The determination of stresses and temperatures in cooling bodies by finite elements. *ASME J. Heat Transfer*, Jan., **76**-HT-YY.

LEWIS R. W. and MORGAN K. (eds) 1979. Numerical methods in thermal problems. *Proc. 1st IC on Numerical Methods in Thermal Problems, Swansea*, Pineridge Press.†

LEWIS R. W. and SZE W. K. 1988. A finite element simulation of frost heave in soils. *Proc. 5th ISGF*, Nottingham, **1**, 73–80.

LI J. C. 1979. Dynamic properties of frozen granular soils. Michigan Univ., *PhD thesis*, 335pp.

LI J. C. and ANDERSLAND O. B. 1980. Creep behaviour of frozen sand under cyclic loading conditions. *Proc. 2nd ISGF*, Trondheim, **1**, 223–234.

LI J. C. et al. 1978. Cyclic triaxial tests on frozen sand. *Proc. 1st ISGF*, Bochum, **1**, 157–172.

LIAN L. and SHI J. 1993. Study of the influence of moisture migration on temperature field during freezing process of pavement proper in cold regions. *J. Glaciol. Geocryology* (Chinese), **15**–3.

LIANDE F. 1986. Applications of the FE method to the problem of heat transfer in a freezing shaft wall. *CRREL 86-08*, 24pp.

LIBERMAN Y. M. 1960. Method of calculation of the thickness of the ice-soil cylinder wall. *Mining Inst, USSR Academy of Science*.

LINNAN Z. 1980. Analysis of the temperature fields of the artificial frozen wall of the deep shaft. *Proc. 2nd ISGF*, Trondheim, **1**, 535–544.

Liu H. 1982. Freezing rate and frost heave of soils. *Proc. 3rd ISGF*, Hanover NH, **1**, 255–260.

Liu J. C. and Andersland O. B. 1980. Creep behaviour of frozen sand under cyclic loading conditions. *Proc. 2nd ISGF*, Trondheim, **1**, 223–234.

Liu M. and Lu J. 1991. Directional drilling and its application. *Proc. 6th ISGF*, Beijing, **1**, 331–336.

Livet J. 1994a. Caractérisation de le sensibilité au gel des sols utilisés en technique de chaussés (in French). *Proc. 7th ISGF*, Nancy, France, **1**, 219–223.

Livet J. 1994b. L'amélioration de la tenue au gel des sols fins ou grenus (in French). *Proc. 7th ISGF*, Nancy, France, **1**, 249–254.

Loch J. P. G. 1980. State of the Art report: Frost action in soils. *Proc. 2nd ISGF*, Trondheim, **1** 581–596.

Loch J. P. G. and Kay B. D. 1978. Water distribution in partially frozen saturated silt under several temperature gradients and overburden loads. *Soil Sci. Soc. of America J.* **42**–3, 400–406.

Loch J. P. G. and Miller R. D. 1975 Tests of the concept of secondary frost heaving. *Soil Sci. Soc. of America J.* **39**–6, 1036–1041.

Lomas K. J. and Jones R. H. 1981. An evaluation of a self refrigerated unit for frost heave testing. *Trnsptn Research Record*, **809**, 6–13.

Losj I. F. *et al.* 1982. Employment of rock freezing techniques for sinking deep mine shafts. *Proc. 3rd ISGF*, Hanover NH, **1**, 337–342.

Lou G. 1991. A finite element mechanical model for shaft freeze wall. *Proc. 6th ISGF*, Beijing, **1**, 251–256.

Lovell C. W. 1957. Temperature effects on phase composition and strength of a partially frozen soil. *Highway Research Record*, **168**, 74–95.

Lovell C. W. and Herrin M. 1953. Review of certain properties and problems of frozen ground, including permafrost. *SIPRE Tech Rpt 9*.

Low G. J. 1960. Soil freezing to reconstruct a railway tunnel. *J. ASCE Construction Div.*, **86**–CO3, 1–12.

Low P. F. *et al.* 1968. Some thermodynamic relationships for soils at or below the freezing point: freezing point depression and heat capacity. *Water Resources Research*, **4**–2, 379–394.

Lowe-Brown W. L. 1933. Shaft sinking by freezing (Swansea). *The Engineer*, **155**, 26 May, 516–519, 526.

Lunardini V. J. 1980a. Neumann solution applied to soil systems. *CRREL Rpt 80-22*, 7pp.

Lunardini V. J. 1980b. Phase change around a circular pipe. *CRREL Rpt 80-27*, 18pp.

Lunardini V. J. 1982. Freezing of soil with surface convection. *Proc. 3rd ISGF*, Hanover NH, **1**, 205–212.

Lunardini V. J. 1987. Exact solution for melting of frozen soil with thaw consolidation. *IS Offshore Mech. and Arctic Engng*, Houston, **4**, 97–102.

Lunardini V. J. 1988. Freezing of soil with an unfrozen water content and variable thermal properties. *CRREL Rpt 88-2*, 23pp.

Lunardini V. J. 1991. *Heat transfer with freezing and thawing*. Elsevier, 437pp.

Lunardini V. J. 1994. Heterogenic and synegetic growth of permafrost. *Proc. 7th ISGF*, Nancy, France, **1**, 361–373.

Lutgendorf H. O. 1986. Design principles of sliding shaft linings for freeze shafts. *Glückauf*, **122**–19, 351–354.

Ma W. and Wu Z. 1991. Elastoplastic calculation of bottom heave in artificially frozen shaft. *Proc. 6th ISGF*, Beijing, **1**, 257–262.

Ma Y. and Wang S. 1985. Shaft sinking in waterbearing non-competent strata in China. *Proc. NC on Shaft Sinking and Tunnelling, Glückauf*, **121**, 1434–1437.

MACFARLANE I. C. 1970. Strength and deformation tests on frozen peat. *NRCC 11340*, 143–149.

MACFARLANE I. M. *et al.* 1988. Application of ground freezing at Rogers Pass, Canada. *Proc. 5th ISGF*, Nottingham, **2**, 563–564.

MAHAR L. J. *et al.* 1982. Effects of salinity on freezing granular soils. *Proc. 3rd ISGF*, Hanover NH, **2***, 77–82.

MAIDL B. *et al.* 1978. Developments in ground freezing techniques. *Tunnels and Tunnelling*, May, 25–26.

MAISHMAN D. 1959. Shaft sinking using the freezing process — Kellingley Colliery. *Iron and Steel Trades Review*, 30 Oct., 707–715.

MAISHMAN D. 1975. Ground freezing. In Bell F. G. (ed) *Methods of treating unstable ground*, Newnes-Butterworth, 159–171.

MAISHMAN D. 1978. Freezing of soils to facilitate construction. *Proc. NS on Improving Poor Soil Conditions*, ASCE, Oct.

MAISHMAN D. 1988. A short tunnel in Seattle frozen using liquid nitrogen cascades. *Proc. 5th ISGF*, Nottingham, **2**, 561–562.

MAISHMAN D. and POWERS J. P. 1982. Ground freezing in tunnels — 3 unusual applications. *Proc. 3rd ISGF*, Hanover NH, **1**, 397–410.

MAISHMAN D. *et al.* 1988. Freezing a temporary roadway for transport of a 3000 ton dragline. *Proc. 5th ISGF*, Nottingham, **1**, 357–366.

MAKELA H. 1979. The laboratory tests on frozen soil of the Kluuvi Cleft. *Finland Geotech. Inst.*†

MAKOWSKI E. 1982. Modelling the interaction between soil freezing system and thermal regime in soils using FEM. *Proc. 3rd ISGF*, Hanover NH, **1**, 157–166.

MAKOWSKI E. and JESSBERGER H. L. 1982. Analysis of heat flow in artificially frozen soils. *Proc. 4th IC on Numerical Methods in Geomechanics*, Edmonton, Alberta, 1211–1220.

MALOWSZEWSKI J. *et al.* 1987. New shaft sinking technologies with low depth strata freezing. *Proc. 13th World Mining Congress*, Stockholm, June, **2**, 959–967.

MALYSHEV M. A. 1973. Deformation of clays during freezing and thawing. *CRREL Rpt TL388*, 9pp.

MANKOVSKY G. I. 1959. Theoretical investigations into the rock freezing process. *Proc. NS on Shaft Sinking*, London, July, 439–456.

MANNING G. P. 1972. Foundations on frozen sites. *Design and construction of foundations*, **17**, 236–263.

MANNING G. P. 1973. In-ground storage of LNG and its effect on the surrounding ground. *Underground Services*, June, 15–23.

MARBACH G. 1934. Stresses in shafts and lining methods for shafts. *Glückauf*, **70**, 321–329.

MARBACH G. 1940. Advantages and disadvantages of pulling freezing pipes after shaft sinking (in German). *Glückauf*, **76**–19, 268–270.

MARCHINA A. R. 1984. *The influence of specimen preparation on the response of rocks and aggregates to freezing*, University of Nottingham, MSc thesis, 181pp.

MARLIN C. and DEVER L. 1994. Geochemical study of carbonate precipitation by soil freezing process in natural environment (Brögger peninsula, Spitsbergen). *Proc. 7th ISGF*, Nancy, France, **1**, 181–187.

MARTAK L. V. 1988. Ground freezing in non-saturated soil conditions using liquid nitrogen. *Proc. 5th ISGF, Nottingham*, **1**, 367–376.

MARTIN R. T. *et al.* 1981. Creep behaviour of frozen sand. *CRREL Rpt*, 238pp.

MATSUOKA N. 1990. Rate of bedrock shattering by frost action: field measurements and predictive model. *Earth Surface Process Land*, Feb., **15**–1, 73–79.

MCCABE E. Y. and KETTLE R. J. 1982. Heaving pressures and frost susceptibility. *Proc. 3rd ISGF*, Hanover NH, **1**, 285–294.

McCabe E. Y. and Kettle R. J. 1983. Frost heave and overburden pressure. *Proc. 1st NSGF Progress in AGF*, BGFS, Nottingham, 31–40.

McCabe E. Y. and Kettle R. J. 1985. Thermal aspects of frost action. *Proc. 4th ISGF*, Sapporo, **1**, 47–54.

McCormick G. 1990. Soil temperatures and freezing indices at depth. *Cndn Geotech. J.*, **27**–6, 749–751.

McGaw R. W. Thermal diffusivity of Missouri clay calculations from measurements of thermal conductivity and heat capacity. *CRREL Rpt.*†

McRoberts E. C. 1975. Field observations of thawing soils. *Cndn Geotech. J.*, **12**.†

McRoberts E. C. 1975. Some aspects of a simple secondary creep model for deformations in permafrost slopes. *Cndn Geotech. J.*, **12**–1, 98–105.

McRoberts E. C. 1978. Creep tests on undisturbed ice-rich silt. *Proc. 3rd IC on Permafrost*, Edmonton, July, 6pp.

Megaw T. M. and Bartlett J. V. 1981. Ground treatment (by freezing). *Ch 5 in Tunnels — planning, design and construction*, **2**, 112–114.

Meissner H. and Eckhardt H. 1976. Deflection of frozen soil beams under constant temperature gradient (in German). *Proc. 6th IC SM & FE*, Vienna, Mar., **1/3**–9, 251–256.

Meissner H. 1985. Bearing behaviour of frost shells in the construction of tunnels. *Proc. 4th ISGF*, Sapporo, **2**, 37–45.

Meissner H. 1988a. Tunnel displacements under freezing and thawing conditions. *Proc. 5th ISGF*, Nottingham, **1**, 285–294.

Meissner H. 1988b. Ground movements by tunnel driving in the protection of frozen shells. *Proc. 6th IC on Numerical Methods in Geomechanics*, Austria.†

Meissner H. and Kroh H. 1994. Plastic and viscous potential of frozen sand. *Proc. 7th ISGF*, Nancy, France, **1**, 181–187.

Meissner H. and Vogt J. 1991. Tunnel construction in the protection of a frost shell in partially saturated soil. *Proc. 6th ISGF*, Beijing, **1**, 337–344.

Meister L. A. and Melnikov 1940. Determination of adfreezing strength of wood and concrete to ground and shear strength of frozen ground under field conditions (in Russian). *Academy of Sciences, Moscow*, Trudy, **10**, 85–107 (Transl Mandel W. Steffansson Library, NY).

Mellor M. 1971. Strength and deformability of rocks at low temperatures. *CRREL Research Rpt 294*, 75pp.

Mellor M. 1973. Mechanical properties of rocks at low temperatures. *Proc. 2nd IC on Permafrost*, Yakutsk, 334–343.

Mellor M. 1974. Cutting ice with continuous jets. *Proc. 2nd IS on Jet Cutting Technology*, G5.†

Mellor M. 1981. Mechanics of cutting and boring. *CRREL Rpt 81-26*, 38pp.

Mellor M. and Cole D. M. 1982. Deformation and failure of ice under constant stress or constant strain rate. *Cold Regions Sci. and Tech.*, **5**, 201–219.

Melnikov P. I. 1985. The ground frost regime regulation at the base of above-mine buildings. *Proc. 4th ISGF*, Sapporo, **2**, 335–340.

Melnikov P. I. *et al.* 1980. Engineering-physical bases of temperature regime regulation of ground massives in northern construction. *Proc. 2nd ISGF*, Trondheim, **1**, 525–534.

Menot J. M. 1978. Equations of frost propagation in unsaturated porous media. *Proc. 1st ISGF*, **1**, 53–60.

Mettier K. 1985. Ground freezing for the construction of the Milchbuck road tunnel in Zurich — an engineering task revolving between theory and practice. *Proc. 4th ISGF*, Sapporo, **2**, 263–269.

Mi H. *et al.* 1991. Experimental study of shear creep of frozen fine sand. *Proc. 6th ISGF*, Beijing, **1**, 163–168.

MI H. *et al.* 1993. Characteristics of shear creep of frozen fine sand. *J. Glacio. Geocryology*, **15**−3.

MI H. *et al.* 1994. A numerical simulation on the forecast of freezing-thawing processes surrounding a mountain tunnel. *Proc. 7th ISGF*, Nancy, France, **1**, 215−218.

MICCHAUD Y. *et al.* 1989. Frost bursting: a violent expression of frost action in rock. *Cndn J. Earth Sci.*, **26**−10, 2075−2080.

MILLER H. W. and GORDON-BROWN T. P. 1967. Recent developments in ground freezing. *Proc. IoRef*, Nov., 8pp.

MILLER R. D. 1963. Phase equilibria and soil freezing. *Proc. 1st IC on Permafrost*, Indiana, 193−197.

MILLER R. D. 1972. Freezing and heaving of saturated and unsaturated soils. *US Highway Research Record*, **393**, 1−11.

MILLER R. D. 1978. Frost heaving in non-colloidal soils. *Proc. 3rd IC on Permafrost*, Edmonton, **1**, 708−713.

MILLER R. D. 1991. Scaling of freezing phenomena in soils. *Proc NS on Scaling in Soil Physics*, SSSA Spl Publication 25, 1−11.

MIN G. and DEWEN D. 1980. Thermodynamic method in the study of frost heave amount in soil. *Proc. 2nd ISGF*, Trondheim, **1**, 670−679.

MINGZHU G. and HUANUANG H. 1985. Calculation of normal frost heave force. *Proc. 4th ISGF*, Sapporo, **1**, 119−122.

MISIONG F. and SZTUKOWSKI B. 1959. An example of overcoming shaft sinking difficulties in Poland, *Proc. IS Shaft Sinking and Tunnelling*, IMM, London, July, 381−393.

MIYATA Y. 1988. A frost heave mechanism model based on energy equilibrium. *Proc. 5th ISGF*, Nottingham, **1**, 91−98.

MIYATA Y. *et al.* 1994. Measuring unfrozen pore water pressure at the ice-lens forming front. *Proc. 7th ISGF*, Nancy, France, **1**, 157−162.

MIYATA Y. and AKAGAWA S. 1991. Factors governing a frost heave ratio. *Proc. 6th ISGF*, Beijing, **1**, 55−64.

MIYOSHI M. *et al.* 1978. Large scale freezing work for subway construction in Japan. *Proc. 1st ISGF*, Bochum, **1**, 255−268.

MIZOGUCHI M. and NAKANO M. 1985. Water content, electrical conductivity and temperature profiles in a partially frozen unsaturated soil. *Proc. 4th ISGF*, Sapporo, **2**, 47−52.

MOHAN A. 1975. Heat transfer in soil-water-ice systems. *J. Geotech. Engng*, Feb., 97−113.

MOREY C. R. 1977. *Some aspects of the strength of frozen Ottawa sand.* Queens University, *MSc thesis*, 112pp.

MORGENSTERN N. R. and NIXON J. F. 1971. One-dimensional consolidation of thawing soils. *Cndn Geotech. J.*, **8**, 558−565.

MOSENBACHER J. 1971. The sinking of the central shaft of Wolkerdorf Colliery by the freezing method (in German). *Montan-Rdsch*, Vienna, Apr., **2**, 128−138.

MUELLER G. 1961. Velocity determinations of elastic waves in frozen rock, and application to frozen mantle of shafts (in German). *Geophysics Prospecting, The Hague*, June, **9**, 276−295.

MUELLER G. 1962. Ultrasonic logging control of freezing in shaft sinking. *Glückauf*, 28 Mar., **98**, 381−387.

MUELLER-KIRCHENBAUER H. *et al.* 1976. Effect of the degree of compaction on the strength relationship of frozen soils. *Tiefbau*, **18**−7, 473−476.

MURAYAMA S. *et al.* 1985. Application of freezing method to construction of tunnel through weathered granite ground. *Proc. 4th ISGF*, Sapporo, **2**, 253−258.

MURAYAMA S. *et al.* 1988. Ground freezing for the construction of a drain pump chamber in gravel between the twin tunnels in Kyoto. *Proc. 5th ISGF*, Nottingham, **1**, 377−382.

MUSSCHE H. E. 1909. Congelation du sol dans les travaux du Metropolitain (in French). *Technique Moderne*, Oct.†

Mussche H. E. 1935. Tir John North power station. *Civ. Engng.*
Mussche H. E. 1937. Ice dam at Grand Coulee. *Engineering News Record*, Feb.†
Mussche H. E. 1938. Millom No. 1 shaft, Workington. *Iron and Coal Trades Review*, Nov.†
Mussche H. E. 1939. Shaft sinking application of freezing and cementation at Workington. *J. King's College Mining Soc.*, July.†
Mussche H. E. 1950. A blast furnace skip/pit constructed by the freezing process. *Civil Engng and Public Works Review*, May, 3pp.
Mussche H. E. 1952. Considerations on freezing holes, their survey, and the freezing process. *Proc. Inst Mining Surveyors.*†
Mussche H. E. and Varty A. 1938. The freezing process as applied to sinking during the past 10 years. *Iron and Coal Trades Review*, **137**, 25 Nov., 877–879.
Mussche H. E. and Waddington J. C. 1946. Applications of the freezing process to civil engineering works. *Proc. ICE*, **14** May, 3–34.
Mussnung G. 1941. Problems and experience in the use of concrete for the lining of shafts sunk by the freezing process. *Glückauf*, **77**, 405–410, 423–428.
Muzás F. 1980a. Thermal calculations in the design of frozen soil structures. *Proc. 2nd ISGF*, Trondheim, **1**, 545–555.
Muzás F. 1980b. Design of circular cylindrical walls of frozen soil. *Proc. 2nd ISGF*, Trondheim, **1**, 880–888.

Nadezhdin A. V. and Sorokin V. A. 1975. Influence of preloading on the strength of frozen soil. *SM & FE*, May/June, 12–3, 185–186.
Nagasawa T. and Umeda Y. 1985. Effects of the freeze–thaw process on soil structure. *Proc. 4th ISGF*, Sapporo, **2**, 219–224.
Nakagawa S. and Fukuda M. 1988. Experimental studies on reducing methods of uplift force to a steel pipe. *Proc. 5th ISGF*, Nottingham, **2**, 565–566.
Nakano Y. 1991. Transport of water through frozen soils. *Proc. 6th ISGF*, Beijing, **1**, 65–70.
Nakano Y. 1994. Dependence of segregation potential on thermal and hydraulic conditions predicted by model M_1. *Proc. 7th ISGF*, Nancy, France, **1**, 25–33.
Neerdael B. *et al.* 1983. Field measurements during construction of an underground laboratory in a deep clay formation. *Proc. IC on Field Measurements in Geomechanics*, Zurich, **2**, 1419–1430.
Neill R. 1951. The freezing process of shaft sinking at Solway Colliery. *Proc. Mining Inst. of Scotland*, 24 Jan., 44pp (U).
Nelson D. W. and Romkens M. J. M. 1972. Suitability of freezing as a method of preserving runoff samples for analysis for soluble sulphate. *J. Environmental Quality*, **1**, 323–324.
Nendza H. *et al.* 1976. Investigations into the freezing of ground with rapid groundwater movement (in German). *Die Bautechnik*, July, 226–232.
Nereseova Z. A. and Tsytovich N. A. 1963. Unfrozen water in frozen soils. *Proc. 1st IC on Permafrost*, Indiana, 230–234.
Neuber H. and Wolters R. 1970. Mechanical behaviour of frozen soils under triaxial compression (in German). *Fortschritte in der Geologie von Rheisland und Westfallen*, **17**, 500–536. (Transl NRCC 1977: NRC-TT-1902, 53pp).
Nishibayashi K. *et al.* 1985. Laboratory performance tests of cryogenic earth pressure cells. *Proc. 4th ISGF*, Sapporo, **1**, 319–326.
Nishimura T. *et al.* 1994. Effective stresses in unsaturated soils after freezing and thawing. *Proc. 7th ISGF*, Nancy, France, **1**, 121–129.
Nixon J. F. 1987. Ground freezing and frost heave -- a review. *Northern Engineer*, **19**–3/4, 8–18.

NIXON J. F. 1991. Discrete ice lens theory for frost in soils. *Cndn Geotech. J.*, **28**, 843–859.

NIXON J. F. *et al.* 1982. In situ frost heave testing using cold plates. *Proc. 4th Cndn C. on Permafrost (1981) (R. J. E. Brown Mem. Vol.)*, Alberta, 466–474.

NIXON J. F. and HANNA A. J. 1979. The undrained strength of some thawed permafrost soils. *Cndn Geotech. J.*, **16**, 420–427.

NIXON J. F. and LADANYI B. 1978. Thaw consolidation. In Andersland O. B. and Anderson D. M. (eds), *Geotechnical Engineering for Cold Regions*, McGraw-Hill, 164–215.

NIXON J. F. and LEM G. 1984. Creep and strength testing of frozen saline fine-grained soils. *Cndn Geotech. J.*, **21**, 518–529.

NIXON J. F. and MORGENSTERN N. R. 1973. The residual stress in thawing soils. *Cndn Geotech. J.*, **10**–4, 571–580.

NIXON J. F. and MORGENSTERN N. R. 1974. Thaw consolidation tests on undisturbed fine-grained permafrost. *Cndn Geotech. J.*, **11**–1, 202–214.

NIXON M. S. and PHARR G. M. 1984. The effects of temperature, stress and salinity on the creep of frozen saline soil. *J. Energy Resources Technology (ASME)*, **106**, 344–348.

NIXON W. A. 1988. Application of fracture mechanics to ice/structure interactions. *J. Cold Regions Engng*, Mar., **2**–1.

NOBLE C. A. and DEMIREL T. 1969. Effect of temperature on strength behaviour of cohesive soil. *US Highway Research Board Special Rpt*, **103**, 204–219.

NOVITOV F. Y. 1978. Pressure of thawing soils on the concrete lining of vertical mine shafts. *Proc. 1st ISGF*, Bochum, **1**, 175–182.

NUMAZAWA K. *et al.* 1988. Application of the freezing method to the undersea connection of a large diameter shield tunnel. *Proc. 5th ISGF*, Nottingham, **1**, 383–388.

O'CONNOR M. J. and MITCHELL R. J. 1978a. Measuring total volumetric strains during triaxial tests on frozen soils. *Cndn Geotech. J.*, 29pp.

O'CONNOR M. J. and MITCHELL R. J. 1978b. The energy surface: a new concept to describe the behaviour of frozen soils. *Proc. 3rd IC on Permafrost*, Edmonton, **1**, 119–126.

O'CONNOR M. J. and MITCHELL R. J. 1982. A comparison of triaxial and plane stress tests on frozen silt. *Proc. 4th Cndn C. on Permafrost (1981) (R. J. E. Brown Mem. Vol.)*, Calgary, 382–386.

O'NEILL K. 1983. The physics of mathematical heave models. *Cold Regions Sci. and Tech.*, **6**, 275–291.

O'NEILL K. and MILLER R. D. 1980. Numerical solutions for rigid-ice model of secondary frost heave. *Proc. 2nd ISGF*, Trondheim, **1**, 656–669, and in *CRREL Rpt 82-13*.†

O'NEILL K. and MILLER R. D. 1985. Explorations of a rigid ice model of frost heave. *Water Resources Research*, **21**, 281–296.

OFFENSEND F. L. 1966. Tensile strength of frozen soils. *CRREL Tech Note*, 21pp.

OGATA N. *et al.* 1982. Salt concentration effects on strength of frozen soils. *Proc. 3rd ISGF*, Hanover NH, **1**, 3–10.

OGATA N. *et al.* 1983. Effects of salt concentration on strength and creep behaviour of artificially frozen soils. *Cold Regions Sci. and Tech.*, **8**, 139–153.

OGATA N. *et al.* 1985. Effect of freezing-thawing on the mechanical properties of soil. *Proc. 4th ISGF*, Sapporo, **1**, 201–207.

OGAWA S. *et al.* 1991. Influence of freezing and thawing on suction of unsaturated soils. *Proc. 6th ISGF*, Beijing, **1**, 71–76.

OHRAI T. *et al.* 1985. Actual results of ground freezing in Japan. *Proc. 4th ISGF*, Sapporo, **2**, 289–294.

OHRAI T. and YAMAMOTO H. 1985. Growth and migration of ice lenses in partially frozen soil. *Proc. 4th ISGF*, Sapporo, **1**, 79−84.

OLIPHANT J. L. et al. 1983. Effect of unconfined loading on the unfrozen water content of Manchester silt. *CRREL Special Rpt* 83-18, 17pp.

OOLBERKINK H. 1951. #4 shaft at Emma Colliery, Schinnen, Holland. *Geologie Mijnbouwm*, Mar., **13**, 97−102.

ORLOV V. O. and KIM V. K. 1988. Method of evaluating the frost-heave pressure of soil against the ice-soil enclosure of an underground structure. *SM & FE N3*, May, **25**, 129−135.

ORTH W. 1985. Deformation behaviour of frozen sand down to cryogenic temperatures. *Proc. 4th ISGF*, Sapporo, **1**, 245−254.

ORTH W. 1987. *Frozen sand as a construction material — fundamental research and a material model* (in German). University of Karlsruhe. PhD thesis.

ORTH W. 1988a. A creep formula for practical application based on crystal mechanics. *Proc. 5th ISGF*, Nottingham, **1**, 205−211.

ORTH W. 1988b. Two practical applications of soil freezing by liquid nitrogen. *Proc. 5th ISGF*, Nottingham, **1**, 389−394.

ORTH W. 1994. Jacking of a railroad overpass with soil stabilization and waterproofing by ground freezing. *Proc. 7th ISGF*, Nancy, France, **1**, 411−412.

ORTH W. and MEISSNER H. 1982. Long-term creep of frozen soil in uniaxial and triaxial tests. *Proc. 3rd ISGF*, Hanover NH, **1**, 81−88.

ORTH W. and MEISSNER H. 1985. Experimental and numerical investigations for frozen tunnel shells. *Proc. 4th ISGF*, Sapporo, **2**, 259−262.

ORZHEHOVSKIY Y. R. et al. 1991. Apparatus and methods for determining of mechanical properties of freezing and thawing soils. *Proc. 6th ISGF*, Beijing, **2**, 529−532.

OSLER J. C. 1966. Studies of engineering properties of frozen soils. *McGill Univ. SM Series*, **18**, 35pp.

OSTERKAMP T. E. 1987. Freezing and thawing of soils and permafrost containing unfrozen water or brine. *Water Resources Research*, Dec., **23**−12, 2279−2285.

OSTROWSKI W. J. S. 1967. Design aspects of ground consolidation by the freezing method for shaft sinking in Saskatchewan. *Cndn Mining and Metallurgical Bulletin*, Oct., 1145−1153.

OSTROWSKI W. J. S. 1985. Industrial tests on application of liquid nitrogen for ground freezing. *Proc. 4th ISGF*, Sapporo, **1**, 265−276.

OUVREY J. F. 1985. Results of triaxial compression tests and triaxial creep tests on an artificially frozen stiff clay. *Proc. 4th ISGF*, Sapporo, **2**, 207−212.

PALMER A. C. 1967. Ice lensing, thermal diffusion and water migration in freezing soil. *J. Glaciology*, **6**−47, 681−694.

PANDAY S. and CORAPCIOGLU M. Y. 1988. Sensitivity of a thaw simulation to model parameters. *Proc. 5th ISGF*, Nottingham, **1**, 99−106.

PANG R. 1991a. The construction of East air shaft, Panji #3 colliery by freezing method 415 m depth. *Proc. 6th ISGF*, Beijing, **1**, 345−350.

PANG R. 1991b. Sinking of main shaft in Chengsilou Mine without freeze-tube breakage. *Proc. 6th ISGF*, Beijing, **2**, 533−536.

PARAMESWARAN V. R. 1982a. Strength and deformation of frozen sand at −30°C. *Cndn Geotech. J.*, **19**−1, 104−107.

PARAMESWARAN V. R. 1982b. Electrical potentials on freezing granular soils. *Proc. 3rd ISGF*, Hanover NH, **2***, 83−89.

PARAMESWARAN V. R. 1985. Cyclic creep of frozen soils. *Proc. 4th ISGF*, Sapporo, **2**, 201−206.

PARAMESWARAN V. R. 1987. Adfreezing strength of ice to model piles. *Cndn Geotech. J.*, Aug., **24**−3, 446−452.

PARAMESWARAN V. R. and JONES S. J. 1981. Triaxial testing of frozen sand. *J. Glaciology*, Feb., **27**, 147−156.
PARAMESWARAN V. R. and ROY M. 1982. Strength and deformation of frozen saturated sand at −30°C. *Cndn Geotech. J.*, **19**−1, 104−107.
PARAMESWARAN V. R. *et al.* 1985. Electrical potentials developed during thawing of frozen ground. *Proc. 4th ISGF*, Sapporo, **1**, 9−16.
PARKHILL S. M. 1960. Cold caisson in Lower Manhattan. *Compressed Air*, Mar., 14−19.
PATTERSON D. E. and SMITH M. W. 1981. The measurement of unfrozen water content by time domain reflectometry: results from laboratory tests. *Cndn Geotech. J.*, **18**, 131−144.
PATTERSON D. E. and SMITH M. W. 1985. Unfrozen water content in saline soils: results using time-domain reflectometry. *Cndn Geotech. J.*, **22**, 95−101.
PAVLOV A. R. and PERMYAKOV P. P. 1980. Numerical determination of thermal characteristics of freezing-thawing soil. *Proc. 2nd ISGF*, Trondheim, **1**, 454−461.
PCHELINTSEV A. M. 1985. Installation for investigation of frost heave forces on foundations. *SM & FE*, **22**−3, 18−19.
PEELE R. 1941 (ed). Shaft sinking in unstable ground — freezing method. *Mining Engineers Handbook*, Wiley, **1**, 8, 20−23.
PEKARSKAYA N. K. 1973. The strain hardening of frozen ground during the creep process. *Proc. 2nd IC Permafrost*, Yakutsk, 841−842.
PELIKAN J. R. 1953. Refrigerated mud supports building during refoundations. *Refrigerating Engineering*, Dec., 511.
PENG X. *et al.* 1991. A model heat, moisture and stress field of saturated soil during freezing. *Proc. 6th ISGF*, Beijing, **1**, 345−350.
PENNER E. 1960. The importance of freezing rate in frost action in soils. *Proc. ASTM*, **60**, 1151−1165.
PENNER E. 1963. Frost heaving in soils. *Proc. 1st IC on Permafrost*, Indiana (Building Res. Advisory Bd NRCC), 197−202.
PENNER E. 1967. Heaving pressure in soils during unidirectional freezing. *Cndn Geotech. J.*, **4**, 398−408.
PENNER E. 1970. Thermal conductivity of frozen soils. *Cndn J. Earth Sci.*, **7**, 982.
PENNER E. 1982. Aspects of ice-lens formation. *Proc. 3rd ISGF*, Hanover NH, **1**, 239−246.
PENNER E. 1986. Ice lensing in layered soils. *Cndn Geotech. J.*, **23**−3, 334−340.
PENNER E. and GOODRICH L. E. 1980. Location of segregated ice in frost susceptible soil. *Proc. 2nd ISGF*, Trondheim, **1**, 626−639.
PENNER E. and UEDA T. 1978. A frost susceptibility test and a basis for interpreting heaving rates. *Proc. 3rd IC on Permafrost*, Edmonton, **1**, 722−727.
PENNER E. and WALTON T. 1978. Effects of temperature and pressure on frost heaving. *Proc. 1st ISGF*, Bochum, **2**, 65−72.
PERFECT E. and WILLIAMS P. J. 1980. Thermally induced water migration in frozen soils. *Cold Regions Sci. and Tech.*, 101−109.
PETROSYAN L. R. and MOSIN V. D. 1985. Failure of frozen soils by high pressure jets in trench and pit construction. *SM & FE*, **22**−5, 194−197.
PFISTER I. *et al.* 1969. Ground freezing for crossing a 60 m triassic crushed zone (in French). *Proc. 7th IC SM & FE*, Mexico, **2**, 449−457.
PHILIPOV P. I. *et al.* 1982. Thermophysical characteristics of frozen, freezing-thawing, and thawed rocks and methods of measurement. *Proc. 3rd ISGF*, Hanover NH, **1**, 151−154.
PHUKAN A. 1980. The strength of frozen fine-grained sands at warm temperatures. *Proc. 2nd ISGF*, Trondheim, **1**, 165−179.
PHUKAN A. (Ed.) 1993. *Proc. 2nd IC on Frost in Geotech. Engng*, Anchorage, Balkema, 191.

PHUKAN A. and ANDERSON D. M. 1978. Foundations for cold regions. In *Andersland O. B. and Anderson D. M.* (eds) *Geotechnical engineering for cold regions*, 276−362.
PHUKAN A. and TAKASUGI S. 1982. Excavation resistance of artificially frozen soils. *Proc. 3rd ISGF*, Hanover NH, **1**, 35−40.
PICKARD F. C. and DUBY G. H. 1958. Freezing for shaft sinking. *Mining Congress J.*, **44**, Dec., 52−55.
PIERRE R. 1903. Sinking a shaft of a coal mine at Eygelshoven, Holland by the freezing process. *Glückauf*. Translation in *Iron and Coal Trades Review*, 5 June.†
PIETRZYK K. 1980. Pressure in the zone of ground freezing. *Proc. 2nd ISGF*, Trondheim, **1**, 702−712.
PIKE D. C. et al. 1990. The BS frost heave test: development of the standard and suggestion for further improvement. *Quarry Management*, Feb., 25−30.
PIPER D. et al. 1988. A mathematical model of frost heave in granular materials. *Proc. 5th IC on Permafrost*, Trondheim, **1**, 370−376, and abbreviated in *Proc. 5th ISGF*, Nottingham, **2**, 569−70.
PIZEY M. H. 1981. *The performance of concrete shaft linings in frozen ground*. University of Newcastle, *MSc Dissertation*, Sept.
PLESNIAK I. and SZCZEPANIAK S. 1988. A modern method of frozen rock mass control in the underground construction enterprise KGHM in Lubin. *Proc. 5th ISGF*, Nottingham, **2**, 571−572.
POETSCH F. H. 1900. Sinking shafts by means of Poetsch freezing process. *Berg U Huttenwesen*, 14 April−5 May.
PONOMARJOV V. D. 1982. Temperature deformations of frozen soils. *Proc. 3rd ISGF*, Hanover NH, **1**, 125−130.
PORCELLINIS P. de and ROJO J. L. 1980. Brine substitute liquids for soil freezing at very low temperatures. *Proc. 2nd ISGF*, Trondheim, **1**, 568−580.
POREBSKA M. and SKARZYNSKA K. M. 1988. Investigations of the frost heave of colliery spoil. *Proc. 5th ISGF*, Nottingham, **1**, 107−114.
PORTURAS F. A. 1980. Geotechnical exploration related to artificial ground freezing.†
POTEVIN G. 1972. La congelation des terrains dans la travaux publics (in French). *Revue de l'industrie Minerale*, **54**, 4, 164−179.
POWERS J. P. and MAISHMAN D. 1981. Ground freezing. In *Construction dewatering: a guide to theory and practice*, Wiley, 20, 349−359.
PUSCH R. 1978. Unfrozen water as a function of clay microstructure. *Proc. 1st ISGF*, Bochum, **1**, 103−108.
PUSCH R. 1980. Creep of frozen soil, a preliminary physical interpretation. *Proc. 2nd ISGF*, Trondheim, **1**, 190−201.

QIU S. and WANG T. 1980. A study of deep shafts using AGF, design of shaft linings and method of preventing seepage. *Proc. NS on Mining*, China, and in *Proc. 3rd ISGF*, Hanover NH, **1**, 363−366.
QUAN X. S. et al. 1989. Viscoelastic constitutive model for creep of frozen soil. *Proc. 3rd Int. Symp. on Numerical Models in Geomechanics*, Niagara Falls, 179−186, Elsevier.
QUINN F. X. et al. 1991. Cryogenic properties of soils and rocks − anomalous behaviour of water. *Géotechnique*, **41**−2, 195−209.

RADD F. J. and WOLFE L. H. 1979. Ice lens structures, compression strengths and creep behaviour of some synthetic frozen silty soils. *Proc. 1st ISGF*, Bochum, **1**, 115−130, and in *Engng Geol.*, **13**−1, 169−183.
RAJALAHTI M. 1987. Ground freezing in full-scale sub-level stoping at the Pyhasalmi mine, Finland. *Proc. 13th World C. on Mining*, Stockholm, June **2**, 937−939.

RAMOS M. et al. 1994. Correlation between heat flux on the ground and permafrost thermal regime near the Spanish antarctic station. *Proc. 7th ISGF*, Nancy, France, **1**, 395–396.

RATKJE S. K. et al. 1982. The hydraulic conductivity of soils during frost heave. *Proc. 3rd ISGF*, Hanover NH, **1**, 131–138.

RAZBEGIN V. N. 1988. Mathematical model for predicting stress–strain behaviour and heat-mass transfer of freezing soils. *Proc. 5th ISGF*, Nottingham, **2**, 419–424.

REBHAN D. 1977. Ground solidification with liquid nitrogen (in German). *Strassen und Tiefbau*, Mar., 23–36.

REBHAN D. 1982. New freeze-pipe systems for nitrogen freezing. *Proc. 3rd ISGF*, Hanover NH, **1**, 429–438.

REBHAN D. 1991. New experience and problems with liquid nitrogen freezing. *Proc. 6th ISGF*, Beijing, **1**, 351–358.

REDFERN A. and PINDER B. F. 1964. Shaft lining and grouting problems associated with Cotgrave and Bevercotes shafts. *Colliery Guardian*, **209**, Dec., 817–826, Jan., 14–21.

REED M. A. et al. 1979. Frost heaving rate predicted from pore size distribution. *Cndn Geotech. J.*, **16**–3.

REIN R. G. and HATHI V. V. 1978. The effect of stress on strain at the onset of tertiary creep of frozen soils. *Cndn Geotech. J.*, **15**–3, 424–425.

REIN R. G. and SLIEPCEVICH C. M. 1975. Rheological properties of frozen soils. *US NTIS*, Feb., AD-A011 459/5SL, 122pp.

REIN R. G. et al. 1975. Creep of sand-ice system. *J. ASCE Geotech. Engng Div.*, Feb., **101**–GT2, 115–128.

REMY J. M. et al. 1994. Relations entre l'endommagement par gel des roches calcaires et la géométrie (in French). *Proc. 7th ISGF*, Nancy, France, **1**, 189–197.

RENO W. H. and WINTERKORN H. F. 1967. Thermal conductivity of kaolinite clay as a function of exchange ion, density and moisture content. *Highway Research Record*, **209**, 79–85.

RESTELLI A. B. et al. 1988. Ground freezing solves a tunnelling problem at Agri Sauro, Italy. *Proc. 5th ISGF*, Nottingham, **1**, 395–402.

RETHY M. A. 1933. L'emploi de la congelation pour les traveaux des tunnels d'Anvers (in French). *Genie Civil*, 26 Aug.†

RETHY M. A. 1936. Le metropolitain de Moscow (in French). *Construction et Traveaux Publics*, May.

RHODE J. K. G. et al. 1987. Ground freezing as water sealing and ground improvement. *Proc. 9th EurC SM & FE*, Dublin, Aug., 235–240.

RICE E. G. and FERRIS W. 1976. Some laboratory tests on creep of steel piling in frozen ground. *Proc. 2nd IS Cold Regions Engng*, Fairbanks, 278pp.

RIEKE R. D. et al. 1983. The role of specific surface area and related index properties in the frost heave susceptibility of soils. *Proc. 4th IC on Permafrost*, Fairbanks, 1066–1071.

RIEMER N. 1905. The latest progress in shaft sinking (transl from German). *Proc. 8th German Mining Congress*, Dortmund.†

RIES A. 1981. Frozen shaft sinkings — past and present. *Proc. NS on Shaft Sinking and Tunnelling*, 22–27. (Glückauf, transl 118)

ROBERTS 1982. Engineering properties of frozen ground. *Applied Geotechnics*, **11**, 204–229.

ROE G. and WEBSTER D. C. 1984. Specification for TRRL frost heave test. *Supp. Rpt 829*, 39pp.

ROJO J. L. and NOVILLO A. 1988. Recent applications of soil freezing techniques in Spanish construction works. *Proc. 5th ISGF*, Nottingham, **2**, 525–532.

ROJO J. L. et al. 1991. Soil freezing for the Valencia underground railway work. *Proc. 6th ISGF*, Beijing, **1**, 359–368.

ROMAN L. T. and ZHU Y. 1991. A new method for predicting long-term strength of frozen soils. *Proc. 6th ISGF*, Beijing, **2**, 437–442.

ROMAN L. T. *et al.* 1994. Using method of temperature-time analogy to determine long-term strength of frozen soil in triaxial compression.†

ROSHENKO V. N. and FEDOSEEV Y. G. 1976. Effect of ice content and cryogenic structure of frozen ground on the bearing capacity of a pile. *SM & FE*, **13**–6, 402–405.

ROTH B. 1981. Tunnelling methods for the Essen underground part 2. *Tunnel*, **2**, 106–114.

RUEDY R. 1955. Shaft sinking by freezing methods. *NRCC Rpt TIS 43*, Aug., 1905–1955.

RUPPRECHT E. 1978. Application of the ground freezing method to penetrate a sequence of water bearing and dry formations — 3 case studies. *Proc. 1st ISGF*, Bochum, **1**, 357–363.

RYDEN C. G. and AXELSSON K. 1988. Laboratory determination of pore pressure during thawing for three different types of soil. *Proc. 5th ISGF*, Nottingham, **1**, 213–218.

RYOKAI K. 1985. Frost heave theory of saturated soil coupling water/heat flow and its application. *Proc. 4th ISGF*, Sapporo, **1**, 101–108.

RYOKAI K. *et al.* 1982. Frost heave susceptibility of saturated soil under constant rate of freezing. *Proc. 3rd ISGF*, Hanover NH, **1**, 269–276.

RYOKAI K. *et al.* 1988. Frost expansion pressure and displacement of saturated soil analysed with coupled heat and water flows. *Proc. 5th ISGF*, Nottingham, **1**, 115–120.

SAARRLAINEN S. and KIVIKOAKI H. 1994. Frost protection of ice arenas. *Proc. 7th ISGF*, Nancy, France, **1**, 237–243.

SACLIER and WAYMEL 1895. The sinking of the Vicq pits of the Anzin Company. *Bulletin Société de L'Industrie Minerale*, **9**, 27–164. (Abstract in *Proc. ICE*, **123**, 541–548)

SADOVSKY A. V. and TICHOMIROV S. M. 1980a. State of the Art report: Artificial freezing and cooling of soils at the construction. *Proc. 2nd ISGF*, Trondheim, **1**, 857–862.

SADOVSKY A. V. and TICHOMIROV S. M. 1980b. Strength and deformability of clays while pressing through them the rigid plates and subsequent freezing. *Proc. 2nd ISGF*, Trondheim, **1**, 120–131.

SADOVSKY A. V. *et al.* 1988. State of the Art: Mechanical properties of frozen soil. *Proc. 5th ISGF*, Nottingham, **2**, 443–464.

SAERTERSDAL R. 1980. Heaving conditions by freezing of soils. *Proc. 2nd ISGF*, Trondheim, **1**, 824–836.

SAGE J. D. and D'ANDREA R. A. 1982. Measurement of soil thaw weakening. *Proc. 3rd ISGF*, Hanover NH, **1**, 105–112.

SAKA F. *et al.* 1994. Application of freezing method to launch a large diameter shield tunnel. *Proc. 7th ISGF*, Nancy, France, **1**, 281–288.

SALTER M. DE G. 1989. Rogers Pass ventilation shaft. *Proc. 9th RETC*, Los Angeles USA, 835–854.

SANGER F. J. 1968. Ground freezing in construction. *J. ASCE*, **94**, Jan., SM1, 131–158.

SANGER F. J. and KAPLAR C. W. 1963. Plastic deformation of frozen soils. *Proc. 1st C on Permafrost*, Indiana, 305–315.

SANGER F. J. and SAYLES F. H. 1979. Thermal and rheological computations for artificially frozen ground construction. *Proc. 1st ISGF*, Bochum, **2**, 95–118, and in *Engng Geol.*, **13**, 311–337.

SARKISYAN R. M. 1973. Evaluation of some physical properties of frozen ground as a function of the frozen water content. *Proc. IC on Permafrost*, Yakutsk, 277–282.

SAUVESTRE L. 1920. Shaft sinking through quicksand under heavy pressure. *Colliery Guardian*, **99**, 1495–1497.

SAVELYEV B. A. *et al.* 1988. The artificial freezing-on of engineering ice−soil structures. *Proc. 5th ISGF*, Nottingham, **2**, 497−504.

SAWADA S. and OHNO T. 1985. Laboratory studies on thermal conductivity of clay, silt, sand in frozen and unfrozen states. *Proc. 4th ISGF*, Sapporo, **2**, 53−58.

SAWADA S. and SUZUKI T. 1991. Experimental studies on frost heaving force and adfreeze frost heaving force on short concrete piles. *Proc. 6th ISGF*, Beijing, **1**, 169−174.

SAWADA S. *et al.* 1994. An inhibition effect of porous, thermal insulating materials against frost heave and frost penetration. *Proc. 7th ISGF*, Nancy, France, **1**, 3−8.

SAYLES F. H. 1966. Low temperature soil mechanics. *CRREL Tech Note*.

SAYLES F. H. 1968. Creep of frozen sands. *CRREL Rpt 190*, Sept., 54pp.

SAYLES F. H. 1969. Stress, strain and time relationships in frozen ground. *MCI Consultants*, Hanover NH, June.†

SAYLES F. H. 1974. Triaxial constant strain rate tests and triaxial creep tests on frozen Ottawa sand. *CRREL Rpt 253*, Aug., 32pp.

SAYLES F. H. 1987. Classification and laboratory testing of artificially frozen ground. *J. Cold Regions Engng*, Mar., 22−48.

SAYLES F. H. 1988. State of the Art: Mechanical properties of frozen soils. *Proc. 5th ISGF*, Nottingham, **1**, 143−165.

SAYLES F. H. *et al.* 1987. Classification and laboratory testing of AGF. *J. Cold Regions Engng*, **1**−1, 22−48.

SAYLES F. H. and CARBEE D. L. 1980. Strength of frozen silt as a function of ice content and dry unit weight. *Proc. 2nd ISGF*, Trondheim, **1**, 109−119.

SAYLES F. H. and EPANCHIN N. V. 1966. Rate of strain compression tests on frozen Ottawa sand and ice. *CRREL Tech. note*, 54pp.

SAYLES F. H. and HAINES D. 1974. Creep of frozen silt and sand. *CRREL Tech. Rpt 252*, July, 54pp.

SCHMIDT F. 1898. Utilisation of the freezing process in the course of mining operations. *Gesanite Kalteindustrie*, **5**−23.

SCHMIDT F. 1917. Shaft sinking by the freezing process. *Proc. IME*, **52**−2. 141−185.

SCHMID G. 1935 Reconstruction of the collapsed Auguste Victoria shaft. *Glückauf*, **71**, 1069−1078.

SCHMID L. 1981. Milchbuck tunnel: application of the freezing method to drive a three-lane highway tunnel close to the surface. *Proc. 5th RETC*, San Francisco, May, **1**, 427−445.

SCHMID W. L. H. 1953. Mining #4 shaft of Emma Colliery, Schinnen, Holland (in French). *Revue Industrie Mining St Ettienne*, Aug., **34**, 561−576.

SCHOLES W. A. 1982. Freezing an island to sink a shaft. *Tunnels and Tunnelling*, Jan, 1pp.

SCHWEITZER R. 1959. Sinking of the Venejoul shaft in waterbearing ground (in French). *Revue Industrie Mining St Ettienne*, Jan., **41**, 17−48.

SCOTT R. F. 1969. The freezing process and mechanics of frozen ground. *CRREL Monograph II-D1*, 20−65.

SCOTT S. A. 1963. Shaft sinking through Blairmore sands and Paleozoic water bearing limestones. *Cndn Mining and Metallurgical Bulletin*, **56**−610, 94−103.

SEGO D. C. *et al.* 1982. Strength and deformation behaviour of frozen saline sand. *Proc. 3rd ISGF*, Hanover NH, **1**, 11−18.

SELVADURIA A. P. S. and SHINDE S. B. 1993. Frost heave induced mechanics of buried pipelines. *J. Geotech Engineering*, **119**−12, 1929−†.

SEMPRICH S. and LOSCH M. 1990. Frozen collar protects tunnel drive beneath rail tracks. *Tunnels and Tunnelling*, **22**−4, 27−29.

SHABABERLE R. *et al.* 1988. Influence of freeze−thaw cycles on clay structures. *Proc. 5th ISGF*, Nottingham, **2**, 573−576.

SHAOXIN X. 1985. The frost heave behaviour of cohesive soils under three kinds of consolidated state. *Proc. 4th ISGF* Sapporo, **2**, 167–169.

SHARBER P. A. 1966. The brineless freezing of soil. *Kholodilnaya Teknika*, **12**, 23.

SHEN J. 1980. Shaft sinking through water-bearing strata. *Proc. 1st IS Mine Planning and Development*, Beijing, Sept.†

SHEN M. and LADANYI B. 1988. Calculation of the stress field in soils during freezing. *Proc. 5th ISGF*, Nottingham, **1**, 121–128.

SHENG Y. and CHEN X. 1991. A numerical simulation of frost heave under overburden stress. *Proc. 6th ISGF*, Beijing, **1**, 175–180.

SHIBATA T. *et al.* 1985. Time-dependence and volumetric change characteristic of frozen sand under triaxial stress condition. *Proc. 4th ISGF*, Sapporo, **1**, 173–180.

SHIH T. S. *et al.* 1988. High strength concrete-steel bond behaviour at low temperature. *J. Cold Regions Engng*, Dec., **2-4**, 157–168.

SHIWU Q. and TIEMENG W. 1980. A study of deep shafts using AGF, design of shaft linings and method of preventing seepage. *Proc. S. on Mining*, China, 363–366.

SHLOIDO G. A. 1968. Determining the tensile strength of frozen ground. *Hydrotechnical Construction*, **3**, 238–240.

SHOCKLEY W. G. and THORBURN T. H. 1978. Suggested practice for description of frozen soils (visual manual procedure). *Geotech. Testing J.*, **1-4**, 228–233.

SHUSHERINA E. P. 1971. Variation of physical, mechanical properties of soils under the action of cyclic freeze–thaw. *CRREL Rpt TL-255* (transl from Russian).

SHUSHERINA E. P. and BOBKOV Y. P. 1969. Effect of moisture content on frozen ground strength (in Russian). *Merzotnye Issledovaniya*, **9**, 122–137 (translator V. Poppe NRCC, NRC-TT-1918).

SHUSHERINA E. P. and VYALOV S. S. 1965a. Triaxial compressive strength of frozen soil in the creep range. *Proc. 6th IC SM & FE*, Montreal, **3**, 348–350.

SHUSHERINA E. P. and VYALOV S. S. 1965b. Study of prolonged bearing strength of frozen soils under uniaxial compression. *CRREL Rpt*, AD 715 056, 36pp.

SHUSTER J. A. 1972. Controlled freezing for temporary ground support. *Proc. 1st RETC*, Chicago, **2**, 863–894.

SHUSTER J. A. 1980. Engineering quality assurance for construction ground freezing. *Proc. 2nd ISGF*, Trondheim, **1**, 863–879.

SHUSTER J. A. 1982a. Ground freezing failures — causes and prevention. *Proc. 3rd ISGF*, Hanover NH, **1**, 315, 10pp.

SHUSTER J. 1982b. Demonstration ground freezing project. *Proc. 3rd ISGF*, Hanover NH, **2***, 103–107.

SHUSTER J. A. 1982c. Ground freezing system at East St pumping station, Tewkesbury, Mass. *Proc. 3rd ISGF*, Hanover NH, **2***, 139–145.

SHUSTER J. A. 1985. Ground freezing for soft ground shaft sinking. *Proc. 7th RETC*, New York, June, 1046–1059.

SHUSTER J. A. and BRAUN B. 1974. Some notes on concreting against frozen earth. *Terrafreeze Corporation*, 5pp.

SHUSTER J. A. and SOPKO J. A. 1989. Ground freezing to control ground water and support deep storm sewer structural excavations. *Proc. 9th RETC*, Los Angeles, 149–155.

SHVETS V. B. *et al.* 1973. Compressometric studies of the strength and deformation of frozen soils during thawing. *Proc. 2nd IC on Permafrost*, Yakutsk.

SIK S. L. 1951. New methods in freezing for shaft sinking. *Glückauf*, **87**, 529–536.

SILINSH C. D. 1960. Freezing keeps shaft dry and holds dirt in place. *Construction methods and equipment*.

SIMONSON E. R. *et al.* 1974. High pressure mechanical properties of three frozen materials. *Proc. 4th IC on High Pressure*, Kyoto, 115–121.

SINIRSYN A. P. 1980. Influence of temperature field on properties of two-layered foundation. *Proc. 2nd ISGF*, Trondheim, **1**, 484–492.

SKARZYNSKA K. M. 1980. Effect of freezing process on selected properties of frost susceptible soils. *Proc. 2nd ISGF*, Trondheim, **1**, 75–84.

SKARZYNSKA K. M. 1985. Formation of soil structure under repeated freezing-thawing conditions. *Proc. 4th ISGF*, Sapporo, **2**, 213–218.

SKIPP B. O. and HALL M. J. 1982. Health and safety aspects of ground treatment materials. *CIRIA Rpt 95*, 32pp.

SLUNGA E. 1989. Determination of frost susceptibility of soil. *Proc. 12th IC SM & FE*, Brazil, **2**, 1465–1468.

SMART P. and HERBERTSON J. G. (ed.) 1993. Ground freezing. In *Drainage Design*, Blackie.

SMITH M. S. *et al.* 1982. Determination of ice/water contents of frozen soils by time domain reflectometry. *Proc. 3rd ISGF*, Hanover NH, **2***, 91–99.

SMITH M. W. and RISEBOROUGH D. W. 1985. The sensitivity of thermal predictions to assumptions in soil properties. *Proc. 4th ISGF*, Sapporo, **1**, 17–24.

SMITH N. 1982. Testing shaped charges in unfrozen and frozen silt in Alaska. *CRREL Special Rpt 82-02*, 10pp.

SMITH W. O. 1939. Thermal conductivities in moist soils. *Proc. Soil Sci. Soc. of America*, **4**, 32–40.

SMITH W. O. 1942. The thermal conductivity of dry soil. *Soil Sci.*, **53**, 435–459.

SOMERVILLE S. H. 1986. Control of groundwater for temporary works. *CIRIA Rpt 113*, 87pp.

SOO S. *et al.* 1985. Finite element models for structural creep problems in frozen ground. *Proc. 4th ISGF*, Sapporo, **2**, 23–28.

SOO S. *et al.* 1986. Flexural behaviour of frozen soil. *Cndn Geotech. J.*, **23**–3, 355–361.

SOPKO J. A. *et al.* 1991. Frozen earth cofferdam design. *Proc. 6th ISGF*, Beijing, **1**, 263–272.

SPIKE I. and WILD W. M. 1958. Shaft sinking at Wearmouth. *Colliery Engng*, **35**, 188–195.

STEGEMAN H. 1901. The sinking of a shaft at Maria Mine, Aix-la-Chapelle. *Glückauf*, 5 Jan.

STEGEMAN H. 1912. Output and costs in shaft sinking by the freezing method. *Glückauf*, 417.

STENBERG L. 1985. Apparatus for determination of frost susceptibility of soils. *Proc. 4th ISGF*, Sapporo, **2**, 141–145.

STEPANOV Y. S. and KHOLIN N. N. 1979. Determination of the stress field during thawing of frozen ground. *SM & FE*, **16**–2, 94–98.

STEVENS H. W. 1975. The response of frozen soils to vibratory loads. *CRREL Tech Rpt 265*, 98pp.

STOSS K. and BRAUN B. 1983. Sinking a freeze shaft with installation of a watertight flexible lining. *Proc. 6th RETC*, Chicago. 513–532.

STOSS K. and OELLERS T. 1985. Influence of freezing on strata behaviour. *Proc. NC on Shaft Sinking and Tunnelling*, Berlin, *Glückauf*, **121**, 1438–1444.

STOSS K. and VALK J. 1978. Chances and limitations of ground freezing with liquid nitrogen. *Proc. 1st ISGF*, Bochum, **1**, 303–312, and in *Engng Geol.*, 1979, **13**, 485–494.

STOSS K. *et al.* 1983. Temperature induced fissuring during the sinking of deep freeze shafts in salt measures (in German). *Proc. NC on Shaft Sinking and Tunnelling*, Berlin, *Glückauf*, **119**–20, 979–984.

STUART J. G. 1964. Consolidation tests on clay subjected to freezing and thawing. *Swedish Geotech. Inst.*, Rpt no 7, 1–9.

Su L. 1988. Some results of in-situ measurement of freezing pressure and earth pressure in frozen shafts. *Proc. 5th ISGF*, Nottingham, **1**, 295–302.

Sui T. L. and Na W. J. 1985. Double-layer grease casting for preventing and treating frost extraction of pile foundation. *Proc. 4th ISGF*, Sapporo, **2**, 301–305.

Sutherland H. B. and Gaskin P. N. 1973. Pore water and heaving pressures developed in partially frozen soils. *Proc. 2nd IC on Permafrost*, Yakutsk, 409–419.

Suzuki S. and Sawada S. 1994. Full-scale test on frost-heaving pressure in reinforced retaining wall. *Proc. 7th ISGF*, Nancy, France, **1**, 311–316.

Suzuki S. et al. 1994. Construction of extension shield tunnel using ground freezing method. *Proc. 7th ISGF*, Nancy, France, **1**, 289–293.

Sweanum S. and Muvdi B. B. 1992. Design method for frozen soil retaining wall. *J. Cold Regions Engng*, **6**–2, 73–89.

Swinzow G. K. 1964. A rheological model for deforming frozen ground. *CRREL Tech note*, 12pp.

Taber S. 1929. Frost heaving. *J. Geol.*, **37**–5, 428–461.

Taber S. 1930. The mechanics of frost heaving. *J. Geol.*, **38**–4, 303–317.

Takagi S. 1978. Segregation freezing as the cause of suction force for ice lens formation. *Proc. 1st ISGF*, Bochum, **1**, 45–52.

Takagi S. 1980. The adsorption force theory of frost heaving. *Cold Regions Sci. and Tech.*, **3**, 57–81.

Takagi S. 1982. Initial stage of the formation of soil-laden ice lenses. *Proc. 3rd ISGF*, Hanover NH, **1**, 223–232.

Takagi S. and Tanaka M. 1980. A model tank test to estimate the additional earth pressure due to freezing of the soil. *Proc. 2nd ISGF*, Trondheim, **1**, 1037–1048.

Takashi T. et al. 1978. Jointing of two tunnel shields under artificial ground freezing. *Proc. 1st ISGF*, Bochum, **1**, 339–348, and in *Engng Geol.*, 1979, **13**, 519–529.

Takashi T. et al. 1978. Effect of penetration rate of freezing and confining stress on the frost heave ratio of soil. *Proc. 3rd IC on Permafrost*, Edmonton.†

Takashi T. et al. 1979. Experimental study on uniaxial compressive strength of frozen sand. *Proc. Jap. Soc. Civil Engng*, 302, 74–88.

Takashi T. et al. 1980. Upper limit of heaving pressure derived by pore water pressure measurements of partially frozen soil. *Proc. 2nd ISGF*, Trondheim, **1**, 713–724.

Takashi T. et al. 1981. Experimental study of uniaxial compressive strength of homogeneously frozen clay (in Japanese). *Proc. Jap. Soc. Civil Engng*, 315, 83–93.

Takashi T. et al. 1982. Artificial ground freezing in shield work. *Proc. 3rd ISGF*, Hanover NH, **1**, 415–422.

Takashi T. et al. 1982. Effect of specimen height on frost heave ratio in uni-directional freezing test of soil. *Proc. 3rd ISGF*, Hanover NH, **1**, 247–254.

Takashi T. 1987. Influence of seepage stream on the joining of frozen zones in artificial soil freezing. *Seiken Company.*†

Takeda K. et al. 1985. Thermal condition for ice lens formation in soil freezing. *Proc. 4th ISGF*, Sapporo, **2**, 89–94.

Takegawa K. et al. 1978. Creep characteristics of frozen soils. *Proc. 1st ISGF*, Bochum, **1**, 133–140, and in *Engng Geol.*, **13**, 1979, 197–205.

Tan A. O. 1985. Laboratory strength and creep testing of frozen Aline sand. *Research Rpt Univ. of Nottingham*, 28pp.

Tanaka M. et al. 1994. Ground freezing application in tunnel constructions. *Proc. 7th ISGF*, Nancy, France, **1**, 405–406.

TAYBASHEV V. N. 1973. Nature of deformation and variation in strength of frozen coarse clactics as a function of their composition and structure. *Proc. 2nd IC on Permafrost*, Yakutsk, 274–276.

THIMUS J. F. 1991. Engineering design: general report. *Proc. 6th ISGF*, Beijing, **2**, 4449–4454.

THIMUS J. F. and GONZE P. Behaviour of two kinds of ground freezing: laboratory tests and case histories. *Proc. 5th ISGF*, Nottingham, **2**, 577–578.

THIMUS J. F. et al. 1991. Determination of unfrozen water content of an overconsolidated clay down to $-160°C$, sonic approaches — comparison with classical methods. *Proc. 6th ISGF*, Beijing, **1**, 83–88.

THIMUS J. F. et al. 1991. Rheological behaviour of overconsolidated clay measured by creep test — application to cryogenic storage. *Proc. 6th ISGF*, Beijing, **1**, 181–188.

THOMPSON E. G. and SAYLES F. H. 1972. In situ creep analysis of a room in frozen soil. *J. ASCE SM & FE Div.*, **98**–SM9, 899–915.

THOMPSON S. and LOBACZ E. F. 1973. Shear strength at a thaw interface in permafrost. *Proc. 2nd IC on Permafrost*, Yakutsk.

TICE A. R. and OLIPHANT J. L. 1984. The effects of magnetic particles on the unfrozen water content of frozen soils determined by nuclear magnetic resonance. *Soil Sci.*, **138**, 63–73.

TICE A. R. et al. 1973. The prediction of unfrozen water contents in frozen soils from liquid limit determinations. *OECD S. on Frost action in soils*, Oslo, **1**, 329–344.

TICE A. R. et al. 1978. Determination of unfrozen water in frozen soil by pulsed nuclear magnetic resonance. *Proc. 3rd IC on Permafrost*, **1**, 149–155.

TICE A. R. et al. 1984. The effects of soluble salts on the unfrozen water content of the Lanzhou PRC silt. *CRREL Rpt 84–16*.

TICE A. R. et al. 1988. Unfrozen water contents of undisturbed and remoulded Alaskan silt as determined by nuclear magnetic response. *CRREL Rpt 88–19*, Nov.

TIEN L. C. and CHURCHILL S. W. 1965. Freezing front motion and heat transfer outside an infinite isothermal cylinder. *AIChem Engng J.*, Sept., 790–793.

TING J. M. 1983a. Tertiary creep model for frozen sands. *J. Geotech. Engng*, **109**–7, 932–945.

TING J. M. 1983b. The creep of frozen sands: qualitative and quantitative models. *US Army Research Rpt 81-5*, 433pp.

TING J. M. and MARTIN R. T. 1979. Application of the Andrade Equation to creep data for ice and frozen soil. *Cold Regions Sci. and Tech.*, **1**–1, 29–36.

TOBE N. et al. 1985. Monitoring the closure of a freeze-wall cofferdam by water level observation. *Proc. 4th ISGF*, Sapporo, **1**, 285–290.

TOGROL E. and ERSOY T. 1977. Stress deformation characteristics of a frozen soil. *Proc. 9th IC SM & FE*, Tokyo, **1**, 333–336.

TOGROL E. et al. 1977. Mechanical behaviour of frozen soil under dynamic stress (in Turkish). *Istanbul Tech. University*, Tech. Rpt 28, 37pp.

TOGROL E. et al. 1982. Influence of repeated loading on the behaviour of frozen silty clay. *Proc. 4th Cndn C. on Permafrost (1981) (R. J. E. Brown Mem. Vol.)*, Alberta, Mar., 440–442.

TONG C. and CHEN E. 1985. Thaw consolidation behaviour of seasonally frozen soils. *Proc. 4th ISGF*, Sapporo, **1**, 159–163.

TONG C. and YU C. 1988. The effect of surcharge on the frost heaving of shallow foundations. *Proc. 5th ISGF*, Nottingham, **1**, 129–134.

TONSCHEIDT H. W. 1991. Rheinberg shaft: planning and execution of freezing methods (in German). *Glückauf*, **127**–11/12, 441–448.

TOULOUKIAN Y. S. et al. 1970. Thermal conductivity. *Thermo-physical properties of matter*, **3**.

TRIMBLE J. R. 1977. *A comparison of the creep deformations of naturally frozen soils under static and repeated loadings*. Queens University, *MSc thesis*, 139pp.
TRIMBLE J. R. and MITCHELL R. J. 1982. A comparison of static and repeated loading tests on natural frozen soils. *Proc. 4th Cndn C. on Permafrost (1981) (R. J. E. Brown Mem. Vol.)*, Calgary, 433–439.
TRUPAK N. G. *et al.* 1982. Contour ground freezing for tunnelling. *Proc. 3rd ISGF*, Hanover NH, **1**, 411–414.
TSYTOVICH N. A. 1937. An investigation of elastic and plastic deformation in frozen ground. *Proc. Academy of Permafrost*, **10**, 120pp.
TSYTOVICH N. A. 1957. The fundamentals of frozen ground mechanics. *Proc. 4th IC SM & FE*, London, **1**, 116–119.
TSYTOVICH N. A. 1960. Bases and foundations on frozen soil. *US Highway Res. Board Special Rpt*, 58, 93pp.
TSYTOVICH N. A. 1963. Instability and mechanical properties of frozen soils. *Proc. 1st IC on Permafrost*, Indiana, 325–330.
TSYTOVICH N. A. 1964. Physico mechanical processes in frozen soils. *SM & FE*, 124–127.
TSYTOVICH N. A. 1975. The mechanics of frozen ground. McGraw-Hill, 426pp.
TSYTOVICH N. A. and KHAKIMOV K. R. 1971. Ground freezing applied to mining and construction. *Proc. 5th IC SM & FE*, Paris, **2**, 737–741.
TSYTOVICH N. A. *et al.* 1959. Physical phenomena and processes in freezing, frozen and thawing soils. *Obruchev Inst. of Permafrost Studies, tech transl*, 1164, Ch. V.
TSYTOVICH N. A. *et al.* 1965. Consolidation of thawing soils. *Proc. 6th ICSMFE*, Montreal, **1**, 390–394.
TSYTOVICH N. A. *et al.* 1973. Physical and mechanical properties of saline soils. *Proc. 2nd IC on Permafrost*, Yakutsk.†
TSYTOVICH N. A. *et al.* 1980. Mechanical properties of frozen coarse-grained soils. *Proc. 2nd ISGF*, Trondheim, **1**, 65–74.
TSYTOVICH N. A. *et al.* 1982. Heaving deformation and thermal creep of frozen ice saturated coarse grained soils. *Proc. 3rd ISGF*, Hanover NH, **1**, 43–52.

UDD J. E. *et al.* 1980a. The strength of a frozen ore in shear. *Proc. 2nd ISGF*, Trondheim, **1**, 297–308.
UDD J. E. *et al.* 1980b. Strength reductions due to the thawing of frozen ores. *Proc. 2nd ISGF*, Trondheim, **1**, 309–324.
UNRUG K. F. 1982. Control of freezing process, an example of deep shafts for polish copper mines. *Proc. 3rd ISGF*, Hanover NH, **1**, 327–336.
UNSWORTH J. F. and SHEPPARD R. P. 1991. Cryogenic properties of soils and rocks — the influence of water types on the uniaxial mechanical behaviour of clays. *Géotechnique*, **41**–2, 211–225.
US Air Force Handbook of Geophysics, 1960. Ch. 12.
US ARMY CORPS OF ENGINEERS 1953. Investigation of the strength properties of frozen soils. *ACFEL*, 300pp.

VAHAAHO I. T. 1988. Soil freezing and thaw consolidation results for a major project in Helsinki. *Proc. 5th ISGF*, Nottingham, **1**, 219–223.
VAHAAHO I. and ERONEN T. 1982. Construction of a tunnel under a major railway with the aid of temporary bridges and V shaped ice-walls. *Proc. 3rd ISGF*, Hanover NH, **1**, 423–428.
VALK J. 1980. The successful application of an unusual ground freezing method to secure tunnel excavation. *Proc. 2nd ISGF*, Trondheim, **2**, 79–93.
VALLEJO L. E. 1982. The effect of freeze–thaw cycles on the structure and the stability of soil slopes. *Proc. 3rd ISGF*, Hanover NH, **1**, 455–461.

VAN ASSCHE R. and DE SMEDT-JANS H. 1989. Road works, hydraulic works and subway works for Place Sainctelette, Brussels, *Tunnelling and Underground Space Tech.*, **4**–3, 293–321.

VAN LOON W. K. P. *et al.* 1988. Thermal and hydraulic conductivity of unsaturated frozen sands. *Proc. 5th ISGF*, Nottingham, **1**, 81–90.

VAN VLIET-LANOË B. and DUPAS A. 1991. Development of soil fabric by freeze/thaw cycles — its effect on frost heave. *Proc. 6th ISGF*, Beijing, **1**, 189–195.

VAN WILK W. R. 1964. Thermal properties of soils. *Proc. I. Study Group on Soils*, Cambridge, 156–167.

VAN WILK W. R. and BRUIJN P. J. 1964. Determination of thermal conductivity and volumetric heat capacity of soils near the surface. *Proc. Soil Sci. Soc.*, 461–464.

VELDEN H. A. V. and SCHAFFERS W. J. 1959. Calculation of the refrigeration required in shaft sinking by the freezing process. *Glückauf*, Sept., **95**, 1237–1244.

VENTER J. 1959. Construction by freezing of a new 50 m length of tubbing while maintaining normal winding. *Proc. IS on Shaft Sinking and Tunnelling*, IMM, London, July, 457–467.

VERANNEMAN G. and REBHAN D. 1978. Ground consolidation with liquid nitrogen. *Proc. 1st ISGF*, Bochum, **1**, 289–302.

VIKLANDER P. and KNUTSON S. 1994. Deformation and compaction of frozen soils. *Proc. 7th ISGF*, Nancy, France, **1**, 109–116.

VINSON T. S. 1975. Cyclic triaxial test equipment to evaluate dynamic properties of frozen soils. *NTIS PB-242 419*, Michigan State Univ., 37pp.

VINSON T. S. 1978. Response of frozen ground to dynamic loading. *Geotech. Engng for Cold Regions*, McGraw-Hill, 405–458.

VINSON T. S. *et al.* 1978. Behaviour of frozen clay under cyclic loading. *ASCE J. Geotech. Engng*, **104**–7, 779–800.

VINSON T. S. et al. 1978. Dynamic testing of frozen soils under simulated earthquake loading conditions. *ASTM ATP 654*, 196–225.

VINSON T. S. *et al.* 1987. Factors important to the development of frost heave susceptibility criterion for coarse-grained soils. *Trnsptn Research Record*, 1089, 124–131.

VJARLOV S. S. *et al.* 1985. Unified laboratory methods for determining strength and deformability properties of frozen soils. *Proc. 4th ISGF*, Sapporo, **2**, 183–187.

VLAD N. V. 1980. The determination of frost susceptibility for grounds using a direct shear method. *Proc. 2nd ISGF*, Trondheim, **1**, 807–814.

VODOLAZKIN V. M. *et al.* 1989. Structural strength of frozen soil during thawing. *SM & FE*, **26**–5, 207–209.

VOVK A. A. *et al.* 1980. Mechanical properties of frozen soils under dynamic loads. *SM & FE*, **17**–2, 64–69.

VOYTKOVSKIY K. F. 1953. Strength and creep of frozen ground. *Academy of Science, USSR*, (transl, 1963, by *US Corps of Engineering*), 185pp.

VRACHEV V. V. and DATSKO P. S. 1980. A differential appraisal of the strength of frozen ground. *Moscow Univ. Tech. Bulletin*, **35**–5, 73–80.

VUORELA M. and ERONEN T. 1980. Driving of metro tunnels with the aid of ground freezing at Helsinki. *2nd ISGF*, Trondheim, **1**, 907–915.

VYALOV S. S. 1955. Creep and long-term strength of frozen soils. *Transl Dok Akad Nauk*, **104**–6, 5–9.

VYALOV S. S. 1959. Rheological properties and bearing capacity of frozen soils. *CRREL, Transl 74*, 120pp.

VYALOV S. S. 1963a. Rheology of frozen soils. *Proc. 1st IC on Permafrost*, Indiana, 332–338.

VYALOV S. S. 1963b. The strength and creep calculation of the barriers made of frozen soil. *Proc. 2nd Asian Regional C. SM & FE*, Japan, **1**, 152–156.

VYALOV S. S. 1966. Methods of determining creep, long term strength and compressibility characteristics of frozen soils. *Tech. transl. from Russian 1364*, NRCC.†

VYALOV S. S. 1973. Continuous frozen soil failure as a thermodynamic process. *Proc. 2nd IC on Permafrost*, Yakutsk, 135–137.

VYALOV S. S. 1978a. Settlements of experimental plates on plastic frozen soils. *SM & FE*, **15**–5, 328–334.

VYALOV S. S. 1978b. Kinetic theory of deformation of frozen soils. *Proc. 3rd IC Permafrost*, Edmonton, 5pp.

VYALOV S. S. and SHUSHERINA E. P. 1970. Resistance of frozen soils to triaxial compression. *Academy for Sci. and Tech. Centre MF*, AD 713 981 36pp.

VYALOV S. S. and TSYTOVICH N. A. 1957. Cohesion of frozen soils. *Cndn Defence Research Board*, T203R, 4pp.

VYALOV S. S. and ZARETSKY Y. K. 1976. Design of ice–soil retaining structures for mine shafts sunk by the freezing method. *Proc. 6th EurC. SM & FE*, Vienna, 1/3–12, 269–274.

VYALOV S. S. *et al.* 1966. Methods of determining creep, long term strength and compressibility characteristics of frozen soils. *NRCC Transl 1364*, Ottawa.†

VYALOV S. S. *et al.* 1978. Stability of mine workings in frozen soils. *Proc. 1st ISGF*, Bochum, Germany, **1**, 339–351, and in *Engng Geol.*, **13**–1/4 (1979), 339–351.

VYALOV S. S. *et al.* 1978. Stability mine working in frozen soils. *Proc. 1st ISGF*, Bochum, **1**, 207–216.

VYALOV S. S. *et al.* 1980. Sinking deep mine shafts by the freezing method. *Proc. 2nd ISGF*, Trondheim, **1**, 980–988.

VYALOV S. S. *et al.* 1982. Artificial freezing of soils in a base of headframes. *Proc. 3rd ISGF*, Hanover NH, **1**, 447–452.

VYALOV S. S. *et al.* 1988. Frozen soil deformation and failure under differential loading. *Proc. 5th ISGF*, Nottingham, **2**, 465–472.

WADSWORTH A. 1957. Recent shaft sinking developments in the East Midlands. *Proc. IME*, **117**, Apr., 397–416.

WALBERG F. C. 1978. Freezing and cyclic triaxial behaviour of sands. *ASCE J. Geotech. Engng*, **104**, GT5, 667–671.

WALBRECKER W. 1910. Tests and studies of the freezing method. *Glückauf*, 1681, 1717.

WALDECK H. 1937. Masonry and concrete linings for shafts in Upper Silesia sunk by the freezing process. *Glückauf*, **73**, 53–58.

WALDECK H. 1948. Safety measures in shaft freezing below the depths of freezing to seal off water flowing from the overlying rock. *Glückauf*, 24 Apr., **84**, 300–302.

WALLI J. R. O. 1963. Application of European shaft sinking techniques to the Blairmore formation. *Cndn Mining and Metallurgical Bulletin*, Feb., **662**, 139–146.

WALMSLEY A. 1962. Shaft and inset work at Cotgrave Colliery. *Trans. IME*, **121**, May, 510–524.

WALSH A. R. *et al.* 1991. Shaft construction by raise boring through artificially frozen ground. *Proc. 6th ISGF*, Beijing, **1**, 369–378.

WANG C. 1991. Deformation of ice-wall in eastern air shaft, Panji #3 mine, and its analysis. *Proc. 6th ISGF*, Beijing, **1**, 279–284.

WANG Z. 1988. The cause and prevention of freeze-tube breakage. *Proc. 5th ISGF*, Nottingham, **1**, 303–310.

WANLOV S. S. and ZARETSKY Y. K. 1966. Design of ice–soil retaining structures for mine shafts sunk by the freezing method. *Proc. 6th EurC SM & FE*, Vienna.†

WARD W. H. 1976. Design and construction of inclined drifts at Gascoigne

Wood using the ground freezing process. *BRS Special Rpt WG2624*, 7pp. (U)

WARDER D. L. and ANDERSLAND O. B. 1972. Soil—ice behaviour in a model retaining structure. *Cndn Geotech. J.*, **8**—1, 46—68.

WASSIF A. A. 1992. Interconnecting tunnel for the Ameria pumping station of the Greater Cairo wastewater project. *Tunnelling and Underground Space*, **7**—2, 145—148.

WATANABE O. and TANAKA M. 1982. Thermal analysis of the position of the freezing front and thermal regime using FEM. *Proc. 3rd ISGF*, Hanover NH, **1**, 177—184.

WEAVER J. S. and MORGENSTERN N. A. 1981. Simple shear creep tests on frozen soils. *Cndn Geotech. J.*, **18**—2, 217—229.

WEEHUIZEN J. M. 1959. New shafts of the Dutch state mines. *Proc. IS on Shaft Sinking and Tunnelling*, London, July, 28—65.

WEILER A. and VAGT J. 1980. The Duisberg method of metro construction — a successful application of the gap freezing method. *Proc. 2nd ISGF*, Trondheim, **1**, 916—927; **2**, 126—128.

WEN Z. and RONGQING P. 1994. Freezing techniques in the deep shaft construction of Chensilou coal mine. *Proc. 7th ISGF*, Nancy, France, **1**, 303—309.

WENG J. 1983. Thermal model and design shaft freezing. *Proc. Soc. Mining Engineers*, **274**, 1898—1903.

WENG J. and ZHANG M. 1991. Back analysis of measured displacements of freeze-wall in shaft modelling. *Proc. 6th ISGF*, Beijing, **1**, 285—290.

WERNET V. C. 1954. Sealing of the Peyerinhof shaft of the Sainte Fontaine Colliery, Porraine. *Bergbau Tech.*, Berlin, 22pp.

WHETTON J. T. 1930. The modern application of the freezing process of sinking. *Colliery Engng*, June, **7**, 228—230, 248—251.

WHITE T. L. and WILLIAMS P. J. 1994. Cryogenic alteration of frost susceptible soils. *Proc. 7th ISGF*, Nancy, France, **1**, 17—24.

WIESE H. 1901. Shaft sinking at Ronnenberg, Hanover. *Glückauf*, 24 Aug.

WIJEWEERA H. and JOSHI R. C. 1992. Temperature independent relationships for frozen soils. *J. Cold Regions Engng*, **6**—1.

WIJEWEERA H. and JOSHI R. C. 1993. Creep behaviour of saline fine-grained frozen soil. *J. Cold Regions Engng*, **7**—3, 77—†.

WILD W. M. and FORREST W. 1981. The application of the freezing process to 10 shafts and 2 drifts at the Selby project. *Proc. IME: The Mining Engineer*, June, 895—904.

WILLIAMS M. and KAPLAR C. W. 1968. Comparison of soil thermal conductivity prediction forumulae with test results. *CRREL Tech note.*†

WILLIAMS P. J. 1962. Specific heats and unfrozen water content of frozen soils. *Proc. 1st Cndn NC on Permafrost*, NRCC TM 76, 109—126.

WILLIAMS P. J. 1964. Unfrozen water content of frozen soils and soil moisture suction. *Géotechnique*, **14**—3, 231—246.

WILLIAMS P. J. 1967. Properties and behaviour of freezing soils. *Norwegian Geotechnical Institute*, **72**, 120pp.

WILLIAMS P. J. 1976. Volume change in frozen soils. *Contribution to Soil Mechanics — Bjerrum Mem. Vol.*, Norwegian Geotechnical Inst., Oslo, 233—246.

WILLIAMS P. J. 1988. Thermodynamic and mechanical conditions within frozen soils and their effects. *Proc. 5th IC on Permafrost*, Trondheim, **1**, 493—498.

WILLIAMS P. J. and MITCHELL R. J. 1969. Isothermal volume change of frozen soils. *Proc. 22nd Cndn C. SM*, Kingston, Ontario, 27pp.

WILLIAMS P. J. and SMITH S. L. 1988. Ice sandwich experiments and the thermodynamic-rheological model of frost heave. *Proc. 5th ISGF*, Nottingham, **2**, 579—82.

WILLIAMS P. J. and WOOD J. A. 1985. Internal stresses in frozen ground. *Cndn Geotech. J.*, **22**—3, 413—416.

WILLS A. J. and DAW G. P. 1988. Temperature monitoring of frozen ground during construction of Gascoigne Wood drifts. *Proc. 5th ISGF*, Nottingham, **2**, 533−540.
WILSON A. H. 1982. Stability of a thick ring of grouted or frozen ground. *Proc. NS on Strata Mechanics*, Newcastle, Apr., 141−146.
WIND H. 1978. The soil freezing method for large tunnel constructions. *Proc. 1st ISGF*, Bochum, **2**, 119−126, and in *Engng Geol.*, 1979, **13**, 417−423.
WINTER H. 1980. Creep of frozen shafts: a semi-analytical model. *Proc. 2nd ISGF*, Trondheim, **1**, 247−261.
WINTER H. 1982. Frozen shafts under time dependent loading. *Proc. 3rd ISGF*, Hanover NH, 79−80.
WINTERMEYER 1922. Recent developments in the freezing process. *Bergbau*, **35**−22, 790−794.
WOLFE L. H. and THIEME J. O. 1963. Physical and thermal properties of frozen soil and ice. *J. Soc. of Petroleum Engineers*, Mar., **4**−1, 67−72.
WOLLERS K. 1985. Causes of fractured freeze pipes in the Hunxe shaft and their restoration. *Proc. C. Shaft Sinking and Tunnelling, Glückauf*, **121**, 1428−1442.
WOLLERS K. 1987. Progressive measurement of ice-wall thickness as a precondition for freeze shaft excavation. *Proc. 13th World C. on Mining*, Stockholm, June, **2**, 953−957.
WOLLERS K. and LUTHE W. 1988. Aspects of modern shaft sinking technology. *Mining Engineer*, **148**−323, 71−75.
WOOD A. F. 1956a. Placing of mass concrete shaft lining in frozen ground. *Proc. IME*, June, **116**, 713−727.
WOOD A. F. 1956b. Sinking at Lea Hall colliery. *Colliery Guardian*, **192**, Mar., 336−340.
WOOD A. F. 1956c. Freezing from an intermediate level in a shaft sinking. *Durham Univ. King's Coll. Mining Bulletin*, **8**, 1 Aug.
WOOD A. F. 1957a. Cotgrave Colliery sinking. *Colliery Guardian*, **194**, Jan., 97−105.
WOOD A. F. 1957b. Placing of mass concrete shaft lining in frozen ground. *Mining Engineer*, **116**, 714−727.
WOOD A. F. 1958. Shaft sinking at Wearmouth. *Colliery Engng*, **35**, May, 188−195; June, 234−238; July, 280−284.
WOOD A. F. 1961. Rapid sinking at British colliery. *Mining Magazine*, **105**, Oct., *213−219.*
WOOD J. A. and WILLIAMS P. J. 1985. Stress distribution in frost heaving soils. *Proc. 4th ISGF*, Sapporo, **1**, 165−172.
WOOD S. E. 1906. Sinking through magnesian limestone and yellow sand by the freezing process at Dawdon Colliery, Co Durham. *Proc. IME, 32, 551−*†
WOON H. K. and DANIEL D. E. 1992. Effects of freezing on hydraulic conductivity of compacted clay. *J. Geotech. Engng*, **118**−7.
WU Q. and TONG C. 1991. Thaw consolidation process and calculation of frozen clayey soil. *Proc. 6th ISGF*, Beijing, **1**, 369−378.
WU Q. and WU Z. 1992. Effect of tunnel wall rock refreeze on lining stability. *J. Glaciology and Geocryology*, **14**−3.
WU Q. et al. 1991. Ground freezing for driving a slope. *Proc. 6th ISGF*, Beijing, **1**, 379−384.
WYLEZOL 1926. Innovations in the construction of concrete shafts by the freezing process. *Braunkohle*, **24**, 923−978.

XIA Z. 1985a. A study of thermal cracks in frozen ground #3. *Proc. 4th ISGF*, Sapporo, **1**, 3−8.
XIA Z. 1985b. A study of frost damage for retaining wall of small-scale hydraulic engineering. *Proc. 4th ISGF*, Sapporo, **2**, 317−322.

XIAOBAI C. et al. 1980. Some characteristics of water saturated gravel during freezing and its application. *Proc. 2nd ISGF*, Trondheim, **1**, 692−701.

XIAOZU X. U. et al. 1980. The thermal properties of the typical soils both in thawed and frozen states. *Proc. 2nd ISGF*, Trondheim, **1**, 413−416.

XIAZOU X. et al. 1985. Experimental study on factors affecting water migration in frozen morin clay. *Proc. 4th ISGF*, Sapporo, **1**, 123−128.

XIE Y. and WANG J. 1985. Effect of saturation level and freeze− thaw cycling on the properties of clayey soil heaving. *Proc. 4th ISGF*, Sapporo, **1**, 197−200.

XU X. and CHUVILIN E. M. 1994. Frost heave experiments with layered porous material. *Proc. 7th ISGF*, Nancy, France, **1**, 399−400.

XU X. et al. 1985. Prediction of unfrozen water contents in frozen soils by a two-point or one-point method. *Proc. 4th ISGF*, Sapporo, **2**, 83−87.

XU X. et al. 1987. Factors affecting water migration in frozen soils. *CRREL Rpt 87-9*.

XU X. et al. 1991. Water and soluble migration of freezing soils in closed system under temperature gradients. *Proc. 6th ISGF*, Beijing, **1**, 93−98.

XU X. et al. 1994. Influence of cooling rate on frost heave of freezing soils in open systems. *Proc. 7th ISGF*, Nancy, France, **1**, 65−68.

YAHAGI H. 1985. On the device for measuring frost penetration. *Proc. 4th ISGF*, Sapporo, **2**, 271−276.

YAMAGUCHI H. et al. 1991. Influence of freezing−thawing on undrained shear behaviour of fibrous peat. *Proc. 6th ISGF*, Beijing, **1**, 197−204.

YAMAMOTO H. et al. 1988. Effect of overconsolidation ratio of saturated soil on frost heave and thaw subsidence. *Proc. 5th IC on Permafrost*, Trondheim, **1**, 522−527.

YAMAMOTO H. et al. 1991. Displacement and increment in earth pressure in unfrozen soil by frost heave. *Proc. 6th ISGF*, Beijing, **1**, 369−378.

YANAGISAWA and YAO Y. J. 1985. Moisture movement in freezing soils under constant temperature condition. *Proc. 4th ISGF*, Sapporo, **1**, 85−92.

YANG Y. 1982. Strength development of concrete placed in frozen soil and its thermal effects. *Proc. 3rd ISGF*, Hanover NH, **1**, 375−382.

YANITSKY P. A. 1991. The formation of heterogeneous geocryological structure while ground freezing. *Proc. 6th ISGF*, Beijing, **1**, 291−298.

YAO L. Y. C. 1964. *Shear strength characteristics of a silty clay subjected to freezing and thawing*. Cornell University, PhD thesis, 187pp.

YARKIN I. G. 1974. Physico-chemical processes in freezing soils and ways of controlling them. *Gersevanov Foundation and Underground Res Inst.*, Collected papers no 64.

YERSHOV E. D. et al. 1988. Thermal characteristics of fine-grained soils. *Proc. 5th ISGF*, Nottingham, **1**, 135−142.

YINQI X. and JIANGUO W. 1985. Frost heave behaviour of cohesive soil due to loading. *Proc. 4th ISGF*, Sapporo, **2**, 153−156.

YONG R. N. 1963. Soil freezing considerations in frozen soil strength. *Proc. 1st IC on Permafrost*, Indiana, 315−319.

YONG R. N. and SERAG-ELDIN N. 1980. Salt treatment effects on frost heave performance. *Proc. 2nd ISGF*, Trondheim, **1**, 680−691.

YONG R. N. et al. 1985. Alteration of soil behaviour after cyclic freezing and thawing. *Proc. 4th ISGF*, Sapporo, **1**, 187−195.

YONG R. N. et al. 1986. Cyclic freeze−thaw influence on frost heaving pressures and thermal conductivities of high water content clays. *Proc. 5th IS on Offshore Mechanics and Arctic Engineering*, Tokyo, **4**, 277−284.

YONGCHENG Z. 1987. New development in sinking deep shafts by special methods in China. *Proc. 13th World Mining Congress*, Stockholm, June, **2**, 969−977.

YOSHISUKE N. 1994. Dependence of segregation potential on the thermal and hydraulic conditions predicted by model M_1. *CRREL Rpt 94−4*.

YOUSSEF H. 1985. Development of a new triaxial cell with self-cooling

system for testing ice and frozen soils. *Proc. 4th ISGF*, Sapporo, **2**, 247–258.
YOUSSEF H. and HANNA A. 1990. Behaviour of frozen and unfrozen sands in triaxial testing. *Trnsptn Research Record*, **1190**, 57–64.
YOUSSEF H. and KUHLEMEYER R. 1981. Dynamic and static creep testing of ice and frozen soils. *Proc. IC POAC*, Quebec, **2**, 726–734.
YOUSSEF H. *et al.* 1982. Development of a testing apparatus for static and dynamic creep of ice and frozen soils. *Proc. 4th Cndn C. on Permafrost (1981) (R. J. E. Brown Mem. Vol.)*, Calgary, 429–432.
YOUSSEF H. *et al.* 1983. Triaxial testing of frozen soils with automatic volume change measurement. *Proc. 4th IC Permafrost*, Fairbanks, 7pp.
YU B. F. and QU X. M. 1985a. Modes of ice-pull action in foundation and its prevention under ice covering. *Proc. 4th ISGF*, Sapporo, **1**, 313–318.
YU B. F. and QU X. M. 1985b. Double layer progressive model and calculation of normal heaving force on base plate. *Proc. 4th ISGF*, Sapporo, **2**, 121–124.
YU B. F. and QU X. M. 1985c. Design of insulating base for culvert sluice. *Proc. 4th ISGF*, Sapporo, **2**, 295–300.
YU C. *et al.* 1991. Simulation tests and studies of triaxial creep in frozen walls. *Proc. 6th ISGF*, Beijing, **1**, 299–305.
YU X. and WANG C. 1988. Structure and stress analysis of seepage resistant linings in shafts sunk with the freezing method. *Proc. 5th ISGF*, Nottingham, **1**, 311–320.
YU X. *et al.* 1991. The influence of waterflow in centre pressure relief hole on ice-wall formation. *Proc. 6th ISGF*, Beijing, **1**, 305–312.
YU X. *et al.* 1994. Study on joint action of freeze-wall and shaft lining. *Proc. 7th ISGF*, Nancy, France, **1**, 271–277.
YUGI Y. 1980. Strength development of concrete placed in frozen soil and thermal effects.†

ZAKHAROV V. A. 1970. Experimental investigations of the mechanical characteristics of frozen ground at different rates of loading. *SM & FE*, Sept.–Oct., 312–319.
ZARETSKY Y. K. 1972. Rheological properties of plastic frozen soils and determination of settlement of a test place with time. *SM & FE*, Mar.–Apr., 81–85.
ZARETSKY Y. K. and GORODETSKII S. E. 1975. Dilatency of frozen soil and development of a strain theory of creep. *Hydrotechnical Construction*, Feb., 127–132.
ZATSARNAYA A. G. 1973. Calculating the settlement of foundations on plastically frozen ground. *Proc. 2nd IC on Permafrost*, Yakutsk, 647, 650.
ZHANG C. and HU D. 1991. Technical innovations in freeze-shaft construction of Jining #2 coal mine, China. *Proc. 6th ISGF*, Beijing, **1**, 385–390.
ZHANG C. *et al.* 1994. Microstructure damage behaviour and change characteristic in the creep process of frozen soil. *Proc. 7th ISGF*, Nancy, France, **1**, 163–167.
ZHANG L. 1991. The law of unfrozen water content change in frozen saline (NaCl) soils. *Proc. 6th ISGF*, Beijing, **1**, 113–119.
ZHANG L. and XU Z. 1994. The influence of freezing–thawing process on the unfrozen water content of frozen saline soil. *Proc. 7th ISGF*, Nancy, France, **1**, 375–377.
ZHANG N. 1991. Application of ground freezing method to two underground projects in water-bearing sand layer. *Proc. 6th ISGF*, Beijing, **1**, 543–546.
ZHANG Y. 1991a. Influence of soil properties on ice-wall formation. *Proc. 6th ISGF*, Beijing, **2**, 463–469.
ZHANG Y. 1991b. Freeze sinking construction of Xie Qiao coal mine. *Proc. 6th ISGF*, Beijing, **2**, 547–552.

ZHANG Y. et al. 1994. Deformation of artificially frozen shafts during excavation. *Proc. 7th ISGF*, Nancy, France, **1**, 225−232.

ZHOU X. et al. 1994. Experimental research for heat preservation of freeze-pipe in the swift water flow. *Proc. 7th ISGF*, Nancy, France, **1**, 35−37.

ZHU Q. and WU F. 1985. An experimental study on the relationship between the frost heave and the water content of the frozen soil. *Proc. 4th ISGF*, Sapporo, **2**, 147−151.

ZHU Y. and CARBEE D. L. 1984. Uniaxial compressive strength of frozen silt under constant deformation rates. *Cold Regions Sci. and Tech.*, **9**, 3−15.

ZHU Y. and CARBEE D. L. 1985. Strain rate effect on the tensile strength of frozen silt. *Proc. 4th ISGF*, Sapporo, **1**, 153−158.

ZHU Y. and CARBEE D. L. 1987a. Creep and strength behaviour of frozen silt in uniaxial compression. *CRREL Rpt 87-10*, 67pp.

ZHU Y. and CARBEE D. L. 1987b. *Tensile strength of frozen silt. CRREL Rpt 87-15.*

ZHU Y. et al. 1991. Constitutive relations of frozen soil in uniaxial compression. *Proc. 6th ISGF*, Beijing, **1**, 211−218.

ZHU Y. et al. 1988. Uniaxial compressive strength of frozen medium sand under constant deformation rates. *Proc. 5th ISGF*, Nottingham, **1**, 25−234.

ZHU Y. et al. 1982. Elastic and compressive deformation of frozen soils. *Proc. 3rd ISGF*, Hanover NH, **1**, 65−78.

ZHU Y. et al. 1994. Creep behaviour of long-term strength of frozen soil under dynamic loading. *Proc. 7th ISGF*, Nancy, France, **1**, 97−102.

ZIENKIEWICZ O. C. et al. 1973. The application of finite elements to heat conduction problems involving latent heat. *Proc. Rock Mechanics 5*, 65−76.

ZYKOV Y. D. et al. 1973. Application of ultrasonics for evaluating the phase composition of water and strength characteristics of frozen soils. *Proc. 2nd IC on Permafrost*, Yakutsk, 335−337.

ZYLINSKI R. 1974. Stability of walls in shafts sunk by freezing methods. *Krakow Univ.* (in Polish).

* *Proc. 3rd ISGF,* Hanover NH, 1982, **2**, is a collection of papers distributed to participants, but not published in bound form. See appendix A: Symposia/Society proceedings.

† Not all details known.

Note: Readers who require further bibliographic information may contact the Author for assistance.

Index

Aberdeen, 140, 183, 190, 200
Acoustic velocity, 23
Alexandria (Egypt), 160, 198
Ammonia, 87, 88, 131, 145, 160, 175, 176–177
Aquifers, 4, 70, 107, 117, 126
Asfordby, 9, 100, 137, 200

Back-wall grouting, 105, 116–117, 132, 188
Base heave, 33, 41, 208
Belle Isle (USA), 108, 128–129, 202
Belmont (Australia), 5, 121–122, 196
Bent sub, 97, 98
Blackpool, 82, 108, 122, 200
Blackwall Tunnel, 144–146, 200
Blasting, 115
Bored piles, 1, 3, 99
Borehole surveying, 102, 189
Born (Switzerland), 153–154, 200
Boulby, 5, 9, 108, 111, 138, 183, 190, 200
Brine, 5, 9, 27, 40, 83, 87, 92–94, 100, 103, 104, 117, 129, 132, 137, 138, 139, 140, 147, 149, 151, 152, 153, 154, 163, 164, 165, 182, 183
  calcium chloride, 87, 88
  circuits, 87, 92–94, 103, 150
  flow, 87, 92, 93, 103
  loss, 88, 94, 100, 103, 139
Bromham Bridge, 82, 108, 167, 190, 200
Brussels metro (Belgium), 82, 108, 170, 196
Burgos (Spain), 5, 82, 108, 168, 200

Cairo (Egypt), 146–147, 198
Caisson sinking, 1, 2, 129
Calcium chloride, 87, 88, 106
Canvey Island, 122–126, 201
Capillary model, 31, 32, 36
Carcroft, 126, 201
Case studies/histories, 121–170
Cheltenham, 165, 201
Chemical injection, 3, 6
Clays, 19, 27, 37, 39, 45, 47, 51, 53, 57, 62, 64, 71, 76, 77, 106, 108, 109, 124, 126, 129, 130, 132, 133, 139, 140, 141, 142, 144, 153, 154, 157, 160, 161, 162, 164, 167, 168, 169, 186, 189
Cofferdam, xix, 2, 4, 17, 70, 71, 82, 99, 105, 107, 108, 141, 153, 165, 167
Compressed air, 4, 6, 129, 149, 155, 156, 160, 163, 169, 176, 183, 210
Compressive strength, xiii, 21, 52, 64, 76, 116
Compressor, 4, 87, 88, 89, 90, 149
Computer program, 79, 83
Concreting, 113, 114–116, 127, 129, 131, 134, 141, 151, 160
Consolidation, xiii, 41, 41–42, 44, 46, 48–50, 51
Contiguous piles, 1, 3
Contract, 5, 62, 89, 96, 121, 129, 140, 155, 179–182, 184, 189, 190
  main, 180
  sub-, 127, 179, 180, 181, 182
Contractors, 2, 5, 6, 89, 113, 121, 127, 129, 181, 183, 184, 189, 204
  main, 121, 140, 164, 179, 180, 181, 182, 183, 186, 187
  specialist sub-, 121, 179, 180, 181, 183
Control, 99, 126, 139, 149, 174, 175, 179, 180, 185
Coolant, 14, 87, 137, 138
  brine, 137, 138
  lost circulation, 88, 94, 100, 103, 139
Cost, xix, 1, 2, 4, 5, 83, 96, 98, 117, 127, 137, 138, 140, 163, 180, 182–183, 184, 185, 186, 186–187, 190
Cracking, 20, 47, 53, 54, 115, 125
Creep, xiv, 19, 20, 21, 22, 23, 54–55, 56–59, 62–64, 109
  equation, 54–55, 56, 57, 58, 59, 60, 61, 62, 76
  strength, 19, 20, 58, 60–61
  test, 19, 20, 21, 55, 56, 59, 60, 62
Crewe, 141, 201
Cryogenics, 87, 94–95, 175
Cut-off, 8, 9, 10, 70, 107–108, 164, 165, 181

Darcy's law, 35
Definitions, 16, 184–185
  frozen soils, 16–17, 71
  risk, 184–185
Deflectometer, 101
Deformation, xiv, 19, 20, 21, 22, 71, 72, 74, 76, 76–80, 82, 101, 109, 117, 185, 187
Dewatering, 2, 4, 6, 107, 129, 139, 142, 163, 169, 186, 187, 190, 204
Diaphragm wall, 1, 2, 164, 165, 168
Directional drilling, xix, 97, 98, 137, 138, 149, 183
Displacement, 79
Domke formula, 72, 74, 75
Drilling, 6, 9–10, 74, 80, 96–99, 99–100, 107, 114, 117, 118, 121, 128, 131, 137, 138, 139, 148, 149, 151, 152–153, 153, 154, 156, 157, 160, 164, 169, 183, 206, 207, 208, 209

261

accuracy, 10, 97, 101, 131, 137, 148, 149
air flush, 17, 96
directional, xix, 97, 98, 137, 138, 149, 183
mud-flush, 74, 96–98
normal, 98, 99
overburden, 9, 74, 80, 98, 99, 128, 155, 160, 190
rotary, 17, 98, 99
shell and auger, 98, 113
stuffing box, 12, 99, 100, 152
targets, 2, 100, 101
Drifts (inclines), 3, 5, 13, 80–82, 101, 108, 132, 141, 154
Du Toits Kloof (South Africa), 116, 147, 190, 200
Duisberg (Germany), 164–165, 198
Dundee, 147, 201

Earth pressure, 94
Edinburgh (Craigentinney), 156, 201
Effective stress, 33, 34, 44–45, 46, 50
Electro-osmosis, 3, 6, 176, 205, 207, 209
Ely Ouse-Essex, 113, 115, 190, 201
Epoxy resin, 99, 152
Excavation, xiv, xix, 1, 2, 3, 4, 5, 6, 7, 8, 9, 10, 12, 14, 17, 24, 46, 53, 70, 71, 75, 76, 77, 78, 79, 80, 96, 100, 105, 106, 107, 108, 109, 113–114, 121, 121–144, 149, 154, 156, 167, 168, 175, 181, 185, 186, 189

Failure, xiv, 52, 53, 55, 93, 94, 141
Finite element method, 78–79
Foundations, xix, 12, 13, 23, 30, 46, 70, 141, 155, 167, 168, 169, 188
Fosdyke, 165, 201
Fotobor, 101–102
Freeze-plant, 167, 173
Freeze-tubes, xiv, 4, 5, 6, 7, 8, 9, 16, 17, 26, 70, 78, 80, 82, 83, 87, 88, 89, 92, 93, 94, 95, 96–118, 121, 123, 125, 129, 130, 131, 132, 133, 135, 136, 137, 138, 139, 140, 141, 142, 151, 152, 153, 156, 157, 160, 161, 164, 165, 169, 176, 189
  accuracy, 100, 137
  angled, 9, 12, 96–99, 101, 138, 153
  assembly, 4, 5, 6, 7, 92, 93
  breakage, 74, 94, 103, 140
  circuits, 5, 6, 8, 88, 93, 94, 95, 103, 104, 125, 145, 149
  configurations, 8, 9, 10, 10–13
  construction, 6, 12, 96–118
  horizontal, 9, 95, 98, 101, 144, 145, 147, 148, 149, 150, 152, 153, 159, 167
  inner tubes, 7, 9, 94, 95, 100, 124, 139, 140
  insulated, 94
  outer tubes, 94, 96, 99, 103, 117, 118, 140
  placement, 9, 13, 96–102, 103, 137, 167
  spacing, 75, 83, 92, 93, 100, 101, 129, 130, 149, 170

  surveying, 100–102
  target, 84, 137
  vertical, 6, 7, 9, 12, 80, 95, 96–99, 132, 142, 148, 154, 160, 161, 163
Freezing process, xix, 26, 91
Freon, 87, 88, 149
Frost,
  adsorbed film models, 36
  discrete ice lens theory, 37–38
  damage, 115
  heave, 16, 18, 22, 23, 27, 30–31, 32, 33, 37, 38, 39, 41, 46, 47, 51, 71, 105, 108–109, 151, 153, 170
  mathematical model, 33–36, 36
  Miller's model, 32–33, 34, 35
  segregation potential theory, 36, 37, 38, 39
  test, 37, 47
  thermo-mechanical theory, 38
Frost susceptibility, 16, 22, 23, 32, 38–39, 46, 51, 71
Frozen fringe, xiv, 30, 32–33, 35, 37, 38
Frozen roadways, xix, 14, 38

Gap freezing, 13, 164–165
Gardner formula, 77
Gascoigne Wood, 113, 132, 202
Geophysics, 132
Gibraltar, 82, 106, 127, 199
Grand Coulee Dam, 13
Gravels, 32, 37, 53, 123, 128, 130, 138, 139, 140, 144, 150, 154, 164, 167, 168, 186
Green River, Kentucky (USA), 5, 14, 167–168
Ground information, 72–73
Groundwater, 1, 2, 3, 4, 6, 7, 13, 16, 83, 84, 104, 105–108, 116, 124, 126, 127, 129, 132, 134, 139, 140, 142, 144, 150, 163, 164, 170, 175, 181, 183, 185, 204, 205, 206, 208
  aquifers, 4, 70, 107, 118, 126
  lowering, 2–3, 144, 187, 190, 205, 206, 208
  movement, 3, 4, 104, 105
  quality, xix, 105
Grouting, 3, 6, 83, 85, 100, 105, 106, 109, 116–117, 118, 128, 132, 138, 140, 149, 160, 161, 163, 183, 186, 187, 190, 204
Gyro, 98, 102, 137

Hampton formula, 77
Hazard, 99, 174, 174–175, 184, 189, 207
Health, 173–175
Heat capacity, xiii, 28–29, 34
Helsinki, 148–149, 198
Hydraulics, xiii, 37, 38

Ice-wall, 4, 6, 9, 12, 14, 16, 17, 23, 26, 29, 71, 72–80, 80, 83, 88, 90, 92, 93, 96, 100, 103, 106, 107, 108, 109, 113, 115, 116, 117, 125, 127, 128, 129, 131, 132,

INDEX

133, 134, 138, 139, 140, 147, 153, 173, 183, 189
growth, 6, 7, 77, 100, 103–105, 107, 125, 138, 141, 151, 157, 189
temperature profile, 7, 8, 9, 14, 72, 75, 80, 106, 140
thickness, 14, 24, 72–75, 78, 80, 83, 87, 88, 93, 103, 104, 105, 106, 124, 129, 131, 134, 148
Inclines, see Drifts
Inclinometer, 101
Injection methods, 3–4, 166, 186, 187
In-situ testing, 18, 23, 24
Instrumentation, 100–102
Insulation, 9, 12, 41, 92, 94, 116, 133, 134
Isle of Grain, 160, 201
Iver, 183

Junctioning, 12

Kelvin equation, 32
Kentwood (USA), 113, 128
Klein formula, 56, 58, 72, 74, 76, 77

Laboratory tests, 16, 17, 18, 19–23, 23, 37, 71, 114, 149, 169
Lamé–Clapeyron formula, 72, 74–75
Linings, xix, 7, 14, 76, 78, 79, 80, 96, 105, 109, 113–114, 114, 115, 116, 117, 126, 128, 129, 131, 132, 135, 139, 149, 181, 185, 189
Liquid nitrogen (LN), 28, 83, 84, 87, 91–92, 95, 126, 138, 140, 147, 153, 156, 157, 160, 161, 163, 164, 165, 167, 170, 175, 177–178, 182, 183
London Chelsea Harbour, 91, 166–167, 201
London St Pauls, 138, 201
Lowestoft, 108, 129, 201

Mannheim (Germany), 108, 149–150
Mathematical model, 33–36, 36
Mechanical properties, 26, 38, 52, 64, 150
Milchbuck (Switzerland), 109, 112, 150–151, 200
Milwaukee (USA), 108, 129–131, 203
Model tests, 33–36, 78
Mohr–Coulomb theory, 21, 46, 61, 72, 76
Moisture content, 5, 45, 47, 83, 84, 108, 113, 149, 150, 153, 170
Monitoring, 12, 14, 102–105, 168, 174, 187, 189

Noise, 2, 89, 161, 175, 210
Numerical methods, 78
Numerical models, 78

Open excavations — see Shafts

Pajari, 100
Peat, 133
Permafrost, xix, 5, 13, 16, 17, 23

Permeability, xii, 2, 3, 4, 6, 8, 32, 35, 36, 37, 42, 48–50, 70, 71, 75, 83, 84, 106, 107, 149, 176, 208
Piezometers, 104
Pimlico, 154, 201
Pits — see Shafts
Poisson's ratio, xiv, 24, 54, 71, 79, 81
Pore pressures, 33, 34, 42, 43, 44
Pressure relief hole, 103–104, 107, 140
Procedure, 14, 16, 17, 22, 41, 70, 78, 134, 174, 189

Ratio of thaw settlement, 42, 43, 44
Refrigerants, 7, 83, 87, 88, 92–95, 138, 156
ammonia, 87, 88, 145, 176–177
freon, 87, 88, 149
liquid nitrogen (LN), 28, 83, 84, 87, 91–92, 95, 126, 140, 147, 153, 156, 157, 160, 161, 163, 164, 165, 167, 170, 177–178, 182
Refrigeration, 5, 7, 9, 13, 14, 20, 21, 22, 71, 82, 83, 87–95, 96, 100, 102–103, 105, 106, 107, 109, 117, 124, 126, 130, 131, 132, 134, 138, 140, 147, 153, 156, 163, 167, 169, 170, 175, 182, 183, 189
plant, 5, 14, 83, 87, 87–91, 92, 102–103, 109, 130, 153
Renfrew, 156
Retaining wall, 13, 108
Reynold's number (Re), xiii, 93
Risk assessment, xix, 14, 173, 184, 185–189, 204–210
River Medway, 156–158, 202
Roadways, 14, 22, 38, 46, 128, 132, 134, 136, 147, 148, 149, 150, 167
Rogers Pass (Canada), 107, 142, 190, 197
Runcorn, 99, 151–153, 190, 202

Salinity, 19, 35, 39–41, 52, 64, 70, 105, 106, 128, 131, 189, 208, 209
depression of freezing point, 39, 41, 70, 105, 106, 189
salt concentration, 39, 41, 64
salt heave, 41
salt redistribution, 41
Samples, 16, 17–19, 20, 21, 22, 23, 79
description, 17
handling, 17, 18
in situ frozen, 18
laboratory frozen, 17, 18, 20, 37
preparation, 17, 18, 18–19, 19, 21, 71
remoulded, 18–19, 19, 22, 77
storage, 17, 21, 71
undisturbed, 17, 18, 19
Sands, 19, 22, 41, 129, 130, 131, 132, 139, 140, 141, 144, 150, 154, 155, 156, 157, 162, 164, 168, 169, 186
São Paulo (Brazil), 5, 82, 169, 197
Saskatchewan (Canada), 5, 131

263

Secant piles, 1, 2, 99, 152
Seepage, 3, 83–84, 150, 176
Segregation potential, xiii, 36–37, 37, 38, 39
Selby, 108, 117, 131–132, 154, 183, 202
Shafts (pits and open excavations), xix, 5, 6, 7, 8, 9, 10, 11, 12, 46, 73–80, 93, 94, 103, 108, 113, 121, 121–144, 149, 151, 161, 166, 168, 186, 190
Shear strength, 46, 64
Sheet piles, 1, 2, 108, 126, 141, 147, 165
Sheffield, 160–161, 202
Shotcrete, 151
Side effects, 105–113, 129, 149, 190
Silts, 3, 39, 47, 53, 57, 59, 71, 108, 123, 126, 129, 130, 141, 150, 153, 158, 162, 164, 167, 169
Site investigation, 70–71, 98, 107, 184
Soil pressure, xiv, 12
Slides/slips, xix, 5, 13
Slopes — see Drifts
Specialists, 83, 126, 179, 187
Standards, 16, 77, 82, 173–178
Stirchley, 161, 200
Stonehouse, 161, 202
Strain rate, xiv, 21, 22, 62
Strength, 4, 14, 19, 20, 23, 46, 50–51, 52, 58, 60–61, 62, 71, 72, 76, 77, 79, 82, 105, 108, 115, 149
Stress–strain relations, 21, 22, 23, 52–54, 76, 79
Structural design, 14, 71, 72
Stuffing box, 12, 99, 100, 152
Suction, xiii, 30, 31, 32, 35, 36, 37, 40, 44, 51
Survey (drilling), 98, 100–102, 137
    deflectometer, 101
    fotobor, 101–102
    gyro single/multi shot, 98, 102, 137
    inclinometer, 101
    pajari, 100

Temperature, xiii, 5, 16, 17, 18, 19, 20, 21, 22, 23, 24, 26, 27, 28, 29, 31, 32, 36, 37, 38, 52, 56, 60, 62, 63, 71, 72, 76, 77, 79, 80, 82, 83, 87, 88, 102–106, 114, 116, 125, 129, 131, 140
    effects, 62
    observation holes, 8, 104, 140
    profile, 7, 8, 14, 29, 82
Temporary works methods, xx, 121–140, 144–155, 184, 185, 206
    compressed air, 4, 6, 129, 149, 155, 156, 160, 163, 169, 176, 183, 210
    diaphragm walls, 1, 2, 164, 165, 168
    electro-osmosis, 3, 6, 176, 205, 210
    ground freezing, xix, 5, 13, 53
    groundwater lowering, 2–3, 144, 187, 190, 205, 206, 208

Tenders, 5, 62, 179, 180, 184, 185, 186, 187, 210
Tensile strength, 75, 82, 115, 116, 125
Testing, 17, 18, 19–24, 38, 46, 54, 62, 79
    frost heave, 16, 18, 20, 22, 23
    in situ, 18, 23–24
    laboratory, 17, 18, 19–23, 23, 37, 38, 169
    penetrometer, 23
    pressuremeter, 23
    seismic refraction, 23
    strength and creep, 19, 20, 21, 55–56, 59, 60, 62
Thamesmead, 162, 201
Thawing, 4, 5, 6, 14, 16, 26–64, 105, 115, 116, 132, 149, 153, 165
    freeze–thaw, 16, 26, 44, 45, 46, 47, 48, 50, 51
    residual stress, 44–45
    settlement, 26, 37, 41–42, 45, 46, 109, 153, 169, 170, 188
    weakening, 26, 39, 46, 47
Thermal analysis, 37, 79
Thermal design, 14, 82–83
Thermal properties, 26–64, 82, 83, 105, 108
    conductivity, xiii, 26–28, 83, 106, 108, 116, 124, 133, 189, 195
    diffusivity, xiv, 29–30
    heat capacity, xiii, 28–29, 83
Three Valleys, 100, 162–164, 202
Timmins (Canada), 132–134, 197
Tokyo, 153
Tooting Bec, 143, 144, 159, 202
Tottenham Hale, 154–155, 201
Trench sheets, 1, 2
Tunnels, xix, 4, 5, 6, 9–10, 11, 46, 80–82, 103, 108, 109, 113, 116, 121, 126, 130, 131, 144–170, 175, 190

Underground structures, 9, 151, 168
Underpinning, 3, 5, 12, 82, 168–170
Unfrozen water content, 27, 31, 39, 52, 53, 61, 62
Uniaxial compression, 53, 64
Unity (Canada), 139, 197

Vauxhall Cross, 154, 201
Vibration, 2, 18
Vienna, 5, 169–170, 196
Voids ratio, 44, 48, 85
Vyalov formula, 74, 76

Water content, xiv
Water movement/seepage, 3, 83–84, 150, 176
Wrexham, 134–136

Young's modulus, 23, 53, 54, 71, 75, 81